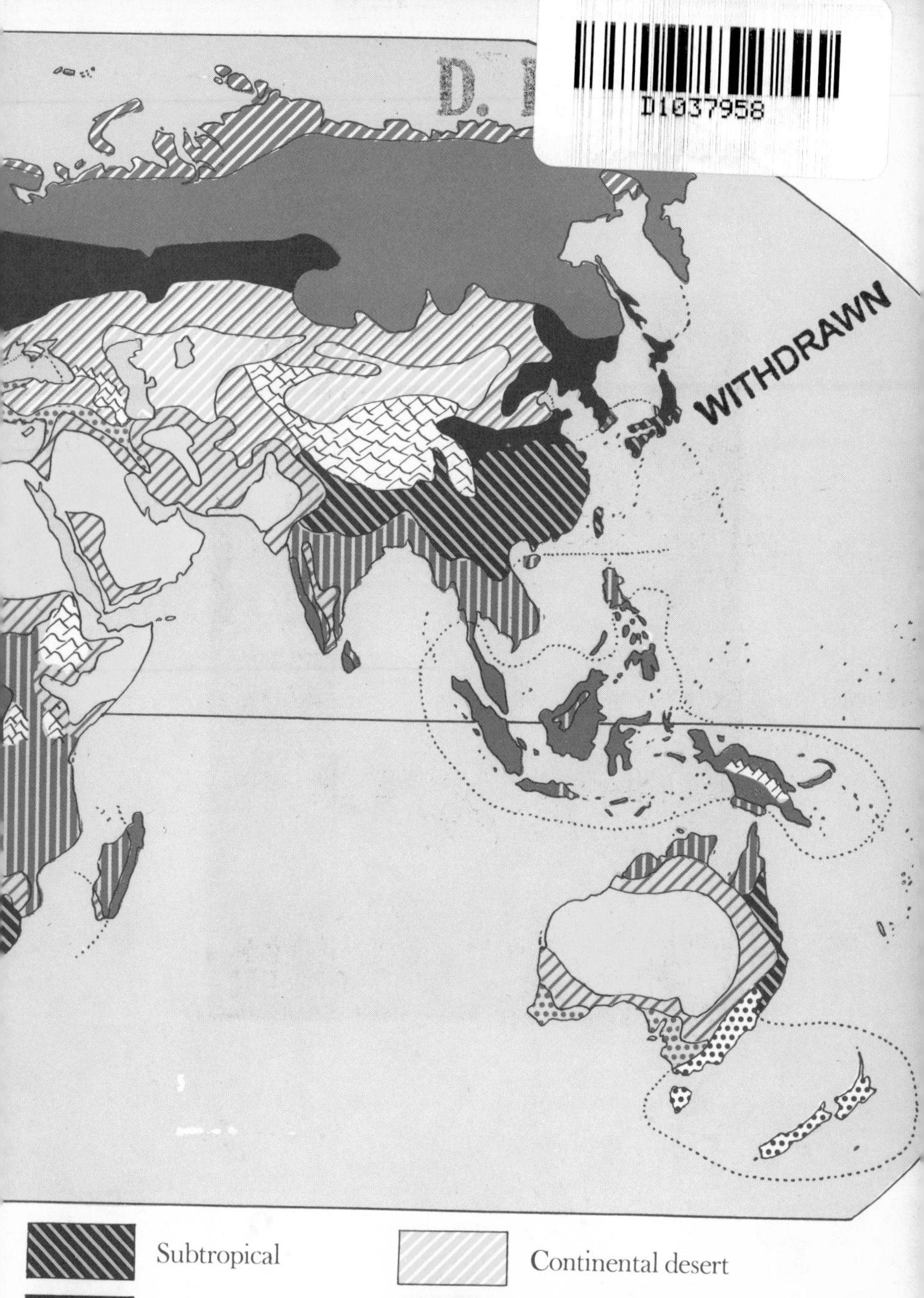

Subtropical

Continental desert

Continental moist

Subpolar

Continental dry

Polar

A.M. JAUSS

Climate, Man, and History

Robert Claiborne

Climate, Man, and History

W · W · NORTON & COMPANY · INC ·

New York

FIRST EDITION

SBN 393 06370 4

Library of Congress Catalog Card No. 68-20815

Published simultaneously in Canada
by George J. McLeod Limited, Toronto

PRINTED IN THE UNITED STATES OF AMERICA

1 2 3 4 5 6 7 8 9 0

To Virginia, who first gave me the taste for wondering about things, and Sybil, who gave me the peace to wonder in.

Any resemblance between the historical situations herein described and more recent predicaments may be coincidental, but is certainly worth thinking about.

CONTENTS

PREFACE

This book will probably annoy quite a number of scientists.

Such was not my original intention, which was merely to make some money in an honest and interesting way. It was only after some weeks, and several false starts, that I realized this book was going to be different from other books I had done.

In ten years of writing professionally about science and medicine, I had, for various persuasive reasons, stuck pretty closely to what scientists thought and said about science. This book, it became apparent, was going to include a good deal of what *I* thought about science—and scientists. And since I am by temperament a skeptic, the book was evidently going to take a skeptical, critical look at the science of climatology.

I think this is a Good Thing, for a number of reasons.

For the last quarter century, ever since mankind was explosively and brutally ushered into the atomic age, science—American science especially—has enjoyed an unprecedented boom in prestige, prosperity, and power. The scientist, once a retiring and underpaid practitioner of arcane mysteries, has been promoted. He now disposes of billions in government and private research funds and is consulted assiduously by our political and corporate rulers; his doings and opinions, both in science and in other areas, are reported respectfully, if sometimes inaccurately, by the mass media. In a world increasingly dominated by experts, he is

The Expert par excellence and, since his field of expertise is likely to be miles away from the experience and training of the ordinary citizen, his opinions and decisions are seldom criticized by outsiders.

This, I think, is a Bad Thing—for science as well as for society. Granting that a complex and highly productive social system like ours cannot be run without the help of many experts in many fields, it is still true that the expert, left too much to himself, can foul up the society he is supposed to be guiding. If war, as recent events have powerfully reminded us, is too important to be left to the generals, then science, it seems to me, is too important to be left to the scientists—and this even though I find scientists, as a class, rather more appealing than generals. Power notoriously tends to corrupt, and in the realm of ideas scientists often enjoy more power than is good for them—or us.

One aim of this book, then, is to depict science as an activity carried on by human beings—men and women who, despite their impressively specialized knowledge and skills, are no more immune to error and carelessness, wishful thinking and personal vanity than the rest of us. The humanity, and therefore the fallibility, of the scientist seem to me worth stressing even at the price of irritating some scientists—in particular those who, like some leaders in other fields, have become victims of their own press notices.

Another aim is to convey the fact that science, despite its often recondite subject matter and forbidding terminology, is frequently not as far beyond the layman's comprehension as some scientists would like to think. Developing a critical understanding of science, like most things worth doing, requires some intelligence, plus the willingness to engage in reasonably hard thinking; it does not require a Ph.D. degree. The theater critic who has never written a play, the dance reviewer who cannot perform an entrechat, can nonetheless put forth opinions valuable to both the artist and the audience. To repeat a trite but true maxim, one need not be able to lay an egg in order to tell a good egg from a bad one.

In trying to distinguish good scientific eggs from dubious

ones, as I many times do in the course of this book, I am not seeking to "debunk" science—which is, as a matter of fact, a good deal less bunk-ridden than such areas of human activity as politics, war, and journalism. I *have* tried to demystify it, to encourage my fellow citizens to look at it seriously, thoughtfully, and critically, with an awareness that the scientist, like any other expert, does not invariably know what he is talking about.

A realistic, adult view of science seems to me especially important at present because science, considered as both a human institution and as a way of thinking, is under attack. Repeatedly, and in a variety of ways, we are being told that "science has failed," and that we must seek answers to the questions which trouble us through mysticism, clairvoyance, or drug-begotten phantasms. Among some young people in particular, antiscience is replacing science as a philosophy; astrology and yoga, Edgar Cayce and Timothy Leary are the new dispensations and prophets. To the extent that this rejection of science represents a reaction to an earlier, uncritical acceptance of science, it is understandable. But understandable or not, all-or-nothing antiscientism is no less foolish than the all-or-nothing scientism it is replacing. Both attitudes arise out of some fundamental misconceptions as to what science is.

In one sense, science is a human institution. Like any institution, it is manned—and financed—by human beings; its activities will necessarily be shaped by the human needs and desires of its personnel and financiers: for prestige, for profit, for power, or sometimes simply for a quiet life. This fact is so obvious that I would not bother to state it except that it is clearly news to many people, including some scientists. The latter, like some other academicians, like to believe that they are engaged in a pure, detached, unflinching search for Truth—which would be true only if the search were conducted by angels and not men. Even then there would remain the rather basic questions of what truths to search for, and what to do with them if and when they were found.

In another sense, science—the scientific method—is a way of thinking about things and as such, it is a tool, with the same limitations as any other tool. No tool can be more useful than the skill of the man using it (watch a novice carpenter trying to hammer a nail); no tool is equally good for every job (don't use a hammer to tune your piano). Science is a tool for answering questions; the quality of the answers depends in the first place on who is asking the questions and what questions he is asking. In science as elsewhere, ask a silly or meaningless question and you get a silly or meaningless answer; as the computermen so elegantly put it, GIGO (Garbage In, Garbage Out). The possession of an advanced scientific degree does not automatically gift its possessor with genius, imagination, or even common sense; there are plenty of hack scientists, just as there are plenty of hack writers.

Even in the hands of the most skilled practitioner, science is still far from infallible. In the first place, it can supply answers only if it has material—data—to work on, just as a hammer is useless without a nail to drive and a board to drive it into. In addition, science requires a conceptual framework which can relate the facts of a problem to one another and thereby allow the scientist to draw conclusions about them; just so, hammer, board, and nail together are still useless unless the carpenter knows enough to set the *point* of the *nail* against the *board* and hit the *head* of the *nail* with the *head* of the *hammer*.

The answers obtained by even the most able scientist will be only as reliable as the data and conceptual framework at his disposal. The astronomer who states that the sun will rise at 5:42 A.M. next Monday is making a statement that for all practical purposes is 100 percent reliable, because he is operating with data and concepts that have been refined over generations (we can ignore the one chance in a billion billion that the sun will rise at 5:43, or not rise at all). In the field of weather and climate, on the other hand, neither the data nor the concepts are anything like adequate; typically and correctly, therefore, the meteorologist couches his conclusions either in vague terms (how cloudy is "partly cloudy"?) or as a probability ("probability of rain, 20 percent tonight, 40 percent tomorrow"). Thus when we say that "science has failed" we may be saying no more than that it cannot, yet, give us answers as reliable as we would like.

A second "failure" of science stems from the obvious fact that it will not answer questions, reliably or otherwise, unless somebody bothers to ask them. DDT, whose effects on our environment are now being viewed with quite justified alarm, represents just such a failure. It was developed in the early 1940s as an answer—an excellent one, so far as it went—to the question: How can we kill insects reliably and cheaply? What the scientists did not ask, at that time, was what DDT might do to other forms of life, such as birds and fish, or whether spreading the chemical abroad was everywhere and always the most effective way of controlling insect pests. Later they began asking these questions, and got some rather alarming answers; still later, voters and legislators became sufficiently aware of the answers to start doing something about DDT. The fact that the questions were not asked sooner can, I think, be fairly described as a failure of scientists—but not of science as such.

Still another "failure" of science really amounts to the complaint that science has succeeded too well. For example, the question of whether a nuclear bomb could be constructed was posed scientifically a quarter century ago; it was answered, horribly, over Hiroshima and Nagasaki. Perhaps the question shouldn't have been asked at all; certainly many people, including me, think the answer should have been left on paper instead of being translated into military hardware. But the question *was* asked. Ask a terrifying question, you get a terrifying answer.

The most conspicuous "failure" of science, however, is the fact that it cannot answer some kinds of questions at all—now or ever. It can tell you what *will* happen, but never what *should* happen.

Every scientific law is in essence a statement of the form:

If A, *then* B. If you construct a square on the hypotenuse of a right triangle, then its area will be equal to that of the squares on the other two sides. If you quickly bring together two masses of uranium-235 of a certain size, then there will be a catastrophic explosion. There is no possible way by which science can tell you whether you should or should not bring those chunks of uranium together. That decision depends, not on "what will happen if?" but on "what do I want to happen?"—and what you want is between you and your conscience, an area in which science is irrelevant. As Martin Gardner and others have pointed out, the dialogue between the scientific method and mankind resembles that between Alice and the Cheshire Cat. Asked, "Which way ought I to go from here?" science will say only, "That depends a good deal on where you want to get to." And if, like Alice, we don't much care where we get to, science will remark with a grin, "It doesn't matter which way you go." If asked, science can usually tell us that if we do A, B will (more or less probably) result. If asked further, as too often we don't, it will even tell us that if B happens, C, D, and E will also happen. But on the question of whether we *should* do A—meaning whether we want B, C, D, and E (and *their* consequences) to happen—science is, and will forever remain, silent.

The scientific way of thinking has told us how to destroy a city in an eye-blink, how to cure pneumonia and tuberculosis, how to raise bountiful crops, move mountains, shift the course of rivers. It has—as we shall see in this book—told us why some parts of the earth are hotter or colder, wetter or dryer, than others, and how (in limited areas) we can alter the temperature or the moisture to better suit our own needs. But it cannot tell us whether to kill or heal, what mountains to move, what crops to raise or who should eat them. It cannot tell us what we want or who we are.

Here, I suspect, is perhaps the major root of antiscientism among young people. To them, the questions of who and what they are loom as the most vital questions in the world

—and if science can't answer them, as it assuredly can't, then science is worthless. A humanly understandable conclusion, perhaps, but a silly one. An automobile will get you from one place to another; a road map will tell you how to get there; do you then downgrade the car or the map because they can't tell you where you want to go?

Who we are, what we want, why we are alive are certainly among the most important questions there are (though we would do well to remember that for hundreds of millions the really pressing question is not "Why am I alive?" but "Will I *be* alive next week, or next year?"). But having answered these questions, at least provisionally, by whatever nonscientific means we choose, we then confront the problem of how to *be* what we are, how to *get* what we want. At this point, if we are not fools, we will turn to science for such answers as it can give us. One truth which writing this book has brought home to me is that people who try to get what they want—an alteration of climate, for instance—by nonscientific, or insufficiently scientific, means generally end up with something they don't want. (This might be noted with special profit by some young "existential" radicals who are seeking to revolutionize society, not by trying to understand it rationally, but simply by feeling like revolutionaries.)

The theme of this book—man's social interaction with his climatic environment—is, I think, interesting and important in itself; I should not otherwise have bothered to write about it. But counterpointing this theme is another even more important: what science is all about. For science, despite its limitations and the limited capacities of its devotees, remains our most useful tool for understanding both our environment and our society—and changing them.

When I set about doing research for this book, I was appalled to discover that there was no single scientific work on the subject, and precious few that dealt with it even peripherally. I have accordingly been forced to piece together information from literally dozens of books, scientific articles, and interviews. To set down the complete list would be a tedious task—even if I could remember them all. Accordingly, I have limited the bibliography at the end of this book to the more useful sources. Courtesy and gratitude, however, bid me list the scientists who have been generous with their time and thought. I have intimated in the text my special debt to Reid Bryson of the University of Wisconsin and Rhodes Fairbridge of Columbia University; other invaluable help has come from Edwin S. Deevey and Elwyn Simons of Yale University, David Ericson and William Donn of the Lamont Geophysical Laboratory, J. Murray Mitchell of the U.S. Environmental Sciences Service Administration, Cesare Emiliani of the University of Miami, Richard S. McNeish of the University of Alberta at Calgary, and Robert L. Braidwood of the University of Chicago. I need hardly say that none of them shares responsibility for my skeptical opinions on science or scientists, or for any of the other opinions expressed herein (apart, of course, from those explicitly ascribed to them), or for any of the book's other deficiencies—which will, I hope, inspire one or more of them to produce a more professionally authoritative work on the subject.

Whatever the errors and omissions of this work, however, I would like to assure my readers, scientists and laymen alike, that none of the facts, however seemingly far-fetched, have been made up—at least not by me. For every factual statement, there is at least one, usually more than one, "reputable" scientific source. Where the facts are controversial, as they often are, I have tried to so indicate, though considerations of readability have inhibited me from inserting such hedges as "may," "probably," and "thought to be" quite as often as the professional academic might deem appropriate.

Truro and New York, 1969

PART ONE

Climate Past and Present

1. TO BEGIN WITH

On Some Books I Would Like to Have Written

There is something seductive about a book that purports to give one the inside track on history. Reading about the fumblings and blunders of past generations, it is very pleasant to be told that it all *really* happened because. . . . To be assured that we understand our ancestors a great deal better than they did themselves provides us with that sense of superiority which is one of the most exhilarating—and insidious—drugs known to science. And if an author's explanation of what happened in history is so resoundingly simple, so transcendently obvious that any fool *ought* to have seen it at the time—so much the better.

For this reason (among, no doubt, many others) there has seldom been a shortage of theories which explain human history in terms of some quick, tidy generalization. When Karl Marx and his sidekick Frederick Engels set about writing what was to become the most famous pamphlet in the world, they began with the declaration: "The history of mankind is the history of class struggles." An older contemporary of Marx, Thomas Carlyle, on the other hand, saw history as shaped, not by classes, but by eminent individuals. Still other writers have explained history in terms of original sin, the rise of Christianity, the decline of the West, the spread of the "Aryan" race (or any other race of which the writer happens to be a member), even the spread of syphilis.

In recent years, the business of interpreting history has become a game at which any number can play. With historical change proceeding at a headlong and often terrifying pace, it is more than pleasant to be told that we really understand what's going on; it is deeply comforting as well. Certainty, even if it is merely the morose conviction that things will get worse before they get better, is a great deal easier to bear than uncertainty.

Thus an American playwright, Robert Ardrey, not long ago produced a lucid and dramatic account of how human history, from Australopithecus to the atom bomb, has been determined by the homicidal and territorial instincts which man allegedly acquired during his African genesis. In sharp contrast is the Canadian "communications specialist" Marshall McLuhan, who, in some of the most opaque prose since hieroglyphics went out, has explained the whole business in terms of media. And while the meaning of this fashionable term seems to shift from chapter to chapter of his interminable works, it appears at any rate to have nothing much to do with instincts of any sort.

Contemplating the escalating sales figures and resplendent public images of these and other purveyors of instant history, I wish profoundly that I could have written a book of that sort. Unfortunately, however, I have an ingrained suspicion of easy answers, in history or any other area. In my time, I have ingested a lot of simple explanations of my fellow men and of myself, and ended up with intellectual indigestion. Thus I feel about instant history much as I do about instant coffee: a thing to be swallowed under protest.

Class struggles have certainly played a major role in history. So has human cussedness, sometimes called original sin, along with human creativity, sometimes called the Divine Spark. So have great men, races, religions, and, I dare say, syphilis as well. And so, without a doubt, has climate, whose influence on the human drama is the subject of this book. But it is not *the* explanation of history, past or present, any more than are the other historical influences I have mentioned. The human drama is as complex and subtle as anything written by Shakespeare or Euripides—and I don't propose to reduce it to the level of a TV skit.

So if what you expected to find in this book was a sort of pipeline to Providence, a tidy explanation of where man has come from and where he is going, I must advise you, with regret, to return it to your friendly neighborhood bookseller and hope he will allow you full credit. What I offer here is not an explanation but a puzzle, for the story of how climate has helped shape human destinies over the centuries and millennia is just that: a complicated and often exasperating jigsaw. It is composed of many oddly shaped pieces—fossil relics of extinct plants and animals, grains of prehistoric pollen dug from Swedish bogs, bits of charcoal from Neanderthal caves, painstaking measurements of the width of tree rings in Arizona, and tabulations of the number of occasions on which the churches of medieval Barcelona offered prayers for rain. It is also a puzzle in which many of the key pieces are still missing, and in which many of the pieces we have seem to fit the others as well (or as badly) when turned upside down. There is plenty of evidence on what the climate was a thousand or even a million years ago, but there are plenty of gaps in the evidence too.

At the moment, there is only one way to fill those gaps: by guesswork. So I have guessed—and with a shamelessness that would give a professional historian or climatologist the green fits. I have picked and chosen among the evidence, speculated about things that the experts can't agree on, and in general conducted myself like the proverbial fool who rushes in. What has emerged is not always the historian's history or the climatologist's climatology, but it has been fun to write and will, I think, be fun to read.

But there is a bit more than fun involved. Climate is a puzzle—and I am admittedly hooked on puzzles. But if the puzzle is worth tackling, surely that is partly because it may help us toward an understanding of an even bigger puzzle: of what man is and how he got that way—in fact, of history.

Edward Gibbon, contemplating the deplorable annals of

the later Roman Empire, could describe history as "little more than the register of the crimes, follies and misfortunes of mankind." But human society is not always in the midst of a decline and fall. There is heroism in history along with crime, ingenuity mixed with folly, and progress mingled with misfortune; there are comedy and tragedy so intertwined that one knows not whether to guffaw hysterically or howl in pity and terror.

Beyond this, history—which is, at any rate, a record of human predicaments—seems to me worth understanding because it may, just possibly, give us some clues as to how to cope with our present predicaments. Several historians (Santayana and Hegel were two of them) have observed that those who do not understand history are doomed to repeat it. And there is a great deal of history, both recent and ancient, that I would just as soon not see repeated in my lifetime. A knowledge of history does not, God knows, supply any royal road to political sagacity, but an ignorance of history is surely a one-way street to political fatuity. Anyone who is curious about how a great democracy can destroy itself by succumbing to the arrogance of power could do a lot worse than read Thucydides on the thirty-years war between Athens and Sparta.

This view, that history is worth bothering about, admittedly runs counter to a very widespread contemporary philosophy: that nothing is worth bothering about. Play it cool, don't get involved, eat, drink and blow your mind, for tomorrow we die. Painters lovingly depict Brillo boxes, as a way of saying that art is bunk; composers produce "music" in which minutes of silence alternate with the blatting of transistor radios, thereby establishing that music is bunk. And there are not a few hip individuals who unwittingly echo the dictum of that arch-square Henry Ford that history is bunk.

Well, if history, the story of the human race, is bunk and not worth bothering with, then human beings are not worth bothering with. As a human being, I find this view obnoxious. The human race is pretty unpleasant company at times, but I know of no way, short of suicide, to get away from it. The games people have been playing since before the dawn of history are often trivial, ludicrous, or vicious—but they are the only games in town.

And climate, as we shall see in a moment, is very much involved in the human game. It sets some of the rules and also helps to shape the field on which the game is played. Let's see, then, if we can make some sense out of the puzzle of climate—and the great game of which it is a part.

2. WHAT IS CLIMATE?

And What Does It Do?

Everyone knows that climate has something to do with weather. And, unlike some things that "everyone knows," this happens to be true. But though climate indeed is akin to weather, it is not quite the same thing.

When I read in my morning paper that temperatures in New York yesterday ranged from a high of 80 degrees to a low of 68, with a mean of 74, and that .34 inches of rain fell, that's weather. When I read, a few lines farther down, that the *normal* mean for yesterday's date is 78 degrees, and that the normal rainfall for the month of July is 3.60 inches, that's climate. Weather is what happened in the atmosphere yesterday, or is happening today, or will happen next week; climate is what has happened and can be expected to happen over the reasonably long run—weather averaged out over fifty or a hundred years. It is the seasonal pattern of heat and cold, of sun, wind and rain, that is—or was—characteristic of a particular place or region.

Differences in these patterns from place to place, and changes in them from time to time, have inevitably exerted a powerful influence on human affairs. They define regions in which man must exert special efforts to avoid freezing or roasting to death. They delimit the sultry, moist areas in which he must coexist with malarial mosquitoes and skin-rotting fungi and the blazing deserts in which he must organize his life around wells and water holes. Climate, along with soil (which is itself heavily influenced by climate), determines what plants can flourish and thus what animals—including man—can survive by eating the plants, or one another. For primitive man (which, up to about ten thousand years ago, included the entire human race), climate set down the list, lengthy or sparse, of herbs, roots, and berries he could gather and animals he could trap or spear. For more advanced cultures, including our own, it still has a weighty voice on what crops can be harvested, which herds and flocks can roam the hills.

Climate has helped determine where man could go on earth. The waning and waxing of ice sheets and deserts has opened or barred the way to the migrations of peoples and ideas. When sailing ships fared forth across the waves, climate, in the shape of prevailing winds, had not a little to do with where they could get to and how quickly, with whether they might have to grope through fog or face the terrors of typhoon or hurricane. Improvements in the climate of Scandinavia helped launch the Vikings west across the Atlantic and south at the throats of Europe; a later climatic change almost literally cooled off their ardor for exploration and conquest.

Climate, and the need for tools to cope with it, has played no small part in the advance of human culture. Fire, the first natural force to be tamed by man, may or may not have found its first application in warding off the cold, but it assuredly became man's first means of manipulating climate, thereby permitting him to spread from the tropics and subtropics into areas where he could not otherwise have survived. Irrigation, man's prime tool for coping with inadequate rainfall, played a leading role in the beginnings of civilization. Today, man still cannot manipulate climate in the large, or at least not deliberately (there is much evidence that he has often done so by accident), but his ability to control the local climate is impressive. With dams and canals he has made the desert to blossom as the rose and, with the aid of air conditioners or heaters, he can survive in comfort amid the Saharan sands or the Antarctic ice.

But, though man's tools for coping with climate have improved along with his tools for coping with other problems, climate is still there to be coped with. Of the millions of tons of coal and oil burned annually in the United States, something like half is used simply for "space heating"—that is, to maintain a tolerable climate inside our homes, offices, and factories. Today's mightiest liners, braving the tempestuous winter climate of the North Atlantic, must, by international law, load less cargo than ships in other seas or seasons. Fog, despite the wonders of radar, can still send ships crashing into each other and draw its clammy, crippling hand across our great airports. More than any other natural force, it is climate that confronts man with his most basic problems—what to eat and drink and how to shelter himself from the elements—even as it supplies him with many of the natural resources he can mobilize to solve these problems.

Climatology, in this sense, is one component—and the most important one—of geography, the study of the planet which for the past million or so years has provided man with the stage for his activities. In thus helping to set the stage for the human drama, it inevitably places important limitations on what sort of drama can be played out at a given time or place. But note well that it does not write the script. Only man does that. Climate may determine what *can* happen, but it does not determine what will happen. Though it poses basic problems which man must solve, how he solves them (or whether he solves them at all) is up to him.

3. CLIMATE PRESENT I

How It Works

For the climatologist, as for all other scientists who study the earth, the present is the key to the past; to understand what happened *then,* they must first understand what is happening *now.* Once they know that the features of today's earth—including those of the atmosphere that beats ceaselessly upon it—are shaped by certain natural processes, they can assume fairly safely that past features, however different these may have been, must have been governed by some rearrangement of the same processes.

The present, after all, can be observed and measured; the past cannot. Statements about what the climate in a particular region may have been a million, or even a thousand, years ago are for the most part conjectures. Some are very probable conjectures indeed, some are merely plausible, but all are based on evidence which is often fragmentary, ambiguous, and at times even contradictory. When we talk about climate today, on the other hand, we stand on a solid foundation of painstaking, worldwide observation extending over more than a century. We need not conjecture; we *know.* The climatologist can stick a pin in the globe at random and, after consulting a few tables, reel off the approximate average temperature and rainfall at that spot in any month of the year. Moreover, he can give a rational explanation of his climatic statistics, joining the innumerable varieties of climate around the world into a reasonably orderly pattern and showing how a desert in North Africa, a rain forest in Brazil, and a glacier in Greenland are all merely different aspects of one grand atmospheric design.

Just how many varieties of climate are there on earth? In one sense, the number is almost infinite. Just as no two people on earth have the same fingerprints, so it can be argued that no two places on earth have precisely the same seasonal pattern of sun and rain, heat and cold. But just as fingerprints can be classified into types, so can climates. One of the best-known systems of classification contains no less than thirty-two basic types—most of which can be further broken down into subtypes. Our own approach is going to be much cruder than that: only twelve kinds of climate, all of which represent variations of only four basic types. Each of these ten is distributed fairly widely around the earth; there is no point in getting detailed about the climates of particular regions until later, when we begin talking about man's activities in those regions.

The most obvious thing about the world's climates is that some places on earth are hotter or colder than others. This obvious fact, in turn, stems from the most obvious fact about the earth itself: its shape. The fact that the earth is round means that some parts of it receive much more heat from the sun than others. At the equator, the noonday sun beats down from almost directly overhead at every season; at the poles, even at midsummer, the sun, though it never sets, hangs low and feeble on the horizon, and for half the year it does not rise at all. Result: the poles receive less than a quarter as much solar heat as the equator. The actual discrepancy is even worse, for since the polar regions are largely covered with permanent snow and ice, the greater part of the incoming radiation is immediately bounced back into space from the brilliantly white surface.

In the first place, then, climate is determined by distance north or south of the equator—by latitude, in fact. We owe this observation to the ancient Greeks, who seem to have been the first people to speculate about climate, as they were the first to speculate about so many other things. Greek merchantmen, sailing north to the Black Sea to barter wine, oil, and pottery for grain, noted that the weather there was noticeably colder than in their home ports around the Aegean. Sailing south to Crete, Egypt, and Libya, they found the weather warmer. Accordingly, they used their word for latitude, *klima,* to mean climate as well.

The actual relationship between latitude and temperature is, as we shall see in the next chapter, a little more complicated than this. The hottest places on earth are not at the equator, nor the coldest at the poles. Nonetheless, the principle that high latitudes are cold and low latitudes are hot is a good rule of thumb for the moment.

When it comes to rainfall, however, there is unfortunately no neat rule of thumb to help us. To be sure, the poles are quite dry and the equatorial regions generally very wet—but in between we find every conceivable variant, from extremely dry to extremely wet. To discover why this is so we must begin by asking why rain falls at all. The basic reason is quite simple (it is, in fact, almost the only simple thing about rainfall): *rain—or snow—falls from air that has been cooled.* To spell it out a bit, when air grows cooler, it grows moister (technically, its relative humidity increases). If it cools enough, the water vapor in it condenses into clouds of water droplets (when you breathe out on a frosty morning, your warm, moist breath, chilled by contact with the outside air, forms tiny clouds). If the cooling process goes far enough, rain or snow will fall from the clouds. Air that is growing warmer, on the other hand, is growing dryer, and produces neither rain nor snow.

To some extent, air is cooled by contact with other, cooler air (as in the breath-on-a-frosty-morning illustration) or with cold land or sea. For the most part, however, the cooling in the earth's atmosphere which produces precipitation occurs because the air, for one reason or another, is moving upward, at a particular place. The essential fact is that *upward moving air grows cooler and moister; downward moving air grows warmer and dryer.* The more upward motion in the air over a particular region, the wetter the climate; the more downward motion, the dryer.

To understand why air moves upward over certain parts of the earth and downward over others we must take a look at something with the rather forbidding name of The General Circulation of the Atmosphere. Considered in detail, the General Circulation is quite as formidable as its name suggests; indeed, even the professional meteorologists, with their weather satellites and superfast computers, do not wholly understand it. In its basic features, however, the General Circulation is hardly more complicated than an old-fashioned steam engine. In fact, it embodies essentially the same working parts as such an engine and operates in much the same way.

A steam engine—such as those which powered the great steam locomotives that are now, regrettably, almost extinct—consists of only four basic parts. First is the firebox, which is simply a source of heat. Second comes the boiler, in which heat is transferred to the engine's "working fluid" (in this case, by turning water into steam and further heating the steam itself). Third comes the cylinder-piston combination in which the working fluid, expanding and cooling, performs work. Fourth and finally we have the condenser, in which the fluid is further cooled and from which it is returned to the boiler to begin its cycle once more.

In the earth's atmospheric engine, the "firebox" is, of course, the sun. This mighty nuclear furnace, though it lies some ninety-three million miles away, still pours some twenty-three trillion horsepower's worth of heat daily onto the earth and its atmosphere—day in, day out. Some of this prodigious heat input is immediately reflected back into space, but most of it remains to keep the engine turning over.

The "boiler" of our engine is the earth's tropical and subtropical areas, where more heat comes in than can be radiated off into space. Most of the surplus goes to heating the atmosphere, and to evaporating water from the oceans and from moist land areas; the resulting water vapor, plus the atmospheric gasses themselves, make up our "working fluid."

The "condenser," on the other hand, is roughly equivalent to the polar and subpolar regions, in which the income of solar heat is much less than the outgo of radiation, so that the atmosphere is chilled and its water vapor reconverted into water (in the form of rain or snow). The movement of the atmosphere from equatorial boiler to polar condenser and back again performs work, precisely as does the moving steam in a steam engine.

What kind of work? Well, the atmosphere itself weighs something over five and a half million *billion* tons—several hundred times the weight of all the cargo moved by all the locomotives, ships, trucks, and planes in the world over an entire year. And nearly all this enormous mass is constantly in motion—parts of it at the walking pace of a gentle breeze, parts at the furious 200-plus knots of a jet stream. In addition to moving its own ponderous weight, the atmosphere, as just noted, is continually picking up water by the trillion tons from land and sea, hoisting it some thousands of feet into the air, and dropping it hundreds or thousands of miles away. Clearly, our atmospheric engine is no puny apparatus. Its work output, in fact, is the General Circulation itself, whose ceaseless shifting of air and water brings different patterns of wind or rain, clouds or sunshine, hot air or cold, to different parts of the earth.

Precisely how and where the atmosphere moves as it circulates is a bit more complicated, but not too hard to understand if we keep in mind a few simple, household scientific principles. The first of these is the same principle on which the old-time hot-air furnace operated: hot air rises, cold air sinks.

Given this single principle, we might expect the atmosphere to behave like a sort of enormous conveyor belt in which air, heated at the equator, would rise high above the earth and move toward the poles. Cooled there, it would sink to the earth's surface and return to the equator again. Unfortunately, this beautifully simple picture is considerably blurred by the fact that the earth revolves.

The atmosphere does indeed move in a north and south direction away from the equator and south and north away from the poles. But while it is doing so, the earth is, of course, spinning from west to east. And some of this spin is transferred to the moving atmosphere. The precise results of this are fairly complicated and need not concern us at this point. The important thing at the moment is that our simple, single conveyor belt is in effect transmuted into three (three in each hemisphere, of course). And it is the actions and interactions of these three belts that create what I have chosen to call the four basic types of climate.

For simplicity, let us trace the course of the belts in the Northern Hemisphere—bearing in mind that three other belts are operating in more or less the same way south of the Equator.

Belt One sweeps toward the Equator along the surface, in the form of the trade winds. In the equatorial regions, its air is heated, rises aloft, and turns north in an upper-atmosphere current sometimes called the antitrades. In the general neighborhood of the Tropic of Cancer (latitude 23½ degrees north), it descends to earth again and turns south to complete its circuit.

Belt Two descends toward the surface around the Tropic of Cancer, just as does the first belt, though its air, originating to the north, is cooler. Some of this air it passes to Belt One, receiving in exchange some of that belt's warm, moist air. Belt Two then moves north along the surface to (very roughly) about 45 degrees north, where it rises aloft and turns southward again.

Belt Three, carries cold, dry air southward from the North pole along the surface until it meets Belt Two. There, after a certain amount of mixing between the air of the two belts, Belt Three rises aloft to turn back toward the pole, where it descends again.

Consider, now, the regions at the *ends* of the belts. Two of these—around the Tropic of Cancer and 45 degrees latitude, respectively—are regions in which different kinds of air meet and interact. And in all of them a great deal of air is moving, not merely along the surface, but also upward or downward.

Upward-moving air grows cooler and moister, remember. Along the Equator, then, the rising air of Belt One (and, of course, its counterpart in the Southern Hemisphere) creates the first of our basic climates: the Equatorial, which we find in the Amazon Valley, central Africa, the islands of Indonesia, and a few other places. There, the cooling of the upward-moving air condenses its moisture into copious clouds and equally copious rain. Things are always damp. Moreover, since these areas are located on or near the equator, they are always hot as well. Not blisteringly so, for the heavy cloud cover blocks out a good deal of the sun's heat, but hot enough so that, given the perpetual high humidity, they are considerably less than comfortable for those unaccustomed to this climate—and often enough even for those who are. In this perpetually damp hothouse, vegetation proliferates furiously into the tropical rain forests of *Green Mansions, The Heart of Darkness,* and *Tarzan of the Apes.* Animal life (for reasons to be explained in Chapter 17) is mostly less prolific; an outstanding exception is the insects, whose ceaseless multiplication helps give the Equatorial climate its nickname of Green Hell.

Moving north to the edge of the tropics, where Belt One meets Belt Two we find a very different situation. Here the air is not rising but falling, and thus growing dryer, not moister. By the time it has descended to the surface, it is dry enough to create our second basic type of climate: the Desert, found in parts of northern Mexico and our own Southwest, and (especially) the great desert belt that stretches across North Africa, Arabia, and on into Iran and Pakistan; counterparts in the Southern Hemisphere cover coastal regions in Peru and Chile, parts of South Africa and a large chunk of Australia. These are the lands of *Death Valley Days, Lawrence of Arabia,* and innumerable films about the French Foreign Legion.

The Desert lands are, of course, dry; one town in Chile recorded 0.02 inches of rain in forty-three *years!* (For comparison, a routine Equatorial downpour can drop this much rain in minutes.) With almost no clouds to block off the sun, they are also searingly hot; daytime temperatures can top 130 degrees in the shade. And there is no natural shade, since there are no trees, or indeed much vegetation of any kind. Such plants and animals as exist there do so only by means of special adaptations for conserving water. A harsh and barren land, in which few people live by choice.

Moving north again out of these insalubrious regions, we come to the meeting place of Belts Two and Three. This region is sometimes called the Temperate Zone—unquestionably one of the major misnomers of history. The dictionary defines "temperate" as "characterized by moderation," but weather in many of these regions is in fact the most immoderate, violent, and capricious in the world—for which reason I prefer to call it the Stormy Zone. Since it is the climate that most of us know first hand, there is no need to say much about it here. Located well north (or south) of the equator, its temperature varies markedly with the seasons. But there is normally a reasonable amount of rain or snow at all seasons, thanks to the interaction of the two belts. Not only does much of their air rise and grow cooler, but the further cooling of the warm, Belt-Two air by chill Belt-Three air produces additional clouds—and more precipitation.

Finally, in the extreme north, where Belt Three descends to earth, we have the Polar Zone—always cold, and, because of the downward movement of air, always quite dry. The dryness may seem unbelievable, since much of the polar region is covered with snow and parts of it with a mile or two of ice. But these great accumulations of frozen water exist not because there is much precipitation but because the little that falls stays put. Because of the year-round cold, almost none of it evaporates or melts to run off. Even in the "warmer," ice-free parts of this zone, runoff is often sluggish;

the deeper layers of the soil remain permanently frozen even in summer ("permafrost"), making for poor drainage.

These facts, by the way, illustrate a point worth bearing in mind, because we shall come back to it repeatedly later on. In most climatic regions, climate depends not merely on the amount of precipitation but on what happens to it once it has fallen—that is, on how quickly it evaporates, runs off, or sinks deep into the earth. The annual precipitation that maintains the mile-thick icecaps of Greenland and Antarctica or the boggy Canadian "muskeg" would, in Brazil or Nigeria, evaporate so quickly as to produce a very fair desert.

I must now regretfully confess that much of what I have written thus far in this chapter is a species of fiction. The three "conveyor belts," in particular, bear rather less relation to the actual complexities of the General Circulation than a child's stick drawing bears to a human being. They are convenient fictions, in that they provide us with a neat, simplified picture of what the atmosphere is doing over fairly broad regions—but fictions nonetheless.

Even more fictitious is the implication that the ends of the belts, and the climatic zones they define, are neatly distributed around the earth along parallel lines of latitude. Much of the Equatorial Zone does, indeed, lie close to the equator, but not all Equatorial lands have Equatorial climates, nor do all areas with this climate lie that close to the equator. The same goes for the Desert Zone and the others; in many places they lie well north or south of where they "should" be, shift around with the seasons, and in general engage in the sort of disorderly conduct that we have learned to expect of natural forces.

Of course there are reasons behind the disorder, some of which we shall go into in the next chapter. Meantime, however, bear in mind that if the conveyor belts I have so carefully depicted are at best a sort of semifiction, the basic climates definitely are not—as anyone who has ever visited the Congo, or Arizona (the name means "arid zone"), or Greenland, or, for that matter, the east-central United States can testify. Moreover, most other varieties of climate on earth—and there are plenty I have not yet mentioned—can be visualized pretty clearly as a variant, or a blend, of one or more of our four basic types: Equatorial (hot and wet); Desert (very hot and dry), Stormy (hot or cold, depending on the season, but fairly moist at all times), and Polar (cold and dry).

4. CLIMATE PRESENT II

Some Complications

The existence of both land and water areas on earth is the major complicating factor in the distribution of climatic patterns: notably, their failure to adhere to tidy belts along the parallels of latitude. This complication is responsible for the fact that London, with a fairly typical Stormy Zone climate, lies in the same latitude as Labrador, which is quasi-Polar; that New York, which spends millions of dollars each winter to clear snow from its streets, is in the latitude of subtropical Naples, where even frost, let alone snow, is almost unknown. It explains how sun-baked Cairo, with a typical Desert Zone climate—about one inch of rain a year—can sit almost astride the thirtieth parallel along with New Orleans, whose 60-odd inches a year give it a climate that at most seasons can be fairly described as steamy.

These peculiar effects of land and water derive mainly from two facts. The first of these is that water—for reasons having to do ultimately with the structure of the H_2O molecule—heats up and cools off much more slowly than land. (On a sunny July day, the sand on a New England beach may be hot enough to make walking uncomfortable, while the water is still cold enough to daunt all but the hardiest swimmers.) This fact is of rather minor importance in the tropics, where neither land nor water cools off very much, or at the poles, where they warm up very little. Not so in the in-between latitudes, where temperatures shift markedly from season to season. There the surface of the sea is notably cooler than the land in summer, warmer than the land in winter. Its *average,* year-round temperature, for a number of reasons, is generally higher than that of the land.

Now it happens that in these same regions, the winds blow predominantly from west to east. The result is some marked differences between climates on the western sides of the continents, with prevailing winds off the ocean, and more easterly regions, where prevailing winds come off the land. On the west, the relatively warm oceans help warm air to move considerably farther toward the poles in winter so that the Stormy Zone (which is, of course, the meeting place of warm, "tropical" air and cold, "polar" air) lies well north, or south, of where it otherwise would. The net result is that northwest Europe and the U.S.-Canadian northwest coast (including the Alaska panhandle) enjoy warmer climates than latitude alone would suggest, as do such regions as New Zealand and southern Chile. Moreover, their climates, though stormy, are more literally "temperate" so far as temperature is concerned. The maritime winds are tempered by the sea, making for relatively mild winters and cool summers.

Matters are quite different in the eastern continental regions. In winter, the chill land helps polar air to thrust itself—and along with it the Stormy Zone—closer to the equator. Thus the "continental" climate of these regions is cooler, on the average, and also more extreme than that of the maritime regions to the west. Continental Boston, though lying well to the *south* of Maritime Seattle, has a *lower* annual average temperature—with colder winters and hotter summers. (Chicago, with an even more continental location than Boston, has still colder winters and still hotter summers.) The same is true, of course, of places on the east and west of the great Eurasian land mass: maritime London *vs*. continental Vladivostok, for example.

The west-east contrast in climate is even more remarkable in the subtropics. In easterly regions, the southward shift of the Stormy Zone (plus other causes too complex for even this difficult section) in effect squeezes the Desert Zone out of existence. As a result, regions such as the southeastern parts of the United States and China are warm and moist rather than hot and dry.

The second great climatic complication produced by the land-water contrast derives from the fact that it is the sea, not the land, which is the ultimate source of every bit of precipitation on earth. To be sure, air passing over land can and does pick up considerable quantities of moisture from the soil and vegetation. But this process itself, plus the sinking of water into the subsoil and its running off through rivers, would eventually dry out the land surface completely were the supply of water not continually renewed by moist winds from the ocean. Since atmospheric moisture ultimately derives from the ocean, it follows that the farther air travels from the ocean, the dryer it is likely to be. This is especially true if it must pass over high mountain ranges on its way inland. Forced upward by its passage across the mountains, the air cools and drops much of its load of moisture, precisely as does the upward-moving air of the Equatorial Zone. Descending on the other side of the range, it grows warmer and therefore dryer, like the descending air over the Desert Zone.

Again the effects are most marked in the temperate latitudes. There, regions remote from the sea or lying in the "rain shadow" of mountain ranges have a far dryer climate than their nominal position within the Stormy Zone would dictate. Hence, such "temperate" deserts as those of Central Asia and the Great Basin region of the U.S.—parts of which are nearly as dry as the Sahara, though, since they lie well to the north of it, they are not as hot.

Unhappily, it must not be assumed that the various climatic zones are each confined to a neatly defined territory. There are innumerable borderline cases: the dry (but not desert) continental climate of our Great Plains, the cold (but not polar) climate of Siberia and Canada, and so on almost ad infinitum. There is no need to burden ourselves with an interminable list of all the climatic variants and subvariants. Two of them, however, are worth brief mention because they have played, at different periods, peculiarly important roles in the development of man and his culture. Man first evolved, as far as we can tell, in the Savanna Zone, and his first civilizations grew up in and around the Mediterranean Zone.

Both these climates are literally borderline climates. They exist because the climatic zones shift with the seasons, pressing toward the poles in summer and drawing back toward the equator in winter. As a result, certain regions, like disputed borderlands alternately controlled by two contending armies, come under the influence of different zones at different seasons. Thus between the Equatorial and the Desert zones we find the Savanna Zone—part of the rainy tropics in summer but an outpost of the desert in winter. The Savanna Zone covers large parts of South and Central America and of tropical Africa, where man first began his upward journey from his apelike forebears.

Between the Desert Zone and the Stormy Zone, we have an opposite situation. There, the Desert rules during the long hot summers, to give way to storms and rain or snow in winter. The Mediterranean Zone is found precisely where its name indicates, the region where the ancient Hebrews, Greeks and Romans laid the foundations of what we now call Western Civilization. It also occurs, in bits and pieces, in California, Chile, South Africa, and Australia.

To help you keep this proliferating list of climates straight, they are set down in a table, with three additions to round out the list.

CLIMATE		DESCRIPTION	MAIN AREAS*
Equatorial		Hot and wet all year	Amazon and Congo valleys; Indonesia; Central America
Savanna		Wet summer, dry winter; hot all year	N and S of Equatorial in S. Amer. and Africa; also India, SE Asia, Australia
Tropical Steppe		Like Savanna but less rain, especially in dry season	Between Savanna and Desert, also parts of Brazil and India
Tropical Desert		Very hot and dry all year	SW U.S., NW Mexico, N. Africa, Mid-East, Pakistan; SW Africa, Cen. Australia
Mediterranean		Hot, dry summer; mild, wet winter	Mediterranean coasts; Portugal; California; Cent. Chile
Stormy Zone	Maritime	Cool summer, mild winter, moist all year	NW Coast of U.S. and Canada; NW Europe; New Zealand and S Chile
Stormy Zone	Continental Moist	Hot summer, cold winter, moist all year	Cent. and NE U.S., SE Canada; Cent. Europe; W U.S.S.R., Cent. China, Japan
Stormy Zone	Continental Dry (Steppe)	Like Continental Moist, but less rain at all seasons	U.S. Great Plains and Basin, S U.S.S.R. NW China, Cent. Argentina
Stormy Zone	Continental Desert	Hot summer, cold winter, dry all year	SW U.S.; S U.S.S.R.; W China
Stormy Zone	Subtropical	Hotsummer, mild winter; moist all year	S U.S.; N India and Burma; S China
Subpolar		Very cold dry winter; cool moist summer	Alaska, Canada, Siberia
Polar		Very cold and dry all year	Greenland; coasts of Arctic Ocean; Antarctica

* A world map, showing the distribution of these climates in detail, will be found on the endpapers. Like all such maps, it is somewhat arbitrary.

5. CLIMATE PAST I

The Evidence and the Problems

Cape Cod is surely one of the pleasantest spots on earth to spend a summer. It has almost everything: the Atlantic surf for skillful swimmers and wave-riders; quieter, bay waters for the more timorous, and for fishermen who troll for bluefish or stripers; tiny, jewellike lakes whose rain-soft water leaves your hair like silk, and above all (to my taste) the solitude of its almost endless beaches and piney woods. Having spent most of my life in New York, I am pretty well attuned to the noise and bustle of the city—but preferably for only ten months a year. Our children, and their children, can be thankful that Congress, under pressure from the more far-seeing inhabitants, has set aside part of this paradise as national property, safe from the spreading blight of motels, pizza parlors, and tawdry souvenir shops.

Cape Cod's climate is superior too. Lying on the east coast of the United States, it of course has the continental climate appropriate to its latitude within the Stormy Zone. There is enough rainfall in an average year to keep its little ponds brimming with water, and set clusters of blueberries to ripening on the hills. Most summer days are hot enough to make swimming a pleasure and often almost a necessity. At the same time, the ocean waters that almost completely surround it give its climate a pleasantly maritime flavor. Its average July temperature is nearly five degrees cooler than that of mainland Boston, only forty-odd miles away as the gull flies—enough to make the difference between a hot summer and a stifling one.

Yet for people who like to roam the countryside on foot there are a number of features about the Cape which point to a climatic past very different from its idyllic present. For one thing, there is no sign of bed-rock. In most areas in this part of the U.S., there are few hills without a ridge or ledge of granite, schist, or gneiss poking up through the soil. In New York's Central Park, my kids climb them on weekends. Driving along any superhighway, you can spot where the bed-rock has been blasted away to let the road cut through the hill crests.

Not so the Cape. Along its eastern, Atlantic shore there are sea cliffs which in places stand over 150 feet high, but they are not the towering granite cliffs of rockbound Maine. Running your eye up the steep slope from beach to cliff top, you see only layer after layer of sand, clay, and gravel. Nor is this merely a surface illusion; the local well-digger, who is putting in a water supply for a new cottage going up down the road, tells me that even two hundred feet down his drills still bring up the same grainy collection of materials.

Not that there aren't rocks on Cape Cod. Wave-rounded pebbles are scattered along its beaches; cobbles and occasional boulders peer out of sandbanks and road cuts. But they are of a remarkably heterogeneous character. Wandering along the beach at low tide, one spots a lump of pink granite next to a bit of whitish quartz. Here is a brownish chunk of sandstone, there a piece of fine-grained gray basalt. A knobbly lump of conglomerate lies next to a smooth, flattened bit of slate—perfect for skipping across the water on a calm day. It is as if some Gargantuan rock-hound had cast a dragnet across thousands of square miles of territory, dredging up for his curiosity specimens of fifty or a hundred minerals.

In fact, this is approximately what happened. From these rocks, from the soil, from half a dozen other clues of geology and topography, scientists know that the "collector" was an enormous sheet of ice which, perhaps 50,000 years ago, drove south from Labrador. Formed by some major alteration of the earth's climate, this great glacier, a mile or more thick, ground forward, century by century, stripping away the soil of the plains, gouging grooves in the rocky ridges and wrenching away chunks of their substance.

Arriving at last at what is now the New England coast, it found itself unable to advance further, for the warmer climate of that region melted the ice as fast as it could push forward. Millennium after millennium the ever-moving ice maintained an uneasy equilibrium with the atmosphere, pushing forward a little in cold years, drawing backward when the weather was warmer, but always adding to its collection of rock and soil. At last, some 15,000 years ago, the forces that had powered the ice sheet waned and it began its final retreat to ultimate extinction. But it left behind the great mounds and ridges of debris which it had collected in the north. Cape Cod is one of these glacial rubbish heaps, shaped and reshaped by the winds, rains, and tides of the intervening centuries.

About 15,000 years ago, then, Cape Cod—though it was not a cape in those days—lay at the very edge of a continental icecap. Its climate, now so salubrious, must have been close to the present climate of the Greenland coast. As a summer resort, its appeal was to the musk-ox and the polar bear.

The climatic history of Cape Cod vividly illustrates one of the central themes of this book: climate, which is the averaged sum of the ceaseless changes of weather, can itself change. Indeed, there are few parts of the earth in which one cannot find evidence that the climate—a thousand, or ten thousand, or ten million years ago—was not different, sometimes radically different, from what it is today. In the remainder of this chapter, and the next, I shall introduce some of the evidence, and some of the problems that climatologists confront in trying to figure out what it means.

As one would expect, the type of evidence varies markedly

depending on how far back in the past one is looking. For the present century, of course, we have actual measurements of temperature, rainfall, and the like for nearly every part of the earth. But as we move backward from 1900, the records thin out until, by the late eighteenth century, they cover only a few places in western Europe. Before that, nothing. At that point, the paleoclimatologist, or specialist in past climates, must begin to work by inference. Lacking evidence of what the climate *was,* he must study what it *did,* and reason from effects to causes. Taking the present as the key to the past, he assumes—and can usually do so quite safely—that the effects of a particular climatic pattern on animals, vegetables, and minerals was the same centuries ago as it is today.

For a good sampling of the kind of evidence used in studying climates of what one might call the recent past, we can do no better than leaf through the minutes of a conference held in the summer of 1961 at Aspen, Colorado, a place whose cool and relatively dry summer climate makes it a favorable resort of scientists, and others, who wish to combine business with pleasure during the hotter months. The aim of this meeting was to bring together all the existing evidence on the climates of two specific periods—the sixteenth and the eleventh centuries—and see whether it would add up to any kind of detailed and consistent picture.

The evidence presented at the conference was later collated conveniently into two large charts, known informally as the "Aspen Diagrams." Running one's eye down the three-foot chart for the sixteenth century, one is struck by the diversity of the evidence. Here is a chart showing the dates on which the spring thaws opened the Baltic port of Riga to navigation; a late thaw, presumably, meant thick ice and, therefore, a severe winter. Lower down, one finds the dates of the French *vendange,* when the vineyards deliver up their ripe grapes for pressing into wine. A late vendange meant a cool summer, an early one, fine, warm weather.

Lower still, we have estimates of summer temperatures in England, culled from manorial records and similar sources, graphs of the price of grain in half a dozen countries, and so on. The overall impression is of an enormous amount of work by a great many people. Closer examination of the graphs, however, suggests that the value of the evidence is not always proportionate to the efforts expended in collecting it.

For example, the archives of Barcelona Cathedral list, year by year, the number of occasions on which public prayers were offered for rain. One would expect that in years when there was lots of praying, there must have been little rain, poor crops, and high prices. And indeed, when we check this graph against another giving the prices of grain in Barcelona, we find that years of much praying were indeed years of high prices more often than not. But there were plenty of exceptions. In 1561, for example, prayers were offered up on more than one hundred occasions, setting a record that was not beaten for more than twenty years—yet grain sold at a price nearly 10 percent below average. That time, presumably, the prayers worked and the rains came after all. (Though one must note, regretfully and with all due respect to the Catalan clergy, that their rogations were fruitless most of the time.) Or perhaps 1561 was a bad year for Catalonia but an unusually good one in nearby regions of Spain, so that grain could be shipped in and the price held down.

The best we can say here is that the years when the prayers and the prices agreed were almost certainly dry years; but where they do not agree, the weather is anyone's guess.

Or consider the climate far across the globe, in sixteenth-century Japan. It happens that Japanese priests, for religious reasons, have for centuries kept careful records of two peripherally climatic events: the date on which Lake Suwa froze over in winter and the date on which the cherry trees bloomed in spring. Clearly, in years when the lake froze late, the winter must have been mild—*up to that point*. But was the rest of the winter equally mild? We have no way of knowing. Judging from the cherry-blossom dates, it often wasn't. In 1547, for instance, the lake froze no less than six weeks behind schedule, but the blossoms the following Spring appeared no earlier than usual. And of course neither the lake nor the flowers tell us anything about an equally important aspect of Japanese climate: were the summers cool or hot, wet or dry?

The problem with these and similar documentary sources is that most men, at most historical periods, have been incurably parochial. The things they bothered to write down bore strictly on their own concerns. The Japanese priests were interested in cherry blossoms; the Barcelona clerks were interested in prayers, or in the price of grain. None of them was in the least concerned with weather, or climate, as such. To be sure, some documents seem to be more consistently reliable, and meaningful, than others. Parchment manor rolls from England of both the sixteenth and the eleventh centuries give what seems to be a pretty reliable picture of the climate, at least in the sense that a particular winter was warmer or colder, a particular summer wetter or dryer than usual. But even here we must reckon with the problem of what seemed "usual" to a manor clerk of the eleventh century.

Small wonder, then, that climatologists, even in dealing with the fairly recent past, turn with some relief from the subjective and fragmentary records left by man to those left by nature. Nature, to be sure, sometimes speaks in parables that require careful interpretation, but her records are at any rate not subject to the fallibilities of human memory or the parochialism of human interests. They are a record of what happened, not what somebody thought happened.

Glaciers provide useful records. To be sure, the great continental ice sheets, of the sort that heaped up the substance of Cape Cod, have been gone for thousands of years. But small glaciers still exist in mountainous areas, such as Switzerland and the northern Rockies. And in their own modest way, they carry out the same business of collecting soil and rock and piling it into the heaps and ridges that the geologist calls moraines.

If a geologist, exploring a glacial valley, spots a moraine a mile or two "downstream" from the glacier's present face, he knows without question that the glacier has retreated and that the climate of that area has therefore changed. Perhaps the area has grown dryer, meaning that less snow has fallen to nourish the glacier. Perhaps the summers have grown warmer, meaning that the face of the glacier has melted back faster than the accumulating snows of winter could push it forward. Without other evidence he cannot tell which change was responsible (both may have been). But change, of some sort, has certainly occurred.

If our geologist finds a grove of spruce growing atop the moraine, the largest of them perhaps two hundred years old, even better. These trees could not have sprung up immediately in front of a glacier face; therefore the retreat of the glacier, and the climatic change that caused it, must have begun more than two centuries ago. From evidence of this sort, we have learned, for example, that in the last decades of the sixteenth century the glaciers of the Alps advanced well beyond their present positions.

Trees can do much more than set a term to glacial advances and retreats; they can provide climatic information in their own right, and in remarkable detail. As every home carpenter knows, the butt end of a board or beam shows a pattern of curved lines. These are, of course, segments of the annual "growth rings" of the tree from which the board was cut. All trees growing in climates with distinct seasons lay down these rings, year by year, and the width of each ring shows how favorable, or unfavorable, that year was for the tree's growth.

In most areas, the chief climatic "growth factor" is rainfall. Not long ago, at the Connecticut home of a friend, I stumbled across the stump of a white oak that had been cut a few weeks before. Just inside the bark, I could trace four or five narrow rings, formed during the drought that had gripped that part of the country for several summers. Working inward, I came to broader rings showing moister years; had I wished, I could have worked out the "growing weather" in that area for fifty or sixty years back.

In Subpolar climates, such as that of central Alaska, the governing factor is not moisture (which, thanks to low temperatures and consequent low evaporation, is seldom in short supply) but temperature—specifically, the length of the summer growing season. The longer the summer, the thicker the rings.

Any stump, or beam from an old house or windfallen log will yield information of this sort, and that information (by methods to be described in Chapter 7) can often be dated to the year. By piecing together such records we can say, for example, that in Lapland, at the top of the Scandinavian peninsula, summers during the sixteenth century were much like they are now, except toward the end, when they grew considerably cooler. And this, taken with other evidence, indicates that narrower tree rings in Lapland and advancing glaciers in Switzerland both reflected a trend to generally cooler climates in Europe. Equally, we can tell that in eleventh-century Alaska the summers were markedly warmer than they are now.

The trouble with these and other natural records of climate is that you have to take them where you find them—and where you find them is, all too often, not where the action is. To an anthropologist studying the migrations of the Eskimos, the climate of eleventh-century Alaska is a useful and indeed almost essential piece of information; the rest of us would be far more interested in the climate of western Europe, where the Normans were preparing their invasion of Saxon England, or of Spain, where the Moslem civilization was entering its brief summer of literary and scientific flowering. Details of the climate in sixteenth-century Lapland are a good deal less enthralling than the same details on

Elizabeth I's England or Michaelangelo's Italy would be.

Where the records are missing, the climatologist must begin a complicated game of drawing inferences from regions on which he happens to have information to more "interesting" regions where he does not—a rather tricky business at best. Moreover, inferences which serve us well under one set of conditions will, under other circumstances, yield conclusions which, when we can check them, turn out to be the opposite of the truth. Consider, for example, the climates of Alaska and Lapland, both of them subpolar. It happens that we can make a direct comparison between these two regions during the sixteenth century, since the Aspen Diagrams give us tree-ring records from both places. It turns out that a warm summer in Alaska was likely (though by no means certain) to be a chilly one in Lapland.

This seems reasonable enough when you think about it. In Subpolar climates, the summer temperature depends largely on how much frigid Arctic air pushes south into the area. And since, other things being equal, there seems to be pretty much the same quantity of such air from one year to another, if the air isn't going to Alaska in a particular year it must be going somewhere else—to Lapland, for instance. Except that other things aren't always equal. The Aspen Diagrams for the eleventh century, plus other evidence not included in them, show pretty conclusively that in that era summers were warmer than the present in both Alaska *and* Lapland—and in Greenland, northwestern Europe and, so far as we can tell, all over the northern part of the Northern Hemisphere. Unlike the sixteenth century, where a warmer year in one place seems to have been balanced by a colder year somewhere else, the eleventh century was warmer overall. Some other factor must have been operating—and the climatologists still don't know what it was.

The farther we try to swing our chain of inference, from "known" to "unknown" regions, the more tangled it is likely to get. One would expect that the eleventh century's marked climatic changes in northerly regions would have been accompanied by equally marked changes to the south. They were, but the changes were not always the expectable ones.

The simplest way of visualizing these eleventh-century changes is to imagine a contraction of the Polar Zone—for unknown reasons. With a smaller Polar Zone, the Subpolar Zone moved northward to fill the gap, and behind it—one would suppose—the Stormy, Mediterranean, and Desert Zones also pressed northward. This, at any rate, is the picture that climatologists visualized for a long time. Up to a point, it seems to be a true picture. Parts of Polar southern Greenland, whose permanently frozen soil cannot support the growth of trees, were then partly forested, and thus must be called Subpolar. Many areas of western Europe, now in the heart of the Stormy Zone, were somewhat dryer then, with a rather more Mediterranean climate.

But when we look farther south, we are in trouble. If western Europe was dryer and more Mediterranean than it now is, surely the Mediterranean lands must have been dryer still. The Desert Zone, which now covers them only during the summer, must, one would think, have dominated them during most of the other seasons as well. In fact, however, such evidence as we have—and it is fairly reliable—indicates that the Mediterranean regions were not dryer than today, but wetter—and wetter in the summer at that.

Some meteorologists explain this by suggesting that since the Desert Zone evidently did not move north during the eleventh century, while the zones to the north of it did, a sort of gap was left, which was filled by a secondary Stormy Zone. This would have brought summer rains to the region which today are nonexistent. What they are saying is that the General Circulation, whose portrait we drew so laboriously in Chapter 3, must have been markedly different from what it is now. This is a rather shocking suggestion, and perhaps it isn't true. All we can say at this point is that drawing inferences about climate is at least as risky as drawing inferences about anything else, and probably riskier.

And all these problems, let me stress, concern the recent

past—a period in which climatic changes were relatively minor and for which we have fairly copious evidence. Just wait! It gets worse the farther back in time you go; the evidence is sparser, the changes are greater, and the dangers of drawing inferences begin to resemble going over Niagara Falls in a barrel. But I, for one, am willing to live dangerously. To the paleoclimatologist, the conflicts and confusions in the evidence of past climates are a challenge, and a headache. To non-scientists, they are an opportunity: when the experts disagree—and in climatology they do so constantly—the nonexpert comes into his own. He can speculate to his heart's content, secure in the knowledge that somewhere there is almost certainly some expert who will agree with him!

6. CLIMATE PAST II

The More Distant View

Because I am so fond of the little ponds that dot Cape Cod, I am mildly concerned that they are contracting. Not quickly—they will last my time, and my children's. But slowly the pond grass, water lilies, and duckweed are moving in from the edges, each year's crop building up the bottom a little more. Land plants, too, make their contribution. On any breezy day, but especially in the fall, the pond receives a rain of leaves from bayberry, oak and black locust, needles and an occasional cone from the pitch pines. Much of this rain of vegetable matter decays, or is consumed by the tiny water animalcules whose myriads nourish the local population of sunfish and bass. But much of it remains. In a few generations, many of these ponds will have become marshes, whose stagnant acid water will slow decay to a crawl; in a few generations more, the marshes will have become wholly filled by layers of vegetation and will then be peat bogs. Not a few ponds have already gone through this metamorphosis; some are now used to grow cranberries.

If the entire Cape were to be wiped out in some unimaginable catastrophe—a volcanic eruption, say—a scientist thousands of years hence could dig through the layers of lava and ash to the peat and from it reconstruct the constellation of plants that flourished in and around the bog. And from the plants he could reconstruct the climate, for the particular combination of species he would find cannot grow in an area much wetter or dryer, hotter or colder, than Cape Cod now is.

Peat bogs are not, of course, found only on Cape Cod. They will form in any place where the climate is wet enough (meaning that precipitation is relatively high and/or evaporation relatively low) and the drainage is poor enough. Indeed if there is enough moisture they will form not only on sodden lowlands but even on hilltops. Peat, and the climatic records "pickled" in the acid water impregnating it, is building up today in the muskeg regions of Canada, in the Florida Everglades and in many other places. In a wide belt across northern Europe, from Ireland to the USSR, bricks of peat, chopped out of the bogs with spades and dried, have contributed a large portion of the local fuel supply; the smoky peat fire is a standard image in any novel about rural Ireland—and, incidentally, imparts to Scotch whisky (made from barley dried over burning peat) its characteristic flavor.

It was peat that gave scientists some of the first, and most reliable, records of climatic changes. And these records, moreover, stretch backward, not for centuries but for thousands of years. Wood, with its tree-rings, rots; even in the dry climate of our own southwest the tree-ring records peter out around the beginning of the Christian era. But peat, provided only that it remains waterlogged or otherwise sealed off from the destructive effects of oxygen in the air, is almost immortal. If it is buried long enough and deep enough it turns into coal.

Nearly a century ago, archeologists and botantists in north Germany and Scandinavia were peering into the deep cuts left in the bogs by peat diggers, sampling the vegetation at different levels, and drawing conclusions about the climate at the time a particular layer of peat was laid down. Pine needles revealed a climate much like the present; oak, elm and linden leaves showed somewhat warmer and dryer conditions, like those of today's Rhineland. Spruce needles, on the other hand, indicated the sort of Subpolar climate now found well to the north, at the top of Scandinavia and in Russia. And the tiny leaves of the dwarf willow showed that the climate must at one time have been Polar, like conditions in Spitsbergen or northern Labrador today.

The scientists observed other interesting things. From time to time, the warm brown of the peat would be broken by a layer of black, sometimes with tree stumps embedded in it. These marked times of notably dryer climate—dry enough for the upper layers of the bog to lose their water so that the air could penetrate them, oxidizing them from brown to black and permitting trees to spring up and flourish. Eventually a shift back to wetter conditions waterlogged the bog once more, killing off the trees and renewing the slow buildup of peat.

From observations like these, climatologists have been able to reconstruct the changing climate of northern Europe and some other regions over a period of better than 10,000 years. No question, then, but that a peat-bog is a valuable tool to a climatologist—but only if you happen to have one handy. In most parts of the earth you don't, because the climate is too dry, or the topography unsuitable, for a long-term buildup of peat. A skilled botanist, however, does not need an oak leaf to infer the presence of an oak tree; he can make the same deduction from a few grains of oak pollen. Under the microscope, pollen grains are quite as distinctive as leaves and, unlike the leaves, which must be pickled in a bog to survive more than a few years, the pollen will survive in almost any situation, wet or dry, hot or cold. The only problem is to find it.

Lakes are a favorite hunting ground. Along with sand and silt washed into the lake by streams, pollen blown in by the wind settles to the bottom, writing a record of climate into the accumulating layers of sediment. But pollen also turns up in the sediments that overflowing streams or rivers spread over their flood plain, in wind-drifted soils and even, in at least one desert area, in the dried dung of animals whose descendants, along with the vegetation they fed on, have been gone from that part of the world for 5,000 years.

Pollen, one would think, is about as close as one can get to an unambiguous record of changes in vegetation—and hence in climate. But there are still skeptics. A leading one is Professor Edward Deevey of Yale University. He is also something of an expert on pollen, thus illustrating the old maxim that the more you know the less you're sure of. Deevey notes, for example, that where the record shows a shift from grass and shrub pollen to, say, pine pollen, the usual interpretation is that pine forest replaced grassland, as a result of increased rainfall. But pines, like most trees, he points out, produce enormous quantities of pollen as compared with the humbler plant species. Thus the jump in pine pollen might result from a single pine tree which happened to have sprung up in a specially moist situation. And one tree does not make a forest. Or again, he says, when pollen records from, say, Minnesota show that trees—their presence confirmed by other evidence—were replaced by grass, does this mean that the climate had grown dryer, or merely that Indians had cut down the trees or burned them off, to clear land for their corn and squash fields?

Human activities, in fact, can do more than merely substitute one type of vegetation for another; Deevey points out that they can actually remove vegetation entirely. If the climate is quite dry to begin with, plowing up the grass to raise crops may allow the topsoil to be stripped off by even sparse showers or by the wind; less topsoil means that the rains run off faster, which means the soil grows dryer, which means less vegetation, and so on in an accelerating decline. And all this can happen even though rainfall has not decreased one bit; in fact it *has* happened in large areas of the Mediterranean and Middle Eastern lands.

In the context of this book, however, such distinctions become something of a quibble. If one is concerned with the impact of climate on man, then a desert climate is a desert climate, regardless of whether man or nature is responsible. The people living in such an area can starve to death quite as effectively, no matter if the rain fails to fall or if, having fallen, it runs off before it can do anyone any good.

Even pollen, for whatever its indications are worth, does not last indefinitely. Beyond a few hundred thousand years, its records vanish, and we are left with whatever evidence of climate can be seen on the face of the earth itself—or in it. If the climatologist is lucky, he may run across fossilized plant remains—leaves or berries which, having fallen into a stream, were buried in silt which was ultimately transformed into shale or slate, bearing in it the impressions of long-dead vegetation. As usual, however, he needs to be careful about jumping to conclusions. The further back he goes, the more likely he is to be dealing with plants which are not the species we know today. And a fossil fragment of a plant that nobody has ever seen "in the flesh" can fool even the experts. Fossil "spruce cones" of perhaps ten million years ago turned out, on later and closer examination, to be the fruits of a relative of the tea plant; some other fruits were actually first taken for insect cocoons.

Fossil animals also provide clues to past climates, but their interpretation is, if anything, even trickier than with fossil plants. The skull of an extinct beaver, to be sure, is pretty conclusive evidence that we are dealing with a climate at least moist enough for a beaver pond. But what about a fossil rhinoceros? Present day rhinos live in both moist and dry habitats, but their homes are without exception hot. Yet some extinct rhinos lived in climates which from every other bit of evidence seem to have been Subpolar. Here we are fortunate enough to have an explanation. For part of that time, there were men about who drew pictures of these creatures on the walls of caves, revealing that they were equipped with a thick coat of hair. In other cases we are not so lucky and must reason very cautiously from the climatic preferences of today's animals to those of their extinct relatives.

Climate leaves its traces not merely on the plants and animals that inhabited a particular area but on the land itself. Glacial moraines have already been mentioned, and there are many other equally useful relics. In many areas of Southern Africa, for example, there are regions in which the excavator's spade turns up what seem unquestionably to be buried sand dunes, now covered by Savanna or even Equatorial forest. These geologic "fossils" leave little doubt that at one time these areas were covered by deserts which now lie scores or even hundreds of miles to the south. Again, geologists exploring the Great Basin area of Nevada and Utah have traced miles of "fossil" beaches along the shore of Lake Bonneville, a vanished body of fresh water almost as big as some of the Great Lakes, which covered a large part of this now desert area (Great Salt Lake is its much-shrunken remnant). Certain types of gravel deposits tell of rushing rivers that long ago sank into the encroaching sands; beds of clay, of great lakes where now the cactus flourishes. Beds of rock salt or gypsum, such as we find beneath well-watered Michigan or central Europe, tell us that those regions were once near-deserts in which inland seas, lacking sufficient replenishment from rain and rivers, dried up to leave their salt for our ultimate exploitation and edification.

Having read so much about indirect indicators of past climates, with all the ifs and buts attached to them, it may come as something of a shock to learn that there are methods by which certain features of ancient climates can be measured directly—or nearly so. The subtle techniques of chemistry and nuclear physics have developed a sort of "thermometer" which can "take the temperature" of oceans a million and even a hundred million years ago.

The story is interesting enough to go into at some length. It begins in 1946 in Switzerland, in a lecture hall at the famous Technical College of Zurich. The speaker was Harold Urey, a distinguished American chemist. His subject was his work on isotopes, which twelve years before had won him a Nobel Prize.

The isotopes of a particular element, Urey pointed out, are usually described as atoms which differ from one another only in their atomic structure and weight, not in their chemical properties. To take a familiar example, uranium-238 and uranium-235 are very different in their atomic properties; a bomb can be made of the latter but not the former. Yet they form precisely the same chemical compounds, a fact that poses serious problems—unfortunately, not insuperable ones—to anyone interested in separating the explosive (fissionable) isotope from the nonfissionable one. In fact, Urey continued, isotopes do *not* behave precisely alike in the test tube. For example, the oxygen in ordinary water (the "O" in H_2O) actually consists of three different isotopes—O-16, O-17 and O-18. And if a glass of water is allowed to evaporate, the water vapor will carry off a slightly higher proportion of the lighter, O-16 atoms, so that the water remaining in the glass will be slightly richer in the other two isotopes than it originally was.

In the discussion that followed Urey's lecture, the Swiss crystallographer, Paul Niggli brought his own specialty to bear on this point. Ocean waters, he noted, ought to contain proportions of isotopes somewhat different from those found in river or lake waters, since the sea is subject to more evaporation than most fresh-water bodies. And this difference ought to show up in minerals that have crystallized in a particular body of water. Thus if one had a bit of limestone, or coral, or the shell of some aquatic animal, the proportion of isotopes in it should show whether it had been formed in fresh or salt water. Urey agreed that this seemed likely, and for the moment the subject was dropped. But back at his University of Chicago laboratory, his mind turned to it once more.

Harold Urey is no narrow specialist. His mind is broad gauge and his curiosity no less so; his studies have ranged from isotopes to the origins of life on earth to the nature of the moon's surface. The notion that one could distinguish fresh-water from salt-water limestone merely by measuring the proportion of isotopes in it tickled his fancy, and he began jotting down some equations to see just how different the proportions would be. He quickly realized that the balance of isotopes in any given case would depend not only on whether the water was salt or fresh, but also *on how warm the water was when the mineral was being formed.* And suddenly, with the kind of imaginative leap that only a great scientist can make, he realized that he had "a geologic thermometer in my hands."

To be precise, Urey had the *idea* of a thermometer, and between the idea and the reality stretched some four years of hard work. Even quite marked changes in temperature produced only tiny variations in the isotope ratios; a shift of fifty degrees in temperature changed the proportion by less than 1 percent. Existing apparatus for separating isotopes was not a great deal more accurate than that and had to be extensively redesigned before it could "read" the thermometer accurately enough to be of any use. To calibrate the instrument, Urey used the shells of mollusks that had been raised under controlled temperature, and preparing these shells for processing raised another set of problems.

At last, in 1950, Urey and his associates were ready to take some temperatures. Their first specimen was the fossilized skeleton of a belemnite, an extinct relative of the squid and octopus which some 150 million years ago had swum about a shallow sea covering the present site of Scotland. Belemnite skeletons are cigar-shaped pieces of limestone which, when sawed across, show rings very like those of a tree stump.

These had long been assumed to be growth-rings, like those of trees, but since nobody had ever examined living belemnites, there was no way of proving this. Urey and his associates shaved off each of the layers and analyzed them one by one. To their delight, they found that the temperature determinations showed a regular fluctuation from layer to layer, which almost certainly reflected seasonal fluctuations in the ocean temperatures. They were able to deduce that this belemnite, dead for 150 million years, had been hatched in early summer, had survived for three more summers, and had died the following spring.

The temperature curves showed something else: the area around Scotland was a good deal warmer then than now. Summer ocean temperatures averaged about 68 degrees, about like those off Lisbon in the same season today; winter temperatures were around 60 degrees, much like those off Morocco. These findings, to be sure, were no great surprise, for a similar picture had long ago been deduced from other evidence. But they confirmed that the thermometer was working. Had they shown a climate resembling that of Scotland today, something would have been very wrong indeed.

Of course Urey's thermometer is subject to the usual ifs and buts which hamper the usefulness of all paleoclimatic evidence. The isotope ratio of a particular specimen depends not merely on the temperature of the waters it was formed in, but on the proportion of isotopes in those waters. And studies have shown that these proportions are far from uniform in different seas. Where evaporation is particularly heavy (as in the tropics), the heavy isotopes are abnormally plentiful; where the influx of fresh water is high (as in the Baltic), they are abnormally sparse. Thus they must be interpreted with the aid of several assumptions—and a sizable addition of metaphorical salt.

Yet even so, there is a certain consistency in the readings from one place to another. For example, a great number of measurements have been made on "cores" from the sea bottom. These are long cylinders of compacted mud and ooze which, with the aid of special apparatus, can be punched out of the sediments that build up beneath the ocean waters. A twenty-foot core can provide a sampling of sediments built up for tens of thousands of years, and, presumably, a record of changing temperatures over the same period. Temperature graphs obtained from cores as far apart as the Mediterranean, the North Atlantic, and the Caribbean show marked differences in detail, but a strong similarity in their major fluctuations. Readings taken at the tops of the cores show temperatures like those of today; farther down one sees a cooling off, evidently reflecting the general refrigeration of the earth as one moves back in time toward the last Ice Age. Similar fluctuations lower down presumably reflect the repeated shifts from warm to cold and back again which characterized the entire Glacial Epoch.

In the next section, there will be much more to say about Urey's "geological thermometer" and what it means. For the moment, it need only be noted that while many climatologists agree that the thermometer gives a reasonably accurate record of changes in climate *as they affected the oceans,* there is an intense disagreement as to what these changes mean *on land,* which, after all, is where man makes his home.

Musing over the apparently interminable controversies, conflicts, and ambiguities in the evidence concerning ancient climates, one may well wonder whether it all adds up to enough tangible information to be worth a book. My own feeling, obviously, is that it does. Generally speaking, I would say that almost every bit of evidence in paleoclimatology is questionable, taken by itself. And all too frequently it must, indeed, be taken by itself. But increasingly the evidence is piling up, at least in certain areas. One matchstick by itself is a fragile thing, but glue enough of them together and they can make a serviceable seat. So when, on certain points, several different kinds of evidence point to one particular climatic conclusion, it is fair to assume that there must be something in it.

7. CLIMATE PAST III

When Did It Happen?

While gathering material for this book, I have several times felt that I would have done better to have tackled the job twenty years ago. In the 1940s, the scientists' evidence on past climates was much sparser than it is now, but they were fairly sure they understood what it meant. Today, with far more extensive knowledge, they are far less certain about what it adds up to.

What has made the difference, probably more than any other single factor, is better methods of dating: the ability to say, with increasing certainty, not simply that certain climatic changes took place but when they took place (and therefore that a particular change in one locality did or did not occur at the same time as another one elsewhere). This new knowledge has exploded many cherished climatological theories and knocked down a lot of easy assumptions. By the same token, the fact that dating methods are still not as good as one would like goes far to account for the confusion and conflict that still remain—and which may well require another twenty years to clear up.

This chapter is devoted to a discussion of dating methods because they will be referred to repeatedly in the remainder of the book. Not too surprisingly, much of the evidence on when the climate has changed comes from the same materials that show how it has changed. There are tree-ring chronologies and pollen chronologies and lake-bottom chronologies. Fossils play their part, so do land forms. By all odds the most useful dates, however, come from techniques quite different from any thus far mentioned. Indeed, the great virtue of these "clocks" for measuring past time is that they do *not* record the vagaries of climate and are not affected by them; in fact, they cannot be speeded or slowed by any natural process known to occur on earth.

The first chronologies of the past employed fossils. The notion of such a chronology, together with the very notion of paleoclimatology itself, seems to have sprung from the mind of a seventeenth-century Englishman, Robert Hooke, a remarkable individual in many ways. Hooke's temper made him a trial even to his friends, of which he had few; his enemies included many of the greatest scientists of his day. He quarreled with the Dutch astronomer Christian Huygens over which of them had invented the pendulum clock (a notable advance in time-keeping); he quarreled with Newton over the law of gravity. But between bouts he found time to discover one of the basic laws of elasticity, invent the balance wheel (and with it, the first accurate pocket watch), and serve for many years as the secretary of Britain's Royal Society, which then as now was one of the most prestigious scientific bodies in the world.

The diversity of Hooke's interests marks him as another of those broad-gauge minds of the kind already encountered in the person of Harold Urey. Among his interests were fossils; he was, in fact, one of the first to insist that fossils were the remains of once-living things. In the year 1686, he was pondering some fossil shells which he had found in the district of Portland. Noting their size—far larger than any contemporary shells found in English waters—and their resemblance to certain tropical shellfish, he asked himself whether England had not at one time "lain within the torrid zone." When this might have been he had no notion; he suspected that the shells themselves might supply a key. "It must be granted," he wrote, "that it is very difficult to read them and to raise a *chronology* out of them, and to state the intervalls of the times wherein such and such catastrophes and mutations have happened; yet 'tis not impossible."

Hooke was quite right. Yet more than a century was to

pass before any progress was made in raising a chronology out of fossils. For most men, the Bible supplied all the chronology they needed. What need to speculate about the ages of fossil shells when Archbishop Ussher, using the most exquisite scholarship, had dated the Creation at 9 A.M., Sunday, October 23, 4003 B.C.? Even educated men were often quite unable to emulate Hooke's simple achievement of believing his own eyes, of accepting fossils as fossils and not as accidental rock formation, "jokes of nature," or anything else except what they were. In the mid-1700s, the governor of Massachusetts, shown some teeth of an extinct form of elephant, proclaimed that they were the remains of a giant drowned in Noah's flood. (About the same time, similar teeth were shown to some Negro slaves in Virginia, who unhesitatingly labeled them as "the grinders of an elephant," such as they had seen in their African homeland.)

There are times when much learning is worse than none. William Smith, the Englishman who devised the first fossil chronology, was not burdened with overmuch learning. Born in 1769 and left fatherless at the age of eight, he learned reading, writing, and ciphering at the village school. For the rest, he was self-educated; having taught himself the rudiments of geometry and surveying, he worked for most of his life as a surveyor and what would now be called a civil engineer. Significantly, even as a boy Smith had collected different kinds of rocks, with a special interest in those containing fossils. Luckily for science, Smith lived in the right time and the right place. In the decades just before and after 1800, England was in the midst of the great "Canal Boom," in which hundreds of miles of artificial waterways were hacked out of the layers of rock, sand and clay that lie beneath that green and pleasant land. And Smith, as a surveyor and construction superintendent, watched the navvies' spades uncover this substructure.

In his business trips about England, sometimes amounting to 10,000 miles a year, Smith made copious notes on the various types of rocks he observed. He soon realized that rock layers (strata) in England did not lie horizontally but were slightly tilted, like "slices of bread and butter" in a dish. Thus even in a few score miles, a canal cut might expose a whole series of strata—sandstone succeeding shale succeeding limestone, layer upon layer. Smith, through his own reading, had already absorbed an obvious fact about rock strata: the ones on top must be younger than the ones beneath. But while this principle could usually give the relative ages of the strata in any locality, it was of little help in relating a layer of limestone in Wales to one along the North Sea.

However, Smith had taken the trouble to note down not merely the distribution of rocks but also the distribution of fossils in them. At last he realized that "each stratum contained organized fossils peculiar to itself, and might in cases otherwise doubtful, be recognized and discriminated from others like it . . . by examination of them." Smith had given the book of the past a set of page numbers by means of which its scattered sheets could be brought into order. With their aid, scientists could examine a fossil-bearing rock anywhere in England (eventually, anywhere in the world) and determine its proper chronological place. They could do this even when, as occasionally happens, the slow folding of rocks had turned the "normal" sequence upside down, so that the oldest rocks lay on the top instead of the bottom.

But Smith's fossil chronology could supply only relative dates, not absolute ones. It could establish that one rock stratum was older or younger than another—but not how old either one was. It took more than a century before reliable absolute dates could be obtained. Meanwhile, the geologists resorted to all sorts of questionable guesses. Their technique was based on a principle we have already met: the present is the key to the past.

Consider a river which leaps down a rocky waterfall and then passes through a narrow gorge. Almost certainly, the fall will be gradually moving upstream as sand and gravel borne by the rushing river slowly erode the rocks.* Observations over some fifty years, let us assume, have established that the present rate of erosion averages a foot every ten years. Then if the gorge is 1,000 feet long, it follows that the river must have begun carving it out some 10,000 years ago. It follows, that is, if conditions in the past were the same as in the present—if the river's flow was not much greater or much less than it now is; if the rock throughout the length of the gorge is not much harder or softer than what the river is now cutting through; and so on.

Too often, conditions have changed. Precisely these methods were used many years back to determine how long ago the Niagara River had begun cutting out its gorge below Niagara Falls, which would in turn determine how long ago the last glaciers departed from that point on the U.S.-Canadian border. The figures, taken with other evidence, gave a date around 25,000 years B.P. (Before Present). But as other evidence on the last Ice Age accumulated, that figure began to seem much too large. And indeed, a much closer examination of the gorge established that a good part of it had been carved out well before the last glaciation and had then been filled with glacial debris—sand, gravel and silt—which of course the river could cut through far faster than solid rock. We now know that the true date of the glaciers' departure is much closer to 10,000 B.P.

Not all estimates of absolute geological dates were this far off. Some of them, we now know, were extraordinarily accurate. But there was no way then that the accurate dates could be singled out. What solved the problem was the discovery of radioactivity. Beginning in 1888, a series of major scientific advances established what is now a schoolboy's commonplace: atoms of certain elements are unstable. Quite spontaneously, they emit radiations of various kinds and are transformed into atoms of other elements.

* Niagara Falls is a notable example of this process at work; indeed there is now some concern that the falls will destroy themselves.

Less than twenty years after the discovery of radioactivity, the physicists had established that atomic transformation is an almost unbelievably constant process. Every radioactive element has its own characteristic rate of breakdown, defined by what is now called the "half-life." For example, uranium (ordinary, nonfissionable uranium) has a half-life of about 4.5 billion years. If you start with a pound of uranium and wait 4.5 billion years, you will have only half a pound left, the rest having been transmuted into something else. In another 4.5 billion years, only half of that half-pound will be left; and so on.

With radioactivity, the present is the most reliable sort of key to the past. There is no need to make assumptions about a river's rate of flow or anything of the sort. Rather, there is every reason to believe that the rate of "decay" of uranium —or any other radioactive element—was precisely the same a billion years ago as it is now, or as it will be a billion years from now.

In 1907, it occurred to the great English physicist Lord Rutherford that radioactivity could be used to measure the absolute ages of rocks. He knew, for example, that uranium, having once decayed, goes through a series of much more rapid transformations; ultimately, each atom of it ends up as a stable atom of lead. And in its progress from uranium to lead, the atom emits eight atoms of helium. Should it not be possible, Rutherford wondered, to take a bit of uranium-bearing rock, measure how much uranium and how much helium it contained, and from the ratio between the two, plus a little algebra, determine how long the uranium had been decaying in the rock? In principle, it was perfectly possible. In practice, it turned out to be a very tricky business. Among other things, helium atoms, being both light and small, can "leak out" even through solid rock. Gradually, however, the problems were solved, or bypassed, so that today uranium dating is a standard and reliable technique for measuring the absolute ages of rocks.

It is, however, of no use in the study of man. Uranium,

because of its very long half-life, gives dates which are accurate to no better than about five million years. To geologists who deal in hundreds of millions of years, this is no great matter. However, to anyone interested in man and his history—a matter of the past two or three million years—using uranium to date happenings would be like using a grandfather clock to time a footrace.

For dating in the human era, an isotope with a markedly shorter half-life is needed. But this creates a dilemma. If the half-life is short enough to yield useful dates in the past million years or so, it will also be short enough so that the isotope will have long since decayed away. The best compromise that the scientists have been able to come up with is potassium-40, with a half-life of about 1.3 billion years. This is a very rare form of potassium indeed, but it has proved to be useful in dating certain types of rock (notably, volcanic lavas) relatively rich in that element. Even so, the really useful dates it yields are measured in millions, not thousands, and are accurate to only about 50,000 years.

There is one other possibility, but a difficult one: find an isotope with a usefully short half-life *which is continuously formed by some natural process.* This proviso is hard enough, but there is a second one. If the isotope is being continuously formed, after all, there is no way of measuring how much of it has decayed, for the supply is being renewed as fast as it is being used up. Unless, that is, the isotope is being segregated by some natural process so that a fixed amount of it can decay in peace, uncontaminated by fresh and undecayed supplies.

Rather surprisingly, scientists have come up thus far with not one but three isotopes that fit these very restrictive conditions. The first was carbon-14. This is continuously formed in the upper atmosphere by the impact of cosmic rays. It combines with oxygen to form carbon dioxide, which is taken in by plants and through them by animals. So long as the plant or animal is alive, its supply of radioactive carbon (about one billionth of its total carbon intake) is being continuously renewed, of course. But once it dies, the carbon intake stops and the carbon-14 clock begins to tick off the centuries.

Carbon-14 has proved enormously useful. For one thing, its half-life is only about 5,000 years, meaning that it gives dates accurate to within a few centuries, sometimes less. For another, carbon is by all odds the commonest element in living things, so that almost any once-living substance will serve: wood, peat, charcoal, shell, or bone. The only hitch is that carbon dates take us back no further than about 50,000 years B.P. In objects that died earlier than that, so much of the carbon has decayed that what's left radiates too feebly to be measured accurately.

The other two "recent" isotope clocks are forms of the radioactive elements protoactinium and thorium. Both are continuously formed by the breakdown of uranium; both of them accumulate in certain sediments on the ocean bottom, while their "parent," uranium, tends to remain dissolved in the sea water. These two isotopes, with half-lives measured in tens of thousands of years, have established dates in ocean cores out to around 300,000 B.P.

Summing it up then, potassium-40 can give dates up to one million years B.P., or perhaps a little more recent. After that, however, its results become increasingly unreliable; a potassium date of 500,000 B.P. actually means somewhere between 400,000 and 600,000. Not until 300,000 B.P. do thorium and protoactinium give reliable dates on the ocean bottom, while on land it is not until about 50,000 B.P. that carbon-14 takes over.

It is not unlikely that the gap between potassium and the short-lived isotopes will be filled, either by more refined techniques or by new isotopes, before too many years have passed. Meantime, however, geologists and climatologists must fill it by estimates—and continue to dispute, sometimes acrimoniously, the validity of one another's estimates.

Before ending this chapter, a word must be said about two other methods of dating climatic changes. Neither of them has anything like the general usefulness of the methods based on radioactivity; they can be applied only in certain special situations and in any case cover only a few hundreds, or at best thousands, of years before the present. In compensation, however, they yield dates which are sometimes accurate to the year.

The first of these techniques, dendrochronology (tree-ring dating; *dendro* means tree), was first practiced during the first decades of this century, but its origins go back for a hundred years before that. The first man to think of using trees for dating was apparently DeWitt Clinton, the governor of New York State who built the Erie Canal. In 1811, Clinton was examining some mounds near Canandaigua, New York. Wondering who had built them, he had workmen cut down some large trees growing atop the mounds, counted the rings, and concluded, quite correctly, that the structure must have been built around 800 A.D., and thus not by Europeans nor even by the local Iroquois, but by some unknown prehistoric people.

The man who really worked out the theory of dendrochronology, however, was an Englishman. Charles Babbage resembled Robert Hooke in the breadth of his interests, though his temper was considerably milder. A gifted mathematician, he also invented the adding machine and, during the 1830s, helped to rejuvenate the Royal Society, which after a brilliant youth had fallen into premature senility. In 1837, Babbage published a detailed description of how tree-rings might be used for "ascertaining the age of submerged forests"; the technique, he said prophetically, "may possibly connect them ultimately with the chronology of man."

The man who made this prophecy come true was an American professor, Andrew E. Douglass. The method, as foreshadowed by Babbage and laboriously worked out by Douglass, is fundamentally quite simple, since it depends on counting tree rings in stumps, logs, or old pieces of lumber. The trick is to connect up wood of unknown age with other wood whose dates are known. This is done by counting backward. One starts with a tree that has just been cut down. Counting rings inward toward the heartwood, one can easily determine that the tree started growing, say, ninety-three years ago. Next, the innermost twenty rings or so are examined. These, like all the rings, will be of variable thickness because of variations in climate from year to year (as pointed out in Chapter 5). A distinctive sequence of rings —something like wide, wide, narrow, medium, narrow, wide and so on—is singled out. Then an old stump must be found in the source area. It is not unlikely that the same sequence will be found repeated toward the stump's outer circumference. By counting stump rings the chronology can be carried back for another century or two. Another distinctive ring sequence from the center of the stump can then be hooked on to the rings in an old beam, for instance, and so on backward. By thus linking together one chunk of wood (or even a large hunk of charcoal) with another, Douglass and his successors have been able to carry back their tree-ring chronology in the American Southwest to the beginning of the Christian era. And of course the chronology, which records the varying widths of the rings, at the same time gives a complete record of rainfall in that area for the same period. Other researchers, using rings from the giant sequoias, have established a dated climate record for the northern California coast stretching back more than 3,200 years. In Arizona, studies of the even more long-lived bristle-cone pine may ultimately set up a record covering the past 8,000 years.

The main limitation of tree-ring chronology is its parochialism. The ring sequences, like the climatic sequences which they record, are different in different regions and must be worked out laboriously for each one. Moreover, climatically, if not chronologically, tree rings are useful chiefly in areas where one climatic factor (rainfall in semidesert areas like the Southwest, summer temperatures in subpolar regions like Lapland) is paramount. In less marginal situations, the width of the rings is likely to reflect both rainfall and temperature and it is hard to separate their influence.

The other year-by-year chronology depends, like dendrochronology, on counting layers. They are not layers of wood in a tree trunk or beam, however, but layers of mud, called varves, at the bottom of a lake.

I first struck up a personal acquaintance with varves some years ago at my sister's home in northwest Massachusetts. Something had gone wrong with the drains and nobody quite knew where the drains were, since the map showing them, if it had ever existed, was lost somewhere in the files of the Water Department. As a result—and to my sister's considerable irritation—several five-foot-deep exploratory trenches had to be dug across the front lawn. Looking into one of these trenches, I noticed that about two feet down the diggers had hit a thick bed of greenish-gray clay. Picking up a lump that had been thrown out of the trench, I saw that it was marked with many narrow stripes, each one darker and coarser on the bottom and lighter on top. That bed of layered clay told me a good deal about the past history of my sister's lawn. At one time—perhaps 8,000 years ago, perhaps more—it had lain beneath a lake or pond at the edge of a retreating glacier. Each spring, water from the melting ice carried down clay into the lake, where the darker, coarser particles settled to the bottom. But the wind, rippling the lake, held the finer particles in suspension, giving the waters a milky look. Then, when winter came, the melting stopped, the lake froze over, and the waters were stilled. The fine white clay slowly drifted down to form a thinner, lighter section. Holding a lump of varve clay in my hand, I could, with a little manipulation, peel each layer away from the one below it, like thin slices of processed cheese. Some were thicker than others, representing, I knew, warmer years in which the clay-bearing runoff from the glacier was heavy. Thinner layers were years in which summer was short and the runoff sparse.

Layered claybanks of this sort are found in many places in northern Europe and the northeastern United States. Any one of them will represent only a few hundred years of glacial runoff, after which either the lake became filled with sediment or, more likely, the glacier drew farther back. But by comparing patterns of striping at the top of one bed with those at the bottom of one lying north of it, the varves can be linked together into sequences like those of the tree-rings. In Sweden these sequences have been carried forward to the present and backward to the time, some nine thousand years ago, when the ice sheet at last drew back to leave the land uncovered. One can sometimes establish the very year when the glacier withdrew from a particular locality. Moreover, as with the tree-rings, the varves record climatic fluctuations even as they establish dates for them—the runs of cold years when the ice sheet paused in its retreat, the hot years when it almost rushed backward (toward the end of its career, as much as 400 yards a year). Varves pinpoint the very year in which the remnants of the ice cap, still clinging to the Scandinavian mountains, split into two parts—thus, according to some authorities, marking the formal end of the Ice Age in that part of the world. The date was 6839 B.C.

Like tree-ring chronologies, varve chronologies must be worked out separately for any given region. The series for the northeastern United States is quite different from that for Sweden, and, unfortunately, also less complete, so that dates obtained from it are a good deal less certain. But for all its problems, varve chronology is a remarkable achievement. Surely it must be an extraordinary experience to be able to stand by a claybank in Sweden and say, with some certainty, "In 8800 B.C. the edge of the glacier lay . . . right . . . here!"

PART TWO

Climate and Emerging Man

8. HUMAN EVOLUTION

The Uses of Prehistory

To understand history, one must first understand prehistory. This is a fine and mellifluous generalization, and I wish I knew whether I believed it. To be perfectly frank, I am devoting this section to prehistory because the subject has fascinated me ever since I was a kid roaming the halls of the American Museum of Natural History on a Saturday afternoon. In those days, the subject was dinosaurs, those massive and grotesque skeletons which are guaranteed to draw the eye and tickle the imagination of any eight-year-old. Later, my interest in that phase of prehistory dwindled. The dinosaurs were still impressive, but became, as soon as I had learned something about evolution, what might be called "expectable." That is to say, once the earliest reptiles had hoisted themselves out of their ancestral amphibian swamps and evolved the biological techniques for living on really dry land, some of them, in the course of time, almost had to become dinosaurs. Just so, the first three-wheeled one-cylinder automobile was bound eventually to evolve into today's overblown Cadillac.

The really puzzling problem about the dinosaurs is not why they appeared but why they disappeared—and rather suddenly, as such things go. The customary explanation is that they were "replaced" by the mammals—but this explains nothing. The mammals already had been hanging about inconspicuously for something like a hundred million years. If the dinosaurs had managed to coexist with them for that long, why not a hundred million years longer? This puzzle, however, has no place in this book, for it lies beyond even the rather elastic boundaries I have set myself.

Even more fascinating than the End of the Dinosaurs is the Beginning of Man. The dinosaurs were at best merely bigger and better land animals; man was something truly new under the sun (our sun, anyway). His appearance on earth was no less revolutionary than the appearance, a few billion years earlier, of life itself.

The dinosaurs and every creature that ever lived, or now lives, on earth, had adapted to their environment—flourishing when the adaptation was adequate to the circumstances, disappearing when it was not. Only man, with the aid of his peculiar brain, can adapt the environment to himself. He is the only creature, for instance, that can manipulate climate, even on a small scale. He is also the only creature that can wonder about it or write books about it. The appearance of this remarkable animal was an event so extraordinary that it is not surprising that a majority of the race professes to see the hand of Divine Providence in the matter. I don't myself believe this notion for a moment, of course—but I don't laugh at it either.

From the climatic standpoint, the evolution of man can be summed up by four statements which are both brief and more certain than almost anything stated up to this point.

First, man—the tool-making, word-mouthing species—evolved from some extinct species of ape during a time when the earth's climate was in a highly unusual state—in most regions, both colder (sometimes much colder) and dryer than it had been through nearly all the earth's history. Because I have previously laid so much stress on the provisional nature of paleoclimatic information, let me say immediately that this statement is as certain as anything, anywhere, in science. Man, the very unusual animal, evolved in a very unusual climatic era, the Glacial Epoch.

The prologue to that evolution occurred before the Glacial Epoch began, in the rich forests of an Equatorial climate or something close to it. In that climatic environment, our remote four-footed ancestors evolved into semierect apes, acquiring in the process a number of characteristics that we ourselves still find remarkably useful.

The next stage in the process took place around the time the Glacial Epoch was getting started, in the hot, tree-dotted grasslands of a savanna climate. There, some apes mutated into the first crude versions of man.

The final stage, in which modern man evolved from his more apish forebears, took place during the Glacial Epoch proper. It was man's evolving intelligence that helped him to spread into a much wider variety of climates than he had hitherto experienced, to cope with far more demanding climatic conditions, and to survive and thrive during a series of almost catastrophic climatic changes which sent dozens of other species into the limbo of extinction.

It is tempting to conclude from all this that the Glacial Epoch caused man's appearance. There is probably some truth in this view, but to define this truth more precisely would involve a long digression on the philosophy of causation. We can at any rate say with some certainty that the Glacial Epoch *accelerated* man's evolution. Man, presumably, would have evolved in any case, but far more slowly than he in fact did. Thus the fact that we exist in our present form can be laid pretty directly to a very peculiar climatic event.

How did that peculiar event come about? To make even a stab at answering this question will require a rather lengthy discussion. The origin of the Glacial Epoch is one of the most fascinating, most puzzling, and most hotly debated in all climatology; there is not only no simple answer to it, there is no answer at all on which climatologists generally agree. But the hunt for an answer has stimulated research in the glacial crevasses of Antarctica, on the lightless abyssal plains of the ocean bottom, and almost everywhere in between. And if the experts have not yet come up with any wholly satisfactory answer, they have at any rate turned up a great deal of intriguing lore about the earth and its history.

9. THE GEOLOGISTS

Enter the Ice Age

It was the Swiss who discovered the Glacial Epoch. This fact is not very surprising if one accepts the idea that the earth's present is the key to its past. By 1820, most naturalists had come to accept it; most naturalists also lived in Europe. And nowhere else in Europe were there so many glaciers present as in Switzerland.

The first clue seems to have come from one J. P. Perraudin, no naturalist but "a skilled hunter of chamois and an amateur in these types of observations." Roaming the high valleys in his native canton of Valais, he noted, well up on the valley sides, heaps of rounded boulders, composed of rock quite different from that of the valley itself. Perraudin had seen rock heaps like these before, along the sides of the glaciers that filled the very highest valleys. He concluded, quite correctly, that the lower rock heaps, too, must have been carried down by glaciers, which therefore must once have reached a much lower altitude.

Perraudin's observations somehow came to the ears of a civil engineer, Ignatz Venetz-Sitten, who, in 1821, presented them, together with similar observations of his own, in a paper before the Helvetic Society. Soon after, a Norwegian, Jens Esmark, reached the same conclusion about the glaciers of his own homeland.

In subsequent years, Venetz and an associate, von Charpentier followed the rocky traces of vanished Swiss glaciers across hill and valley. They found rocks from the Central Alps scattered all the way to the Jura mountains, scores of miles away.

Louis Agassiz, a bright young naturalist at the University of Neuchatel, didn't believe it. Glaciers, he was convinced, could not move as far or as fast as the theories would require. To test his belief, he drove stakes into several glaciers and over a period of months carefully measured their rate of progress. He found they moved far faster than he had expected. Converted from a skeptic into a believer, Agassiz soon leaped far beyond the theorizing of his predecessors. Pulling together the observations of half a dozen glacier buffs, he reached a startling conclusion. The ice, he said in 1837, had not advanced from the Alps to the adjacent plains; rather, it had once covered all the low ground and had subsequently retreated to the mountains.

Three years later, Agassiz traveled to Great Britain where glaciers, thanks to the maritime climate and low mountains, were, and are, unknown. Nonetheless, in Scotland, northern England, and Ireland he found evidence "that not only glaciers once existed in the British Islands but that large sheets of ice covered all the surface." Here we have the first working definition of the Glacial Epoch.

And here, too, our neat theory of Swiss priority in glacial studies breaks down. Much later, it became known that Agassiz' theory of "sheets of ice" had been anticipated as early as 1832 by A. Bernardi, an obscure German professor at a school of forestry in the small Thuringian town of Dreissigacker. But nobody was listening. In science, as elsewhere, the race is not always to the swift. While Bernardi vegetated among the Thuringian backwoods, Agassiz was roaming about Europe, collecting additional data to support his theory and wrangling with its critics.

At first there were plenty of critics. The generation gap existed then as now, and older geologists had scant respect for Agassiz' youthful heresies. Leopold von Buch, a pillar of the geological Establishment, "could hardly contain his indignation, mingled with contempt, for what seemed to him the view of a youthful and inexperienced observer." But the facts kept piling up, and before long even the stuffiest critics had to concede that the heretic was right.

Once the geologists agreed on what they were looking for, they began piecing together a picture of what had happened. The basic pattern was awesomely simple. At some time in the past, the earth's climate had grown colder. As a result, winter snow all over such mountainous northerly regions as Scandinavia and Scotland no longer melted in summer but instead accumulated from year to year. Gradually, the deeper snow layers became compacted by pressure into glacial ice which, as it grew thicker and heavier, began to flow sluggishly downward into the lowlands, accumulating ever more snow on its surface as it moved. This process continued until enormous sheets of ice, a mile or more thick, covered most of northern Europe—and, it soon became clear, large parts of North America as well. (Agassiz, who had emigrated to the United States, carried with him to his new home the new science of glaciology.)

From these centers, the ice stretched south from Scotland and Scandinavia to cover, at one time or another, the future sites of Dublin, the northern suburbs of London, Amsterdam, Berlin, Warsaw, Kiev, Moscow, and Leningrad. Farther south, the extended Alpine glaciers covered nearly all of Switzerland and bits of the adjacent countries.

North America was worse off. With a more continental, and therefore colder, climate, most of Canada lay beneath the ice. To the south, the glaciers reached St. Louis and covered Chicago, Cleveland, and part of New York City. (Much of that city, indeed, is built on the hills of debris heaped up directly in front of the ice sheet.) To the west, the Rockies generated their own ice sheets, which at various times linked up with the bigger sheets to the east, forming a mighty wall of ice from coast to coast.

Northern Asia, with an even more continental climate than North America, would, one might suppose, be even more completely covered. In fact, apart from the northern edges of Siberia and a few high mountain ranges, it remained ice free. For glaciation you need not merely a cold climate, but a reasonable abundance of snow as well. Siberia was cold—probably colder than today (and even today it can hit 100 degrees or more below zero), but it was also dry. North America gets much of its moisture from the Gulf of Mexico, whence winds pass up through the Mississippi Valley to deposit rain or snow well into Canada. In Siberia, however, southerly moisture is blocked off by the massive wall of the Himalayan mountain system.

Even so, there are oddities about the distribution of ice around the Northern Pacific. Eastern Siberia lies in the same latitudes as Labrador and Newfoundland, which were one of the centers of the North American ice sheet. Like them, it is fairly mountainous and, like them, can draw abundant moisture from the southeast. Yet glaciation there was patchy. Large parts of Alaska, too, were never iced over. We shall come back to this point later on, because it has some importance for the migrations of primitive man.

Where the ice came, it reshaped the land almost beyond recognition. In places it stripped off the soil down to bedrock (as in much of Canada north of the St. Lawrence). In exchange, of course, it deposited the same soil elsewhere, but mixed with cobbles and boulders which, as in New England, would one day bedevil generations of farmers. The ice, or its heaps of debris, blocked rivers and forced them into new channels; old lakes were obliterated and new ones gouged out.* The debris heaps themselves, of diverse shapes, have contributed such picturesque names as moraine, esker, and drumlin to the lexicon of topography. Not merely Cape Cod, but also Nantucket, Martha's Vineyard, and the one-hundred-mile length of Long Island—to say nothing of thousands of other less conspicuous topographical features—are souvenirs of the Ice Age.

Note that nothing has been said about the Ice Age in the Southern Hemisphere. The reason is that nothing much

* Notably in Wisconsin, Minnesota and the adjacent parts of Canada, and in Finland and Sweden, where thousands of them remain today.

happened. In what one might call the "ice belt" of that hemisphere, there is almost no land for snow and ice to accumulate on—barring, of course, Antarctica. That grim continent was probably even more ice-covered than it is now, but that isn't much of a change. For the rest, there were mountain glaciers in the Andes, Australia, and southern New Zealand, some lowland glaciers at the tip of South America, and that was about it. Nonetheless, though the Ice Age brought about few significant direct changes in the Southern Hemisphere, it produced a number of indirect changes that were world wide. This came about through its effects on the oceans.

A few years ago, I was poring over a marine chart of the Virgin Islands. This thickly bunched collection of islands, islets and rocks, with one exception (St. Croix), rests on an underwater plateau no more than forty fathoms deep—and often a good deal shallower than that. Glancing at the southern edge of that plateau, just before the bottom drops off sharply to some 2,500 fathoms, I spotted a curious ridge. Along almost the entire edge of the plateau the bottom rose suddenly from a depth of some thirty fathoms to twenty or less, and with equal suddenness dropped off again. Following the carefully plotted soundings, I was able to trace the narrow ridges for fifteen miles or more. Looking further, I spotted other, shallower ridges closer inshore. They were, I suddenly realized, drowned coral reefs. At one time, the waters around St. Thomas must have been considerably shallower, to the point where the reefs which now lie beneath 120 feet of water were almost at the surface. The plateau, now featureless ocean, must have been a broad lagoon, fringed by a barrier reef like those surrounding many Pacific islands today. Later, the rising water covered the reef too deeply for the coral to grow, but other reefs formed in shallower water farther inshore, to be inundated in their turn.

Those submarine ridges, like Cape Cod, are relics of the Ice Age. The water that made up the great ice sheets—some millions of cubic miles of it—obviously came from somewhere, and where it came from was, of course, the oceans. Result: "sea level," which we are accustomed to think of as a fixture from which all other heights are measured, dropped—according to most estimates by over 300 feet when the glaciers were at their height.

The most obvious result of this world-wide oceanic recession was an "expansion" of the land wherever the ocean bottom was shallow enough to be exposed. Fishing boats 150 miles off our east coast have dredged up mastodon teeth—evidence that these mighty beasts once roamed where now the hake and flounder play. Bears and bison wandered about the basin of the North Sea. Perhaps the biggest single addition to the land was the broad plain formed from the bed of the shallow Bering Sea; this formed a bridge from Siberia to Alaska that carried a steady traffic of plants, animals and, ultimately, men between Asia and the Americas. Other land bridges helped carry other men through the East Indies and into Australia.

The combined effects of glacial ice and lowered sea level much later inspired one of the more charming legends of natural history: the expulsion by St. Patrick of the snakes from Ireland. With all respect to that eminent missionary, the job was done some thousands of years before he, or even Christianity, was born. At the height of glaciation, Ireland was almost totally covered by ice, and what land remained was far too cold for snakes or any such cold-blooded creature. But by the time the melting of the glaciers had left Ireland fit for serpents, the rising sea had inundated the valley between Ireland and England, forming a barrier that snakes (and a number of other creatures) found impossible to cross.

Along innumerable coastlines around the world, geologists and oceanographers have traced miles of submerged beaches, marking the lowered shores of glacial times. These beaches are of special value to students of the Glacial Epoch; because they are world-wide, they can help link events in unglaciated parts of the world to those in regions where the ice sheets left their own characteristic traces. In particular, as we shall see later, they have helped climatologists to figure out what was happening in tropical Africa when more northerly regions were ice-covered.

All these, and many other bits of evidence, have helped the scientists to build up the fascinating history, climatological and biological, of the Glacial Epoch. One of the earliest and most fundamental features of that history was the finding that there was not one single ice age, but many. For example, geologists examining a German deposit of glacial "till" (the same conglomeration of sand, silt and gravel that we find on Cape Cod) would discover beneath it traces of temperate zone, and perhaps even subtropical, vegetation —and beneath that another layer of till. In that region, clearly, there must have been at least two glaciations, separated by an interval with a far milder climate.

By the beginning of the twentieth century, it had become clear that the Glacial Epoch had encompassed four main Ice Ages (plus, it has subsequently been established, two or more less severe glaciations). In 1906 these were named, by two German geologists, after four Swiss valleys: Günz, Mindel, Riss, and Würm.

(These four names are a perfect example of what scientific nomenclature should be, but all too rarely is. They were picked carefully, so as to place the four successive Ice Ages in alphabetical order—and to leave room within, before, and after the sequence in case additional ice ages should turn up. Thus the minor ice ages which preceded Günz are often referred to as the Donau glaciation. The scientists who named the corresponding ice ages on the American continent would have benefited by some Germanic thoroughness. The names they picked—Nebraskan, Kansan, Illinoisan, and Wisconsin —form no rational sequence whatever.)

All four glaciations were, we now know, separated by periods in which the climate was quite as warm as it now is, and sometimes warmer. During at least one interglacial period, hippopotami wallowed in swamps where London stands today. In addition, the Wisconsin (Würm) Age at least (and probably the others, though the evidence is less clear) was interrupted by one or two *interstadials*—periods in which the ice sheets retreated some hundreds of miles but did not disappear entirely. The difference between an interglacial and an interstadial is approximately the same as that between a depression and a recession.

How long did all this coming and going of the ice last? The answer is still debated, estimates ranging from 300,000 to more than two billion years. Some of the differing answers, and the reasons for them, will emerge in the course of the following chapters, in which we enquire why the Glacial Epoch happened.

10. THE CLIMATOLOGISTS I

A Plethora of Theories

In *The Deep and the Past,* one of the best recent books on the Glacial Epoch, David B. Ericson and Goesta Wollin estimate that a new theory of glaciation "has been published for every year since the first recognition of the evidence for past glaciation. . . ." The authors may be exaggerating a bit for rhetorical effect, but even if they are stretching the truth, they aren't stretching it much. Explanations of the glacial epoch resemble the demon inhabiting the Gadarene swine: their name is Legion. They range, as Ericson and Wollin put it, "from the possible but unprovable to the internally contradictory and the palpably inadequate."

The basic problem stems from the fact that the Glacial Epoch was, as already noted, a highly unusual event during the earth's history. For as far back as the evidence of past climates can be traced—some hundreds of millions of years

—we find that at most times the earth was not merely unglaciated, but much warmer than it is today. In Antarctica, geologists have turned up beds of coal, evidence that a rich vegetation flourished where there is now perpetual ice and snow. In barren Spitzbergen, half way between the top of Norway and the North Pole, rock layers show tracks which Iguanodon, a 40-foot plant-eating dinosaur, left there some 120 million years ago. Unless Iguanodon had a metabolism totally different from the large reptiles of today (and there is not the faintest reason to suppose that it did), the climate of Spitzbergen then must have resembled that of Miami now. Even fifty million years ago, palm trees grew on the site of Paris, while breadfruit, cinnamon, and mango flourished in the regions around Leipzig. No matter what period is examined, the evidence adds up to much the same picture: the poles were temperate, the temperate zones were subtropical or tropical, and the tropic regions were . . . tropical.

Then, perhaps thirty-five or forty million years ago, things began changing. The tropical vegetation of Europe and North America began yielding to subtropical species like cypress and magnolia; by the time we reach five or ten million years B.P., these species in turn had given place to temperate vegetation much like that of today. After that, the glacial deluge. What happened?

One of the earliest theories of glaciation has the distinction of being one of the few—apart from the outright fantasies—which has been conclusively disproved. A century or so ago, it was generally believed that the earth had begun its life as a ball of molten material torn from the sun, which in the course of time had cooled to the point where it could form a solid crust and, eventually, to the point where life could exist on that crust. If the cooling continued, as it presumably had, the logical outcome would be glaciation.

We now know, from half a dozen lines of evidence, that the earth is not cooling off and has not done so appreciably for several billion years. Even more conclusive is the evidence that "our" Glacial Epoch was not the first. Geologists have turned up deposits of a rock called tillite, which as its name suggests is compacted and solidified glacial till. Often the tillite beds lie on other rocks that have been scratched and polished precisely like those we find in the paths of recent glaciers. These and other findings show clearly that there was a glacial epoch some 280 million years ago and another perhaps 600 million years back. Before that, the evidence is much less conclusive, but it is suggestive enough to raise the total of known or suspected glacial epochs to half a dozen, spread out over the past billion and a half years. Any theory that seeks to explain the great ice sheets must reckon with the fact that, though they were unusual events, they were by no means unique ones.

One way of explaining these rare but repeated deep-freezes is to assume that the earth or, more likely, the entire solar system has from time to time passed through a cloud of interplanetary dust which, by blocking off a portion of the sun's radiation, would cool the earth enough to get a glacial epoch underway. This notion is in scarcely better repute nowadays than the cooling-earth theory. There are, indeed, plenty of clouds of interplanetary dust around, but they are far distant from us. We know quite definitely that the last retreat of the ice began some eighteen thousand years ago, and if any dust cloud had been in our vicinity that recently, it would still be close enough to be easily spotted by modern telescopes.

Somewhat similar in its reasoning is a theory that the sun's radiation decreases from time to time because of events within the sun itself. This one was put forward by a distinguished astrophysicist, Ernst Öpik. Put very crudely, his theory amounts to saying that the sun, through the accumulation of waste products formed by the nuclear processes in its interior, develops periodic indigestion and, as it were, belches. The energy that produces the belch is subtracted from the sun's radiation output. Few astrophysicists, however, agree with Öpik that the sun does in fact undergo these bouts of cosmic dyspepsia. Nor do all meteorologists agree that a decrease in solar output would necessarily produce an ice age. Some of them, in fact, believe that the ice ages resulted from an *increase* in the sun's radiation. This theory, put forward by Sir George Simpson, who for many years headed the Royal Meteorological Office (the British equivalent of the U.S. Weather Bureau) is not quite so lunatic as it looks. Simpson reasons as follows:

For an ice age, you need above all much more snow in high latitudes. As we have already seen, snow (or rain) ultimately is drawn from the oceans—primarily, the warm oceans within the Tropic zones, whence moisture is transported to the temperate and higher latitudes. As Simpson sees it, an ice age begins with a slight increase in solar radiation. This naturally heats up the oceans and thereby steps up evaporation of water vapor. It also speeds up the "heat engine" which is our General Circulation. An elementary principle of thermodynamics states that the work output of any heat engine depends in the first place on the difference in temperature between the beginning of the heat cycle (the boiler, or the equatorial regions) and the end (the condenser, or the polar regions). It is for this reason that modern power plants use "superheated" steam, at a temperature around 500° F., rather than "normal" steam at 212° F. (the boiling point of water).

Now an increase in solar radiation would raise the temperature of the entire earth somewhat. But because of the earth's spherical shape, the tropics would be heated more than the poles. Result: a greater contrast in heat between the two regions and thus more work done by the atmosphere—meaning that air and water vapor would be transported more quickly and copiously to the Polar and Subpolar regions. Ultimate result: ice sheets.

Beyond a certain point, of course, the increase in temperature would reverse the process, for the precipitation around the poles would begin to fall as rain rather than snow. Now the ice sheets begin to shrink and eventually disappear and the earth goes through a warm and very moist interglacial.

When the radiation curve turns down again, the whole process is repeated in reverse. As the Polar regions cool, ice sheets form again. Then, with further cooling, the evaporation and transport of water vapor is no longer sufficient to maintain the ice, which once more disappears to begin another interglacial—this one cool and dry. Two cycles of this sort, each generating two ice ages, would, Simpson believes, account both for the four main glaciations and for the interglacials that separated them.

Despite a penchant for paradoxes, and a consequent wish to believe that a hotter sun could in fact produce a colder earth, I must report that thus far this notion seems to run counter to several kinds of evidence. First, there is the matter of ocean temperatures. Simpson's theory states that during ice ages the oceans around the equator would be *warmer* than they are now. So far as the Equatorial Atlantic is concerned, this is simply not so. Estimates of ice-age ocean temperatures, made by both the isotope technique already mentioned and by other methods to be described later, agree that since the end of the last ice age, the surface waters have grown warmer, not cooler, and the same is true for the Indian Ocean. For the Equatorial Pacific, the evidence is less clear. Some methods give ambiguous readings, others indicate a slight warming during the ice age—but not much.

Second, there is the matter of the land. If one believes Simpson, then the Equatorial regions must have been both warmer and moister during the ice ages. Again, what evidence there is runs the other way. To be sure, when Simpson first evolved his theory, it was generally assumed that the glacial ages had been accompanied by rainy—"pluvial"—ages around the equator. But as the evidence accumulates, it begins to look as if this was a notable scientific bum guess. The full story here will have to wait until Chapter 16; suffice it to say meanwhile that some very convincing facts show that during the last (Wisconsin) glaciation, most of Africa was distinctly dryer, and probably cooler, than it now is.

Simpson hasn't yet been positively proved wrong, but if

he is ever proved right, a lot of other scientists will have to eat a lot of learned journals.

Quite apart from this, I have a certain aesthetic objection to Simpson's theory, or any other (the dust-cloud hypothesis, for instance) that explains glaciation by external and apparently fortuitous processes. It has too much of the flavor of the god from the machine—that clumsy device employed by Greek playwrights with third-act trouble, when they lowered a god on a pulley to clean up the dramatic mess they had gotten themselves into.

If our earth had undergone only one Glacial Epoch, I could believe in almost any kind of explanation, probable or improbable. After all, *something* must have caused it! But since we have to explain not a single glacial epoch but at least three (and quite possibly six), I am more disposed to look for the cause, not in three (or six) fortuitous events but in the character of the earth itself.

11. THE CLIMATOLOGISTS II

What on Earth?

Before digging further into what—on earth—might have caused the great glaciations, let us refresh our memories by summing up what we are trying to explain.

In the first place, the Glacial Epoch did not arrive suddenly. The evidence of fossil plants and animals is quite clear on this; the climate of the temperate and polar regions had been growing slowly cooler for at least thirty-five, and perhaps fifty million years before the ice moved in. At the culmination of this process, came a series of relatively abrupt changes. By the most generous current chronology, the four glacials and three interglacials occupied no more than two million years altogether.

Putting it this way, it seems that what we have been looking at may be the result of at least two different processes: one very slow, the other relatively fast. In fact, when surveying the many unsatisfactory explanations of glaciation, it seems likely that a great deal of the trouble may have come from asking the wrong question: What is the cause of ice sheets? More and more, scientists are beginning to think that the right question is: What are the causes? What sort of events could cool the earth slowly over many millions of years—and then what other events could make the climate swing wildly back and forth from glacial to interglacial and back again?

Taking the first question first, what we are looking for is some long-term change in the earth which is known to have occurred not once but several times. One very eligible candidate is the changes which are known to have taken place in the extent and height of the earth's land areas.

Several times in its history, the earth has gone through geological "revolutions," world-wide epochs in which large areas of the crust were folded, or lifted, into great mountain ranges by processes deep in the earth's interior. During these same epochs, sea levels tended to drop. Apparently a rise in the earth's crust in one place tends to be "balanced" by a fall elsewhere, so that if the land is rising, the ocean bottom is sinking. Moreover, the same processes that form mountains are thought to form deep ocean trenches, such as we now find in many parts of the Pacific and a few other places as well.* As the result of one or more of these processes, the water of the oceans (which seems to be pretty constant in volume) then collects in the deep basins, so that a good deal of once shallowly submerged land is left dry, and sometimes high as well. Eventually, of course, the mountains are eroded away

* A more recent theory suggests that the squeezing together of the continents into mountains reduces their overall area, thereby correspondingly increasing the area (and therefore the volume) of the deep ocean basins.

by the ceaseless beating of wind and rain; the sediments formed by erosion are washed into the sea, where they raise the level of the ocean bottom, thus causing the oceans to "overflow" back on to the land. At the end of the cycle, land levels are generally low, and large parts of the continental areas are covered by shallow seas.

For a closer look at this process, let us consider the most recent revolution, which began some fifty million years ago. At that time, there were few of the high mountain ranges that are today such a conspicuous feature of the earth's topography. Much of what is now land was covered by water. A good part of the southeastern United States was submerged and so was the isthmus of Panama; western Europe consisted of several large islands; nearly the entire Middle East, including Arabia and the Iranian plateau, was under water; India was an island lying off the Asian coast, and a broad arm of the sea separated northern Europe from Siberia.

The subsequent period saw the gradual elevation of all the earth's great mountain systems: the Alps, the Rockies, the Himalayas, the Andes, the Caucasus, and (somewhat later) the Coast Ranges of California, Oregon, and Washington. Simultaneously, world shorelines advanced to something like their present positions. The earlier revolutions, though they are not known in such detail, must almost certainly have followed a similar course.

The first thing to be said about geological revolutions as a possible cause of glaciation is that their pattern is right. Though a limited amount of mountain building has gone on even during the "quiet" periods in the earth's history, there have been only a few revolutions; these have been separated by long periods in which nothing very radical happened. (This is rather what one would expect. Mountains and highlands are, if we take the long geological view, inherently transient things. The higher the land, the faster the streams flow—and the faster they cut away the land's substance. But once the land has been leveled down, it will stay that

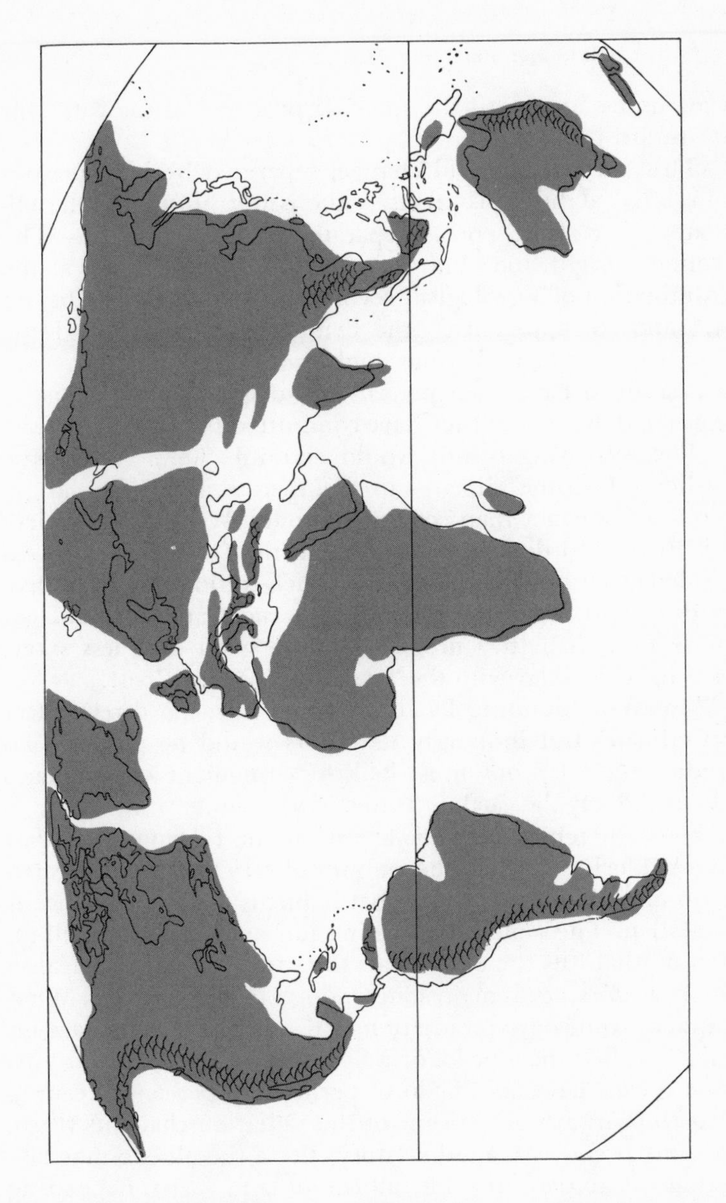

Some 50 million years ago, earth had more extensive oceans, fewer and lower mountains. Climates were mild everywhere.

way, unless and until the internal processes of the earth lift it upward.)

This pattern of rapid change, separated by long periods of relative stability, is precisely the pattern of the earth's climate: a few glacial epochs separated by long, mild eras. The timing is right, too. The recent Glacial Epoch came at the culmination of a geological revolution; so did the one before it,* and so, apparently—the evidence is sparser—did the one before that. Each time, high mountains and contracted seas seem to have been present, so to speak, at the scene of the crime. But could they have committed the crime itself?

The retreating oceans would certainly help. As we saw earlier, Maritime climates in Temperate and Subpolar regions are warmer than continental ones. Winters in particular are a good deal warmer. Thus if the Gulf of Mexico extended as far north as St. Louis (which it did some thirty-five million years ago) the central U.S. and Canada would get more rain than they now do—and a good deal less snow, as is the case today with the "maritime" Pacific Coast.

Mountain building by itself would have no direct effect on climate. But indirectly its effects would be marked. To understand why, one must look for a moment at some facts about the way the earth is heated by the sun.

You will remember that at one point I emphasized that in considering rainfall, the amount of rain that hits the earth is no more important than what happens to that rain once it has fallen. The same is true of the sun's radiation. Not all radiation that hits the earth heats the earth, and a good deal of the heat does not remain long enough to do any good. What happens to incoming radiation depends in the first place on what is called the albedo, or reflectivity, of the surface it hits. The oceans have an albedo of perhaps 10 percent, meaning that they absorb 90 percent of the radiation that hits them. (Stormy oceans are another story; there the albedo may rise as high as 40 percent.) The albedo of land varies, depending

* Though some geologists dispute this.

on the vegetation and whether it is moist or dry; it runs from about 8 percent for forest to 15 percent for grassland and as much as 20 percent for sand or bare rock.* Clouds have a high albedo—about 70 percent, and ice or snow the highest of all, with up to 90 percent of the radiation bouncing back into the sky.

Now mountains are, as everyone knows, cooler than lowlands, and because they are cooler, they receive more precipitation (which comes, remember, from rising, cooling air). If they are high enough, the precipitation will fall as snow. And once snow has fallen, it immediately starts sending nine tenths of the incoming radiation back into space—meaning that there is little radiation available for melting the snow. Mountains that are high enough will be permanently covered by snow even at the equator, as we can see in the Andes, in Central New Guinea, and as (Hemingway fans know) on Kilimanjaro in Africa. And even mountains that are not snow-covered will often be cloud-covered—meaning again a loss in "available" radiation.

The first effect of mountain building, then, will be more snow- and cloud-cover and less solar radiation available to heat the mountain areas. But the mountains also affect, indirectly, the fate of the radiation that does turn into heat—specifically, how quickly the heat itself is radiated back into space. Heat radiation from the earth's surface depends less on the albedo than on the atmosphere. This involves what is known as the greenhouse effect. Every motorist knows that if you leave a parked car with its windows shut on a sunny day, the interior will quickly reach an uncomfortable temperature, even in cool weather. The reason is that the windows are transparent to incoming sunlight, but less transparent to the outgoing heat radiation (which is not visible light at all but longer-wave infrared rays). Gardeners take advantage of

* Sand and rock *feel* hotter (e.g., to bare feet) than grass, but this does not mean that they absorb more of the sun's light; rather, for a number of reasons, our feet absorb more heat more quickly from them. In addition, the temperature of grass is held down by biological processes.

this "insulating" effect of glass when they build greenhouses.

The atmosphere, too, acts as a greenhouse. But not all the atmosphere. The oxygen and nitrogen of which it is chiefly composed are pretty uniformly transparent to both visible light and infrared. But some minor constituents of the atmosphere are not—notably, water vapor. And, of course the amount of water vapor in the atmosphere varies greatly from place to place.

Consider, now, a mass of air which is being forced over a mountain range. It will be cooled as it rises, and if the mountains are high enough it will be cooled enough for any moisture in it to fall as rain or snow. The Sierra Nevada of California gets over 40 *feet* of snow a year in some places ("nevada," in fact, is Spanish for "snowcovered"). As the air descends the other side of the mountains, it will regain the heat it lost in rising. But it won't regain the lost moisture. The result is the "rain shadow" desert, with air that may be hot but is also very dry, so that there is little water vapor to block the escape of heat radiation. If the desert itself is at a relatively high altitude (like our Great Basin) the effect becomes even more marked. Not only is there less water vapor in the air above it, there is less air altogether, since it is thinned out as the altitude rises. The effect is to somewhat increase the incoming radiation, but to increase the outgo of radiation even more. Thus if the earth's atmosphere is a greenhouse, then mountains, high plateaus and deserts are, as one scientist has put it, "holes in the glass."

Let us now consider the past 35 million years of the earth's history and see how all these processes together might have altered the climate from its then balmy state.

What we are starting from, remember, is a world with more sea and less land, few mountains and almost no high ones, no permanent snow cover (let alone glaciers) on land, no ice on any of the oceans. There is probably some winter snow in the Polar regions, but none anywhere else. Because of the lessened contrast in heat between poles and equator, the atmospheric heat engine moves sluggishly; winds are moderate and, though there are plenty of showers in most places (and doubtless downpours in the Equatorial belt, even as today), there are few severe storms. Most of the land is covered by forest or well-watered savannas. Into this Eden, then, enters the serpent of mountain building.

As the mountains push up, they begin to trap moisture wherever they lie across moist ocean winds—all across central Asia and along much of the western Americas, North and South. At first, the moisture they trap falls as rain, but the rain-shadow deserts that result begin to punch holes in the greenhouse glass. The climate grows colder. The Himalayas begin to block off Siberia from the warm winds of the Indian Ocean; the mountains of Alaska and the Canadian Northwest begin to wall off parts of North America from the Pacific, and the interiors of North America and Asia grow even colder. As the mountains continue to push upward, the seas draw back. No longer is the Mississippi Valley half inundated; the seaway to the west of Siberia has dried up. The continental interiors grow more continental—and colder still.

On the dryer highland areas, forest gives way to grassland or even desert (this, by the way, we know from independent botanical evidence). Result: a small but significant rise in albedo—and a still cooler climate. Already there is more "contrast" in the earth's climate—between continents and oceans, between highlands and lowlands. And more contrast makes for more turbulence, meaning more clouds, more and bigger storms. Though we cannot be certain, it seems probable that the main belts of cloud and storm would first have appeared far north or south of their present locations. In any case, they must have brought both more winter snow and still cooler climate because of the greater cloudiness and the higher albedo of storm-whipped oceans.

By this time the winter snows are falling in the Subpolar as well as the Polar regions. They are staying on the ground longer too; near the poles there may be only a few months in which the lowlands are free of snow and can absorb any sizable quantity of the sun's rays. The still-rising mountains of Alaska, Antarctica, and Greenland now have permanent snow caps which, before long, will begin to pack into glaciers that will fill the higher valleys. Still colder. Now the glaciers begin to reach sea level and, if their ends are close to the ocean, to split into icebergs that further chill the surface of the far northern and southern seas. Meantime, the accumulating glacier ice hastens the retreat of the oceans, making for still more "continental" climates on land.

The cooled Polar seas can no longer supply as much moisture as they did. But the loss is made good from southerly regions, for the sharpened contrast between poles and equator speeds up the atmospheric heat engine, transporting more moisture north and south from the tropics—and dropping more snow to cover more land for more months of the year, and to swell the size and number of the glaciers, which by now are appearing even in some mountainous regions of the Temperate Zone.

At this point—a few million years before the present—the world's climate is close to what it is today. The stage is set for an Ice Age.

Is this in fact what happened? The climatologists are still not sure. Few of them doubt that something of the sort occurred, the only question is "how much" it occurred. There seems little doubt that mountain building would cool the earth in the manner just described, but no one is sure whether it would have cooled it enough to make an ice age.

This problem of how much is, in fact, the stumbling block of every attempt to arrive at a "hard" explanation of the Glacial Epoch. We can assume any qualitative change we like in the earth or the sun—both those which might have happened but almost certainly didn't (like the dust cloud) and those which unquestionably did happen (like the geological revolution of the past fifty-odd million years). Putting these changes together with our knowledge, still woefully incomplete, of the atmospheric circulation, we can determine with considerable certainty what *kind* of alterations a given change, or set of changes, would produce in the earth's climate. What we cannot do is tell *how far* the alterations would go. Qualitatively, the Glacial Epoch problem has been "solved" in half a dozen ways; quantitatively, it hasn't been solved yet.

For all that, the geological theory of glaciation is in perhaps the strongest position of any. For one thing, we know that geological revolutions are a fact, while major changes in the sun's radiation, for example, are only conjecture. For another, the revolutions happened at about the right time and in the right manner to produce the glaciations which we know occurred. Finally, the revolutions were, until fairly recently, the *only* long-term changes in the earth that were known to occur periodically. Only within the past twenty-five years has another candidate come forward for the role of long-term climatic controller.

12. THE GEOPHYSICISTS I

Wandering Continents and Poles

Those who remember high-school algebra will recall the problem of equations with more than one variable. If a single equation contained only a single variable, x, it could be solved. If there were two variables, x and y, a single equation was not enough; two were required for a solution—and if there were three variables, three equations were needed.

The trouble with theories of glaciation has always been that there were more variables than equations. And over the past twenty-five years, things have been getting worse, not better. Several things which were thought to be constants seem to be turning into variables.

In the last chapter, we watched mountain ranges rise and fall and continents expand and contract as the seas retreated or advanced. But we assumed all along that the *locations* of the continental masses—flat or mountainous, large or small—remained roughly constant. Thus in speaking of Antarctica, for instance, we were talking about a land mass more or less centered on the South Pole; Africa was assumed to straddle the equator as it does today, and so on.

In fact, however, there is now much evidence that the continents, and perhaps the poles and equator as well, have moved around during the earth's history. The evidence for these startling possibilities derives in part from the study of glaciation; in turn, theories derived from the evidence have supplied some possible new causes for the glacial epochs.

No one knows who first conceived the notion that the continents might not always have occupied their present positions. It surely dates from at least as early as the beginning of the nineteenth century when the first really accurate maps of the South Atlantic became available. Some sharp-eyed explorer or ship's captain may well have noticed the remarkable mirror-resemblance between the east coast of South America and the west coast of Africa. Over a distance of many thousand miles, the two coasts fit together almost like two pieces of a jigsaw puzzle. It is not hard to imagine that the two continents were once a single land mass which at some point broke apart.

Studies of the last Ice Age turned up another interesting fact. Measurements of ancient shorelines established that in glaciated areas, such as Scandinavia and northern Canada, the enormous weight of the ice had pressed down the land beneath it; with the retreat of the glaciers, the land slowly rose again—parts of northern Sweden have pushed up more than 700 feet since the glaciers vanished.

Motions of this magnitude imply quite clearly that most of the earth is not solid. Rather, it seems, the only really solid part is a crust perhaps sixty miles thick; below the crust is a far thicker layer of liquidlike rock which can flow out of the way when pressure is applied from above and flow back when the pressure is removed. The rock must be enormously viscous, moving far slower than molasses in January—but it is a sort of liquid for all that.

Over fifty years ago, a German meteorologist named Alfred Wegener, contemplating this fact and a number of other bits of evidence, evolved the theory called Continental Drift. If the continents can move upward and downward on their liquid rock foundations, he asked, why can't they move sideways as well? Wegener believed that they not only could but did. Perhaps 200 million years ago, he suggested, there had been only a single continent—"Pangea"—which had broken into several large pieces. Gradually, these had drifted apart, opening up the present Atlantic and Indian oceans. Continental Drift, Wegener believed, would explain not only the jigsaw fit of Africa and South America but also a number of curious similarities among rock formations and fossils in regions now separated by thousands of miles of ocean. Most geologists, however, took a dim view of his conjecture. Many of the geological similarities, they declared, were not nearly as similar as he thought. For the others, coincidence.

Nonetheless, troublesome facts kept piling up. There was the matter of fossils. In both South America and Africa, for instance, paleontologists have unearthed the bones of Mesosaurus, a primitive reptile vaguely resembling a two-foot crocodile. From the nature of the rocks its fossils are found in, we know that it inhabited shallow waters, river mouths, and estuaries; it was not a strong swimmer. How could it have crossed three thousand miles of open ocean? Characteristic groupings of extinct plants have turned up in South America, Africa, Australia, Antarctica, and India—but not elsewhere. How could they have spread around the oceans to these widely-separated regions without leaving traces in land areas in between?

Some paleontologists sought to explain the fossil evidence

by inventing land bridges, or even "lost continents," over which plants and animals could have traveled (as we know they later did across the Bering Sea bridge). But growing knowledge of the earth's crust has made these bridges increasingly tenuous. It is one thing to reconstruct a bridge between Alaska and Siberia, which are separated by only a few dozen miles at their closest and by a sea only a couple of hundred feet deep. It is quite another to put a land bridge where now there is an ocean thousands of miles wide and thousands of fathoms deep. If such a bridge had indeed existed and then sunk, it would, from all we now know about the earth's crust, have left unmistakable evidence of its existence in the rocks beneath the ocean bottom. No such evidence has been found.

Among the few geologists to take Wegener seriously were those working in the Southern Hemisphere. They found his theory essential to explain another bit of troublesome evidence: the Permo-Carboniferous Glaciation.

This was the last-but-one Glacial Epoch, beginning some 280 million years ago. Even in Wegener's day, it was becoming clear that it had produced ice sheets in a number of southern continents. We now know that it occurred more or less simultaneously in Australia, South America, Africa, India, and Antarctica. Tillites and other deposits marking the presence of ancient ice sheets have turned up in subtropical and even tropical regions—southern Brazil, South Africa, and equatorial Africa as well.

Now if ice sheets occurred on the equator, surely the Northern Hemisphere must have been almost totally ice-covered. Only it wasn't. In fact, there is no convincing evidence of any ice sheets north of the equator—except in India, where the troublesome fossils also occur.

Remember that the most recent Glacial Epoch produced no ice sheets in the Southern Hemisphere, outside Antarctica, because that hemisphere is mostly sea. During the Permo-Carboniferous Glaciation, then, the *Northern* Hemisphere must have been mostly sea and the Southern mostly land. But this could have happened only if the continents were situated quite differently from where they are now.

From all this, and much other evidence, it has become almost certain that Wegener was basically right. Most scientists today would agree with geologist Rhodes Fairbridge of Columbia University. "It's not a question any more of whether the continents moved," he says, "but of how fast." To be sure, Wegener's concept of "Pangea" seems to have been overambitious. Asia (except for India), Europe, and North America were never merged with the southern continents, at least not at any time of which we have any record —though they seem to have done some drifting on their own. But all the southern continents, plus the Indian peninsula and Madagascar, do seem to have once formed a supercontinent (as large as present-day Eurasia) * which, some 280 million years ago, underwent its own "private" Glacial Epoch. Clearly this supercontinent must have been centered somewhere near the South Pole. And in fact there is evidence, quite apart from climatic evidence, that it was. This has to do with what is called paleomagnetism (*paleo* means "ancient").

Every Boy Scout knows that the earth is a magnet which makes the compass needle point north and south. Less widely known is that many rocks, too, are slightly magnetic. Their magnetism is not strong enough to swing a compass needle, but it can be measured with sensitive instruments. And it is "ancient," not modern magnetism, derived from the influence of the earth's magnetic field when the rocks were formed. Thus it can tell us—at least in theory—where the magnetic poles lay 100 or 200 million years ago, and presumably the geographic poles as well. For though the magnetic poles today are hundreds of miles away from the geographic poles (which is why navigators must allow for

* The clinching evidence came in 1969, when a species of fossil reptile, previously found in South Africa, was unearthed in Antarctica. There is no conceivable way in which this creature could have gotten from one place to the other across an ocean.

"compass variation" in plotting their courses), it is believed that over some thousands of years they average out the same. So the paleomagnetic poles are deemed to have been the same as the paleogeographic poles.

Now paleomagnetism is at present a highly controversial subject. Some of the conclusions drawn from it, climatologic and otherwise, seem to me rather far-fetched, viewed in the light of current knowledge. But for what it is worth, the paleomagnetic evidence does indicate that during the great "southern" glaciation, both Australia and Africa lay far closer to the South Pole than they now do. Some geophysicists have gone much further. Paleomagnetism, they believe, shows that not merely the continents but the poles themselves have been moving about; that is, while the continents were moving relative to one another, the entire crust—continents and ocean bottom alike—was slowly sliding around on the underlying "liquid" material. It is this "polar wandering" which, some believe, partially or totally explains the earth's great glaciations. One of these theories will be explained in some detail in the next chapter, but all of them center on the fact that the poles—the ends of the axis around which the earth rotates—are, as we know, the areas of the earth that receive the least solar radiation. For this reason, any change in the position of the poles with respect to the earth's major land masses and ocean basins (or, for that matter, vice versa) should mean a change in the climates of at least some regions.

Now from what we have already said about the plastic underpinnings of the earth's crust, it seems likely that the poles could have wandered in this way. But it is a long and tortuous way from saying that they *could* to proving that they *did*—and a still longer way to proving that the wandering caused the great glaciations.

The paleomagnetic evidence of polar wandering is reasonably unambiguous, but even so is by no means watertight. Readings on the poles' positions at a given period show a certain rough consistency from one place to another. But they also vary a good deal, especially from one continent to another. In part, at least, this could be explained by continental drift, but there are also cases where rocks of the same age, only a few hundred miles apart, show polar positions varying by several thousand miles. It is hard to chalk this up to continental drift, and, in fact, hard to separate the paleomagnetic effects of continental drift from those of presumed polar wandering. Have the continents been moving relative to the poles (and to one another), or have the poles been moving relative to the continents? Or does this amount to the same thing?

There are other problems. Although the paleomagnetic evidence squares pretty well with what is known about the great southern glaciation, when one turns back to the glaciation before that—some 600 million years ago—something seems to be very wrong. If the magnetic findings are to be believed, nearly all the continents at that time were closer to the equator than they now are, and the bulk of the glaciation occurred in what were, geographically, the Temperate and Tropic Zones. According to present knowledge, this is next to impossible.

There is also geological evidence of polar wandering, but its significance seems to me a good deal less clear. Much of this has to do with the distribution of "evaporites"—rocks that are formed when seas, or other mineral-bearing waters, evaporate and lay down deposits of substances such as rock salt or gypsum. Now it is clear that evaporites can be formed only in regions where the climate is reasonably hot (so that evaporation proceeds vigorously) and quite dry (so that the evaporated water is not replaced by rainfall and river flow). In fact, evaporites are forming today in desert and semidesert areas—around Great Salt Lake (the Bonneville Salt Flats), the Dead Sea, the USSR's Aral Sea, and so on. Logically, then, past deposits of evaporites must have been formed in or close to what were then the earth's Desert Zones.

The German geologist Franz Lotze, of the University of

Münster, has plotted the distribution of known evaporite deposits in the Northern Hemisphere for various geological periods and believes that they fall into belts, which shift with time. Some 500 million years ago, the belt crossed the North Pole; subsequently it moved slowly southward to its present location. If we assume that the evaporite belts represent past Desert Zones (which seems logical), and that these past zones lay in roughly the same position relative to the North Pole as does the present one (which seems at least plausible), one has to conclude that the pole has shifted. Five hundred million years ago it must have been located in the mid-Pacific, somewhere near Midway Island. Interestingly, this is roughly the same story told by the paleomagnetic evidence.

There are lots of things that can be said for or against Lotze's findings. One possibility is that his evaporite belts, like beauty, may lie in the eye of the beholder. The evaporites are real enough, to be sure. But the belts drawn around them are partly creatures of the imagination—especially for geological periods in which the known evaporite deposits are sparse and scattered. On the other hand, some of the deposits themselves lie only a few hundred miles from the present North Pole, a region which it is hard to imagine as ever having had a hot, dry climate unless the pole—or the continents, or both—has moved.

Other studies of evaporites are even more puzzling. Two Australian geophysicists, J. C. Briden and E. Irving, instead of trying to reconstruct evaporite belts have drawn graphs of evaporite deposits for different periods, first according to their present latitude and then according to their paleomagnetic latitude. The present-latitude graphs agree pretty well in showing two groups of deposits, one lying north and the other south of the Equator, corresponding more or less to the two desert zones we know today. The paleolatitude graphs, however, show the deposits lying closer and closer to the Equator the further back we go; from about 300 to 500 million years ago, they are actually centered on the Equator.

This doesn't seem to bother Briden and Irving much—

certainly not to the point of rejecting the paleomagnetic evidence. But perhaps it should. Almost the only constant feature of the earth's past climates appears to be the warm wet Equatorial Belt. Occasionally that belt has been narrower than at present; at most periods, it seems to have been considerably wider. But it probably has always lain about where it does today—nor, given the dynamics of climate as we know them, could it have been otherwise. The Equatorial regions receive the most solar radiation, the Equatorial seas, therefore, generate the most water vapor and the Equatorial lands receive the most rainfall. Unless you are willing to imagine the continents as strung out along the Equator, thus largely eliminating the Equatorial seas, an Equatorial desert belt becomes unthinkable (and even then would be barely thinkable). And while almost everybody concedes that the continents have moved around, nobody is willing to claim that their motions placed all of them on the equator at the time that Briden and Irving are talking about.

But it is when we examine the biological evidence bearing on polar wandering that the real trouble starts. Climatic zones, present or past, are defined not only by evaporite belts but also by biological belts; the animals and plants of Canada and Siberia are obviously different both in kind and in number from those of Brazil or the Congo.

As has been seen, for the fairly recent geological past—from eighty million years ago to the present—plants are a pretty good guide to climate, since they are usually similar enough to present species to judge their climatic preferences. Ralph Chaney, of the University of California, has plotted out the distribution of fossil plants over the Northern Hemisphere sixty million years ago. Since this was a benign, nonglacial period, the various plant species are found well north of where their relatives grow today—temperate forests in central Alaska and Spitsbergen, cool-temperate trees in the now near-barren Canadian archipelago. But allowing for this northward shift, the distribution of the plant zones around the North Pole is remarkably like today's. This, in turn, means that the pole must have been very close to its present position, instead of having shifted a thousand miles or so, as paleomagnetic evidence indicates. Chaney himself is quite definite: sixty million years ago, he believes, "Eurasia and North America . . . occupied essentially the same positions with relation . . . to the North Pole that they do today."

Another biologist who takes a poor view of polar wandering is F. G. Stehli of Western Reserve University. He has concentrated on the period around 250 million years ago, and studied the distribution of certain types of fossils. These are of animals—primarily large reptiles and certain species of the mollusklike marine creatures known as brachipods—which, from our knowledge of their present relatives, seem restricted to quite warm environments. He finds that their distribution around the globe in those far-off times forms a "warm zone" centered on today's poles—rather than on the poles delineated by paleomagnetism.

Stehli has also tackled the problem from a somewhat different angle, using what he calls "diversity gradients." This depends on the well-known rule that groups of animals tend to be less diverse (i.e., to include fewer species) in cooler climates. A good example is the grazing animals of today. In tropical Africa there are literally dozens of species, ranging from rhinoceroses to antelopes. In the north Polar regions, there are only two: the musk-ox and the caribou (reindeer). Again, Stehli finds that diversity 250 million years ago decreases the nearer you get to today's poles—meaning that the position of the poles was apparently the same then as now. He himself concedes that his findings are based on very limited amounts of data (some of his critics use much sharper language, one geologist of my acquaintance calls him an idiot). My own feeling, indeed, is that in some respects—notably, in casting doubt on continental drift as well as polar wandering—he is arguing ahead of his facts. But for whatever his findings are worth, they seem to conflict sharply with the paleomagnetic evidence.

So there we have it: geophysicists on one side, biologists on the other, with geologists in between but leaning toward the geophysical side. These conflicts of evidence, as well as conflicts in the paleomagnetic findings themselves, have caused some geophysicists to wonder whether the earth's magnetic field may not have changed radically in the course of geologic time. Conceivably, they believe, it may in the past have had not two poles but four—or even eight!

The plain fact is that we really know very little about terrestrial magnetism. The earth's magnetic field has, indeed, been mapped in great detail (largely because of its importance to navigation). But as to what creates that field, our knowledge is still rudimentary. Current theory (which is extremely complex) accounts pretty well for the known facts. But it is not detailed enough to let us predict what the intensity of the field should be, or whether its intensity is stable or fluctuating. Rather recently, geophysicists discovered that the field has a habit of reversing itself from time to time, so that the North Pole becomes the South Pole, and vice versa! This fact, as we shall see in Chapter 15, has turned out to be of great importance in the study of the Glacial Epoch, but it, too, was not predicted by current theory. Clearly, we will need a clearer understanding of both magnetism and paleomagnetism before we can be sure that the paleomagnetic data means what it seems to.

13. THE GEOPHYSICISTS II

The Pitfalls of Elegance

Since I have been fascinated by science since childhood (to the point where I began my higher education at MIT) I have sometimes wondered how I happened to end up as a writer rather than a scientist. The answer, I think, is a kind of laziness.

To be a scientist requires, above all, what Carlyle claimed was the prime attribute of genius: "a transcendent capacity for taking trouble." The scientist must perform innumerable measurements (and then repeat them down to the smallest detail to make sure there were no mistakes), engage in tortuous calculations, and rummage through stacks of papers describing other scientists' work—many of them written in a language which only by courtesy can be called English. For all of this I lack the patience. What intrigues me about science is what is sometimes called the aesthetic element—the putting together of the facts and the spinning of theories to explain them. For science, make no mistake about it, is more than the mere piling up of data; it is also an exercise in applied aesthetics—at least, to any scientist who is more than a hack. Having gotten his facts, he tries to account for them by a theory that is aesthetically "elegant," that will account for the maximum number of facts with the minimum of assumptions.

It was this impulse toward elegance, I suspect, that led William Donn and Maurice Ewing, of the Lamont Geological Observatory, to put forward still another theory of the glacial epochs. For their explanation, whatever else it may be, is elegant. It explains the entire process—the long preglacial cooling, the glaciations and the interglacials alike—as the result of one single terrestrial change: polar wandering. Though their theory has undergone several revisions, it has retained this element of fundamental simplicity. In its most recent version (1966), it goes like this:

Not all the heat transferred from the Equatorial to the Polar regions is carried by the atmosphere. A sizable portion of it is borne by the oceans. Warm currents, like the Gulf Stream and Japan Current, move toward the poles, just as warm air masses do; cold currents, like the Labrador and Humboldt currents (the latter off the west coast of South America) of course move in the opposite direction. It follows, then, that if the poles are located in regions where warm currents can move toward them more or less freely, they will receive more heat than when they lie in regions of "thermal isolation."

When the earth's climate began to cool some thirty-five million years ago, the South Pole, according to paleomagnetic readings from Antarctica, was not located in the center of that continent as it now is, but in the Southern Ocean just off the Antarctic coast. Similarly, the North Pole did not lie in the center of the Arctic Ocean (which, because it is almost totally landlocked, receives few warm currents) but off the northern coast of eastern Siberia (or, possibly, northwest of Alaska—it depends on whose paleomagnetic readings one believes).

As the poles moved toward their present thermally isolated positions, the polar regions grew steadily colder, so that eventually the stage was set for the Glacial Epoch. The glaciation probably began in Antarctica. Once that continent was largely ice- or snow-covered, its high albedo would help cool the atmosphere still further. Before long, glaciers would begin to form in the far north of Canada, Greenland and Siberia.

Here we run into an apparent paradox. Today these areas, though bitterly cold, are also very dry—too dry, it would seem, for glaciers to form. But at the beginning of the glaciation, Ewing and Donn believe, the Arctic Ocean, though certainly cold, was not ice-covered as it now is, so that evaporation from it would supply adequate if not copious quantities of moisture to keep snow falling and ice accumulating in the far northern lands.

Eventually, the Arctic Ocean froze over. But by this time, glaciers were, so to speak, off and running. They had already pushed far enough south so that they no longer needed moisture from the north; the North Atlantic and the Gulf of Mexico supplied plenty of snowfall to keep them grinding on in North America and Europe. In Siberia, as we have already seen, the barrier of the Himalayas cut off moisture from the south, so that the glaciers ceased to grow once the Arctic Ocean had frozen. Eventually, the southward-moving ice sheets reached the point of balance, having pushed into warmer regions where melting canceled out their forward movement. At the same time, however, the North Atlantic was growing cooler, from the icebergs shoved into it and from the cold air off the ice sheets. And the cooler it got, the less moisture evaporated from it. Eventually evaporation dropped to the point where the great ice sheets became starved for moisture and began to retreat. And the retreat, so it is held, was bound to continue until they had vanished, for the still-frozen Arctic Ocean could not serve as a "reserve supply" of moisture.

With the ice sheets gone, however, the Arctic Ocean itself must have thawed—and the whole process began all over again, with the ice sheets advancing and retreating three more times. Thus in the Ewing-Donn view, the four Ice Ages resulted essentially from a self-maintaining cycle which began with glaciers in the far north, nourished by an ice-free Arctic Ocean, and ended with extensive glaciation that starved for moisture when the Arctic Ocean froze and the North Atlantic cooled.

The Ewing-Donn theory has generated plenty of controversy, notably with Cesare Emiliani of the University of Florida. Emiliani has his own theory of the ice ages, which includes the "geological revolution" theory described in Chapter 11. Writing in the august pages of *Science,* a journal that does not encourage strong language, Donn has called Emiliani's work "inaccurate" and his theories "untenable." Emiliani has struck back with "wholly uncritical evaluation" and "questionable statistical arguments." What they say about each other in private had best be left to the imagination.

My own feeling about the Ewing-Donn theory is that it is a bit too elegant to be true. In their published work, at least, they simply ignore the whole question of mountain building and land elevation. Personally, I find it hard to believe that polar wandering could have produced the Antarctic or

Greenland ice caps without the mountains that rose in those areas during the same period. Polar wandering may have contributed to the Glacial Epoch, but I can't see it as the sole cause. And even the case for it as a contributing cause seems to me a long way from being proved. So far as the South Pole is concerned, the shift into regions of ever greater thermal isolation seems pretty clear—always assuming we can believe the paleomagnetic evidence. But for the North Pole, even paleomagnetism shows the pole on the edge of the Arctic Ocean at the very beginning of the long preglacial cooling. With that land-locked ocean on one side and a sizable chunk of Siberia on the other, it must have been pretty thermally isolated even then. Moreover (still according to paleomagnetism), the North Pole reached its present position ten to fifteen million years ago. Yet the earliest hard evidence of glaciation anywhere outside Antarctica takes us back no more than some three million years. Why the long wait?

Ewing and Donn try to get around these problems by saying that the paleomagnetic evidence isn't really accurate; it may, in fact, be out by as much as 1,200 miles. This strikes me as something of a scientific fast shuffle. It is a little like a lawyer trying to prove an embezzlement by pointing out discrepancies in a company's books—while admitting that the bookkeeper wasn't very good at arithmetic. If the magnetic evidence is less than accurate—and it certainly is—surely it is just as likely to be off in either direction. An error of 1,200 miles could place the North Pole of fifty million years ago in two locations: the North Pacific Ocean, or almost in its present position. And without further evidence, nobody can say which, if either, was actually the case.

Concerning the Ewing-Donn explanation of the advance and retreat of the glaciers, I shall have more to say in the next chapter, on glacial chronology, which is as important for understanding the comings and goings of ice sheets as it is for understanding the evolution of man. But so far as the causes of the Glacial Epoch as a whole are concerned, I must for the present give Ewing and Donn credit only for an elegant try.

As a radical contrast to the stark elegance of Ewing and Donn, let us close this chapter with a quick look at one more theory. Its author, Rhodes Fairbridge calls it "an eclectic theory of ice ages." Eclectic it certainly is. Fairbridge explains the advance and retreat of the ice sheets by a combination of a freezing and thawing Arctic Ocean (à la Ewing-Donn), changes in solar radiation (of which we will hear more in the next chapter), and variations in the number of sun-spots. The Glacial Epoch itself he explains by a combination of polar wandering, mountain building, and shifts in ocean currents. On this last, he loses me, since the changes he describes, though apparently real enough, seem likely to make the polar regions warmer, not colder. But this doesn't really matter. A scientist with an eclectic theory is like a man who wears a belt, suspenders and an elastic waist band. If one, or even two, of his assumptions gives way, he is still not caught with his theoretical pants down.

14. ICE AGE CHRONOLOGY I

The All-too-human Scientists

On the record, mankind seems to have an inveterate need for supermankind. Many of us seem to derive some comfort from the notion that there is somewhere, on earth or in heaven, a being that transcends man's limitations, who is all-seeing and all-knowing, who does not make mistakes.

In earlier epochs gods and kings (sometimes the same individual) filled this role. More recently, fuehrers of various sorts have joined the superman club. But now God is rumored to be dead—or at least ailing—and kings are out of style (as I write this, Constantine of Greece has just joined the Association of Unemployed Royalty). And, for the moment, the memory of the havoc wrought by dictators has made them unfashionable in the Western world.* The most recent candidate for superman-in-residence has become the Scientist. He may be a Good Scientist—the kind who brings us wonder drugs and atomic power—or a Bad Scientist—the kind who brings us germ warfare and atomic bombs (sometimes, the same individual). But good or bad, he is Super. He may, indeed, have a few endearingly human personal traits—absentmindedness, for instance. But once he is in business within the walls of his laboratory, he becomes superhuman: a cold detached intellect reasoning flawlessly from premise to conclusion, as incapable of error as the computer whose keys he punches.

This view of the scientist (or of computers!) is of course, nonsense. Many years of experience with scientists has shown me that though they are, as a class, considerably brighter than average, they are not a whit less human than other people. On occasion they are blind to things they do not choose to see, deaf to merited criticism, and perfectly capable of selecting and twisting facts to support their own prejudices.

The long controversy over the chronology of the Ice Ages is a case in point. The scientific record—an almost interminable one—is replete with fragmentary facts, dubious data, and the kind of logic heard in a husband-wife argument over who forgot to put out the cat. The ambiguities and contradictions within it are by themselves enough to give employment to a whole generation of paleoclimatologists.

The first reasonably scholarly estimate of ice-age chronologies comes from Penck and Brückner, the same two systematic Germans who originated the classic Günz-Mindel-Riss-

* Unless, of course, they are "anti-Communist" dictators, in which case they become respectable if not fashionable.

Würm terminology. (Not a few climatologists nowadays consider this something of a Greek gift, by the way. The four-ice-age theory, they feel, is a Procrustean bed into which the facts can only be fitted by mangling them. But of that more later.) The two Germans faced a forbidding job even in trying to piece together what had happened, let alone how long it had taken. Ice sheets have an infuriating habit of scraping away the rocks and soil which lie in their paths—including, of course, the characteristic deposits left by earlier ice sheets. It is partly for this reason that both the extent and the chronology of the last (Würm-Wisconsin) Ice Age are far better understood than those of the preceding ones. Much of the earlier evidence was destroyed (or sometimes buried) by the later.

Glaciologists have therefore been forced to concentrate their efforts around the edges of the ice sheets, examining moraines and similar formations of glacial debris (like those that make up Cape Cod), deposits of "loess"—the wind-blown soil deposited by the fierce, dry winds that blew along the ice edges—and so on. It was by examining geological features of this sort that Penck and Brückner worked out the scheme of four main ice ages, each composed of two or (for Würm) three substages ("stadials").

To estimate the ages of the various stages, they sought to determine how extensively the different collections of glacial relics had been altered by erosion and by the chemical "weathering" which helps to change the characteristics of soils. One clue, for instance, came from a type of topography very common around glacial edges: the so-called knob-and-kettle-hole terrain. Once again, Cape Cod is a good example. The land is not dominated by any major system of ridges carved out by streams, but instead is covered with hundreds of small hills ("knobs"), interspersed with steep hollows ("kettle holes") which, if they are deep enough, form the little ponds that make the terrain so attractive. These are relics of the retreating glacier. As the ice drew back, enormous blocks broke off and were left behind; meltwater piled debris on top of them (insulating them from the sun) and around them. In the course of time, the blocks gradually melted, leaving their imprints as kettle holes, with the surrounding glacial rubbish forming the knobs. Thanks to this process, the secondary roads of the Cape that weave their way through the pock-marked terrain often resemble the track of a drunken python. The knob-and-kettle-hole pattern is so distinct on the Cape that we know it must have been formed relatively recently. Elsewhere, one finds similar patterns which have been blurred by erosion; these must be of an earlier date. In still other places, one sees a "normal" terrain with only the main ridge of the glacial moraine as witness that the ice was once nearby.

One can also examine the terraced banks of rivers. During an ice age, there is little water in nearby river-beds; at the same time, there is plenty of geological debris, supplied both by the glacier itself and by the splitting action of frost on rock surfaces. As a result, there is more debris than the river can handle. Instead of carrying its load of rubbish well downstream, it will drop much of it, gradually elevating its bed and building up its banks into terraces. During an interglacial stage, by contrast, the river flows more freely and has less junk to carry, so that instead of building up its bed it cuts down into the glacial deposit it had previously built up.

By estimating how long these and other processes had taken, Penck and Brückner arrived at an educated guess for Ice Age chronology. In round figures, they said, each Ice Age had lasted 60,000 years. The first two and last two were separated by interglacials of the same length; between these two pairs came "The Great Interglacial," lasting nearly a quarter of a million years. Grand total for the whole process: 600,000 years.

These figures were, by any standard, based on the most shaky kind of evidence. It was therefore something of a surprise when a second chronology, based on quite different evidence, apparently confirmed the Penck-Brückner estimates. This time scheme was the work of many hands, but most particularly the Serbian mathematician Milan Milankovich. It was derived not from the results of the ice ages but from their presumed cause.

According to the Milankovich theory, ice ages are caused by variations in solar radiation. Put this way, it sounds very much like Sir George Simpson's rather dubious theory, but in fact it is very different. First, the variations it deals in are real, not conjectural. Second, they are not variations in the total radiation falling on the earth (i.e., in the output of the sun), but merely in the distribution of radiation from one place to another and from one season to another. These shifts in distribution result essentially from imperfections in the earth's motions. For example, our orbit around the sun is, as most people know, not a perfect circle but is slightly flattened into an ellipse. Moreover, due to the gravitational influence of the other planets (especially giant Jupiter) that ellipse is a trifle flatter at some times than others.

For somewhat similar reasons, the tilt of the earth's axis, which causes the seasons, is not always precisely the same. At present it is about 23½°, but it can increase to as much as 24½° and diminish to as little as 21½°. There are other changes in the earth's motion that modify the effect of these two changes, but they need not concern us. Nor, for that matter, need we cudgel our brains over the precise way in which all these factors work. The essential thing to remember is their result, which is twofold: (1) A redistribution of radiation between summer and winter—meaning that winters can average either colder or milder than at present while the summers, to balance things, are hotter or cooler, respectively. (2) A redistribution of radiation among the zones of latitude, so that the regions between Lat. 45° and the poles would receive more (or less) radiation while the regions between that latitude and the Equator would receive less (or more).

Let me repeat: The earth as a whole does not receive a bit more (or less) radiation—but some places receive more (or less) at some seasons.

Now the various changes in the earth's motion, which in turn produce the radiation changes, are all roughly cyclical, but they are by no means "in phase"; that is, one set of changes sometimes reinforces another and sometimes cancels it out, in whole or in part. The net result is that the radiation changes for a particular latitude belt will go through a sort of cycle of their own, swinging from maximum to minimum, or vice versa, in something like 22,000 years. But this figure itself is only approximate; the actual swing may take a good deal more or less time. Moreover, some swings are much bigger than others. The result is that a graph of the radiation changes looks rather like one of those Cape Cod roads just mentioned—only more so.

It took Milankovich nearly twenty years to work out his radiation tables. The math is not particularly difficult, but the figuring—this was before computers or even rapid desk calculators—was laborious. Painstakingly, he worked out graphs for half a dozen different latitudes and then picked out one of them (Lat. 65° N.) as the rationale for his theory. Milankovich's stated reason for picking this particular graph was that 65° N. was where the glacial action was; that is, the various centers from which the great ice sheets had expanded lay at about that latitude. In fact, however, this was true of only the one in Scandinavia. Another, in Scotland, lay at about 58° N., while the two main centers in North America (where much more ice was) were at around 55° and 70°, respectively. And, quite apart from this factual inaccuracy, I wonder why Milankovich, if he knew all along that 65° N. was the key latitude, wasted years in figuring out graphs for all the other latitudes. My suspicion is that he worked backward; that is, having first plotted out all his graphs, he then picked out the one that best fitted the traditional glacial chronology of Penck and Brückner. In so doing, the Serbian savant was employing a time-hallowed, though very unofficial, scientific concept: Finagle's Constant—sometimes crudely known as The Bugger Factor.

Finagle's (or Phenagle's or Murphy's—the exact name

of its discoverer is lost to history) Constant is a mathematical curiosity because it is a variable constant. As its vulgar name suggests, it is used to bugger up the results of computations —in a constructive way. Technically, it is defined as that number which, when added to, substracted from, multiplied by, or divided into the answer you've got, gives you the answer you want. In my MIT undergraduate days, a proper understanding of bugger factors was absolutely essential if you wanted to pass Physics I; I have no doubt that it still is.

(If this sounds cynical, consider that the proper bugger factor can not only convert a wrong answer to a right one, it can also help you discover why you got the wrong answer.)

Milankovich's first bugger factor, then, was the choice of Lat. 65°. For this latitude, he plotted out the curve of radiation changes for the summer half of the year. Troughs in this curve represented times of cooler summers and milder winters; peaks represented periods of hotter summers and colder winters. These, he believed, represented, respectively, times of glacial advance and glacial retreat.

The reasoning was as follows: When summers are cool, the winter snows will stay longer; eventually they will stay all year and will begin to build up into ice sheets. The fact that during the same period the winters will be relatively mild is unimportant, since in the higher latitudes they won't be mild enough to reduce the snowfall. Average winter temperatures may be higher, but not high enough to top the freezing mark. Conversely, when summers are especially hot, the ice (or snow) will melt more rapidly than it can be replaced during the following winter. Winters will be extra cold, to be sure, but this will not increase the snowfall; indeed, it may actually reduce it (most of us are familiar with the really bitter winter weather in which it is "too cold to snow").

Though something less than watertight, this theory seemed plausible enough. It seemed even more plausible when all the "troughs" of Milankovich's graph—the presumed cool-summer, ice-building periods—turned out to fall just about where Penck and Brückner's chronology had placed the ice ages. Well, almost all. There were, as a matter of fact, several troughs which did *not* correspond to ice ages at all; Penck and Brückner's "Great Interglacial" included no less than five of them. To get around this, Milankovich's partisans (of which there were by this time quite a number) introduced a second bugger factor: they simply disregarded all the troughs that failed to reach a certain depth, assuming that these periods of cool summer had not been quite cool enough for an ice buildup.

The next difficulty was raised by Sir George Simpson, the veteran meteorologist whom we have already met as the author of a competing ice-age theory. He pointed out, in effect, that Milankovich has been using still another bugger factor —and it was the wrong one. It isn't the drop in solar radiation itself that would produce an ice age, Simpson pointed out, but the fall in summer temperatures presumably produced by the drop in radiation. Now, as we have already seen, temperatures anywhere on earth only in part reflect the radiation at that point; they also depend heavily on how much heat is carried to, or from, the region by the atmosphere and oceans. Milankovich, said Simpson, had underweighted this factor, so that his estimates of the drop in summer temperatures were four times too large. The actual drop would have been quite insufficient to build ice sheets.

Another blow came in 1953. Forty years of astronomical observations had yielded new and more accurate figures on the irregularities of the earth's motions. Using these new figures, and a computer, a Dutch-American astronomer, A. J. J. van Woerkom, recalculated Milankovich's curve and found a deep "glacial" trough smack in the middle of the Great Interglacial. Clearly something was very wrong, with either the Milankovich theory or the Penck-Brückner chronology— or both.

One of the reactions to this embarrassing fact throws an embarrassing light on the vaunted objectivity of science. The late Frederick E. Zeuner, one of the great names in geochronology, and a strong partisan of the Penck-Brückner time scheme, described the differences between van Woerkom's and Milankovich's curves—including that troublesome interglacial "glaciation"—as "negligible." This was precisely the opposite of what van Woerkom had said. Some of the Milankovich partisans, however, were less impervious to facts than Zeuner. Recognizing that something had to give, they simply jettisoned the Penck-Brückner chronology, at the same time restating the Milankovich theory so as to take account of Simpson's criticisms.

Perhaps the most attractive of these "revisionist" views was put forward by Cesare Emiliani and Johannes Geiss. Their version is clear, well thought out, and attempts to take account of as many facts as possible. It may or may not be true, but it certainly tries to cover all the bases.

Because the Emiliani-Geiss theory is a fine example of a well-constructed scientific hypothesis, it will be described here in some detail—the more so in that the description will help to pull together a lot of relevant facts about the ice ages themselves. I shall also, however, inject my own comments (*in italics*) as a way of indicating how many assumptions, conjectures and other types of bugger factors even a good climatic theory must employ.

Emiliani and Geiss begin by conceding Simpson's point that changes in solar radiation by themselves would produce only insignificant effects on world climate, and cite several kinds of evidence in support. For example, they note, measurements of ancient ocean temperatures (by the oxygen-isotope method) show that, say, twenty million years ago, well before the Glacial Epoch began, there were no cyclical temperature fluctuations corresponding to the fluctuations in radiation. Thus, they say, a radiation change "may produce important climatic effects only if it may act as a trigger to produce glaciation. It is glaciation [*not radiation*] that affects climate directly." *This amounts to saying that a gentle push may have little effect on a man walking along the street but very serious effects on the same man standing at the edge of a cliff. As a principle it is obviously sound.*

But if radiation is the climatic trigger, what loads the gun? Here, Emiliani and Geiss favor the theory of rising mountains and contracting oceans already set forth in Chapter 11. It is this long-term process, they believe, that sets the stage for glaciation.

Their story proper begins with the stage set—world temperatures considerably lower than their "normal" level, glaciers in the mountains and ice caps in Antarctica and Greenland—in fact, with conditions much like those of today. At that point, the next radiation minimum "made it possible for permanent snow to persist over some areas of the high latitudes," especially in Scandinavia, where the eastern part of the North Atlantic, warmed by the Gulf Stream, "supplied an ample source of moisture." Before many centuries had passed, the accumulating snow would have begun to pack into ice caps, not only in Scandinavia but also in Siberia and North America. The latter, the biggest of all, "were fed mostly from the Gulf of Mexico and the Eastern Pacific." *Not quite right. The main North American ice caps were centered in the northeastern part of the continent, and must therefore have gotten a good deal of their moisture from the northwest Atlantic; the Pacific contributed little, except to the Rocky Mountain ice sheet.*

"After having reached a certain size, the young ice-caps expanded mainly because of their own effect" on the climate. *This is the "trigger effect"—a key point in the argument.* "Air temperatures above them became very low during the whole year" *because their high albedo reflected most of the sunlight back into space.* A permanent, sharp difference in temperature developed "between the growing ice caps and the adjacent oceanic areas." *Perfectly sound.* "Atmospheric turbulence increased and the North Atlantic . . . and Gulf of Mexico were swept by increasingly swifter, colder and dryer winds." *Equally sound. As already noted, the atmospheric circulation is powered by temperature differences, and the greater the difference, the more vigorous and turbu-*

lent the circulation. "Evaporation increased, even though surface waters were beginning to cool, resulting in greater average precipitation." *Not so sound. Swifter, dryer winds would increase evaporation, but cooler winds and oceans would reduce it. Our authors are guessing here—but it isn't a key point.*

"The increase of albedo and decrease of temperature were furthered by the expansion of the ice caps [*Obviously*] . . . and by the increase in cloudiness over areas which are now arid or semi-arid, caused by the shift of the climatic belts toward the equator." *Some effect of this sort is probable, but how big it was is conjectural.*

"The decrease in sea level [*from the locking up of water in the ice caps*] and piling up of ice on land increased the average . . . height of the continents by more than 300 meters [*about 1,000 feet*] at the last glacial maximum."

A good point. Mountain building is mountain building, whether the mountain is made of rock or of ice. And the higher the land the greater the probable snowfall. Our authors go on to suggest that the higher land would further lower temperatures by the "holes in the greenhouse" mechanism depicted in Chapter 11. This would be true only if you accept that particular assumption. The figure of 300 meters is of course an estimate, but if anything it is on the conservative side.

"Vigorous thickening and expansion of the ice caps continued" as long as the adjacent oceans remained relatively warm, with resulting heavy evaporation. *This is an important point—and a very uncertain one. It is likely that, for a while at least, the amount of evaporation and precipitation may have changed little, but this is not to say that the location of precipitation remained constant. In fact, it almost certainly did not. There must have been an accumulation of very cold masses of air over the ice sheets, which would have had the effect of pushing the Stormy Zone well to the south of where it now is—and in fact we know from other evidence that the Mediterranean and other areas south of the ice were wetter than they are now. Thus precipitation over the ice sheets must have dropped off considerably. But the ice sheets could still have continued to expand, since even the lessened precipitation would mostly have failed to melt —i.e., would have stayed to form more ice. We don't really know.*

"Cooling of the surface waters, however, was continually reducing . . . evaporation." *This, at least, is certain.* "When the water masses of the Atlantic Ocean were sufficiently cooled, the northern surface water began to freeze. Eventually [a large area] became largely covered with pack and drift ice." *Having gotten the glaciers going, our authors now have to work out some way of stopping them. The frozen North Atlantic is probable—even today sizable areas of the seas around Greenland are largely ice-covered in winter —but not certain. The reader will note the similarity to the Ewing-Donn frozen Polar Sea theory; both might be true.* "The process of freezing was undoubtedly favored . . . by the intense discharge of ice-bergs to the sea, which produced a cold surface layer of relatively low salinity." *Fresh water is lighter than salt and therefore tends to float on it; fresh water also freezes at higher temperatures than salt water—hence the use of rock salt to melt sidewalk ice and snow in winter. But note that that "undoubtedly" only applies if in fact the freezing did take place.* "Cooling and eventual freezing of the northern North Atlantic greatly reduced evaporation," and the Scandinavian and western Siberian ice caps began losing more water than they gained from precipitation. *Again, a plausible consequence of what has gone before, but far from certain. How greatly evaporation was reduced would depend on whether the ocean actually froze (see above) and also on air temperatures, which in turn would depend partly on the atmospheric circulation. This last is heavily conjectural. Note also that this process, plausible enough in Europe, is much less plausible in North America. The North Atlantic, major source of European precipitation, may well have frozen; the Gulf of Mexico, major source of eastern and central North American precipitation, certainly did not, nor did the northeast Pacific, which provided snow for the western glaciers. The general cooling of air and ocean would certainly have made for some drop in precipitation, but whether this would have been sufficient to tip the balance toward glacial decline in North America is a very dodgy question.*

"The onset of these conditions . . . should not have resulted in an immediate stop of the ice advance. Ice, in fact, should have kept advancing for some time under its own weight." *As pancake batter poured on a griddle will continue expanding for a moment after the pouring stops. But for the ice, this is a wide-open question. If precipitation dropped off quite suddenly, there would presumably be a "bulge" of ice that would gradually flatten out, in the process pushing the ice-front farther south. But if the precipitation tapered off gradually, this would happen only slightly, or not at all. There's no evidence on which actually occurred.*

"Advancing ice [*if it did advance*] means further increase of albedo as well as of rates of melting." *Because the ice is now farther south, in warmer areas. But here our authors may be trying to melt their ice-cake and have it. The greater the advance, the greater the melting—but the greater the melting the less the (net) advance!*

"Therefore [*that is, assuming all these assumptions are true*], the northern North Atlantic will remain largely frozen and surface water elsewhere should [remain quite cold] while the ice-caps flatten under their own weights [*if they did flatten*] and while the ice [then melts] at least as far back as it was when the northern North Atlantic began to freeze." *Assuming it had in fact spread beyond this line.*

"This mechanism," say the authors, seems to describe the observed facts with no straining. On the one hand, isotope measurements from deep sea cores show "nearly constant low temperatures between 25,000 and 11,000 years ago." *That is, if you accept these temperature readings as accurate. Not all oceanographers do—see next chapter.* On the other hand, the ice sheets reached their maximum area 18,000 years ago. *This date is derived from carbon-14 techniques and is pretty reliable. Our authors are saying that the northern North Atlantic may have frozen 25,000 years ago, while the ice continued to advance under its own weight for another 7,000 years.* From then on, the ice "kept decreasing, in a process of flattening [thinning] rather than retreat. Thus, albedo was decreasing only slowly, and low temperatures were maintained." *It depends what you mean by "slowly." The ice sheets certainly backed up considerably between 18,000 and 11,000 years ago.* The thawing of the frozen North Atlantic may not have occurred until about 11,000 years ago—and in fact there is considerable evidence of a marked temperature rise at about this date. *There is. But equally reliable evidence shows a somewhat less marked* fall *in temperature a few centuries later.*

With the North Atlantic now unfrozen and warming up, there ought to have been a further advance of the ice. But the enormous weight of the ice had depressed the earth's crust in the ice-covered areas. *No doubt about this.* Thus the tops of the remaining ice sheets probably lay not much above present-day sea level. *An estimate, probably not too wide of the mark.* And it remained at this level because the "rebound" of the depressed crust takes far longer than the disappearance of the ice holding it down—something like 10,000 years. *This estimate, very plausible, is based on the fact that parts of Scandinavia which became ice-free a little less than 10,000 years ago are still rebounding.*

Thus the ice, which had begun by building up the mountains, ended by mashing them down almost to sea level. As a result, the incoming ocean air, though it will be moist, will not be forced very high; precipitation will remain fairly sparse and will also fall at lower altitudes—i.e., as less snow and more rain. *No doubt about this.* A further result would be continued deglaciation, possibly now helped along by an increase in summer radiation, until the ice sheets have vanished.

Thus as Emiliani and Geiss see it, the radiation changes

merely serve to start the ice sheets going, and perhaps to preside over their final demise. In between, the ice creates the conditions of its own advance (more storminess and precipitation, greater land elevation, greater albedo, and thus lower temperatures) and eventually the conditions of its decline (cooler and partially frozen oceans, lower land elevation).

For all its assumptions, this theory strikes me as a reasonably plausible description of what *might* have happened. But the whole thing stands or falls on the chronology. If the ice ages did in fact occur in step with the troughs in the radiation curve, then Emiliani and Geiss are almost certainly right; if not, they're not. And the question of just when the ice ages occurred is far from settled. In fact, it is one of the most hotly debated questions in climatology.

15. ICE AGE CHRONOLOGY II

The Oceanographers

David Ericson's business is with the sea, and he looks it. Broad-shouldered, with ice-blue eyes and a grizzled fringe of beard, he resembles nothing so much as one of those old New England skippers who sailed clipper ships around the Horn or hunted whales across the Pacific. Like Melville's Ahab, Ericson has spent some years traversing the seas in search of his own white whale—in this case, a complete climatic record and chronology of the glacial epoch. What's more, he has found it and caught it, at least to his own satisfaction. Some of his oceanographic colleagues are less impressed with the animal.

Ericson began his professional life as a petroleum geologist. This may seem an odd route to oceanography and paleoclimatology, but in fact it is quite logical. Any competent oil geologist must be an expert in what is called micropaleontology. This science, like paleontology, is concerned with fossils, but not with the massive bones which that term usually conjures up in our minds. Instead, it deals in *micro*fossils—the tiny shells or skeletons of prehistoric marine organisms. For hundreds of millions of years, untold myriads of these minute creatures have inhabited the oceans. When they die, their shells sift down on to the bottom, piling up layers of oozy sediment tens and even hundreds of feet thick. Ultimately, these are transformed by pressure and/or heat into rock.

Microfossils are no bigger than a sand grain and to the naked eye about as distinctive as one. Under a low-power microscope, however, they are seen to possess rather elaborate and often beautiful structures by which they can be classified into families, genera, and species. And, as with other fossils, the species change with time, so that the age of many rocks originally deposited on the ocean bottom can be accurately estimated by noting which microfossils are present in it.

As a petroleum geologist, then, Ericson found microfossils essential in dating the rock "cores" brought up by oil rigs. As an oceanographer, he finds them equally essential in deducing the climatic changes reflected in the mud cores brought up from the ocean bottom.

On land, as already noted, the climatic record of the earlier ice ages is blurred by the action of later ones and by the erosion of wind and water; it is, in Ericson's words, "a tattered old book in which many pages, even entire chapters, are missing." Beneath the ocean waves, however, where there is neither wind nor rain nor grinding ice sheets, the climatic record of the Glacial Epoch should lie complete and in order. Such, at least, was the theory with which Ericson began his hunt. Fortunately, it turned out to be false.

The first cruise in search of the glacial Moby Dick was less than successful. The research vessel, *Atlantis,* after refueling at the Cape Verde islands and picking up equipment at Dakar, set out across the Atlantic. At the first attempt to secure a sediment core, the wire rope holding the large coring apparatus broke, and the device sank to its final resting place more than 2,000 fathoms down.

Next day the mainsail ripped. It was the only one aboard.

That same evening a crewman came down with acute abdominal pains, and the *Atlantis* was forced to make for Barbados, the nearest land.

The following night the propeller dropped off.

Finally, by "alternate sewing and sailing" on the flimsy mainsail, the *Atlantis* made her way to Barbados. The seaman was put in hospital, where he shortly "recovered as mysteriously as he had fallen ill."

By this time the *Atlantis* had to turn homeward to New London, Connecticut, which she fortunately reached without further catastrophe. The cruise, as Ericson observed, provided an elegant proof of what is known as Murphy's First Law: *If, under a given set of conditions, anything can go wrong, it will.* (This is the scientific formulation. Many nonscientists know it better as the Buttered Side Principle, from the rhyme "I never had a piece of bread/ Particularly large and wide/ But when it fell upon the floor/ It fell upon the buttered side.") Apart from supplying further proof of Murphy's Law—proof which nobody really needed—the expedition collected a good deal of valuable information on undersea topography, ocean temperatures, and the like. Thanks to the early loss of the large coring device, however, it did poorly in that department. With the remaining equipment, no cores longer than 10 feet could be obtained—not nearly long enough to produce useful climatic records. One of these cores, however, turned out to be of peculiar interest.

A typical undersea core is a long cylinder about two inches in diameter. Colors vary both from one layer to another and from one core to another. Some are rather prettily marbled, but these are of little value since the marbling shows that the sediments have been disturbed. The useful ones range from a whitish putty color to an unpleasant brown. The core in question had two distinct layers: the upper one sandy, the lower a fine white mud. Both contained microfossils of foraminifera—one-celled organisms whose tiny, snaillike shells make up well over half of all microfossils. Because of their ubiquity, they are of enormous use in dating geological deposits and hence have been extensively studied; the present catalogue of species, living and extinct, has reached some seventy volumes.

Back at the lab, the core was sliced open for examination and samples of the two layers were washed. Under the microscope, the foraminifera ("forams," for short) in the upper layer turned out to be all of species now living in the North Atlantic; that is, the upper layer must be of quite recent vintage. The lower layer, however, contained only species which, as Ericson well knew from his oil-prospecting days, dated from the Eocene epoch, more than forty million years ago. Clearly, something was missing—specifically, sediments representing forty million years worth of deposits!

The first implication of this fact was all too obvious: if forty million years could simply "disappear" from a portion of the ocean bottom, the marine records of glacial climates (or of anything else) must be far less undisturbed than had been thought. Without the closest examination, it would be unsafe to assume that any core actually represented an unbroken series of sediments; Ericson's whale was turning out to be as elusive as its Melvillean prototype.

In fact, subsequent work proved that the ocean bottoms are far from the placid unchanging localities they were once thought to be. Deepsea currents—notably the so-called turbidity currents of sediment-laden waters—can rush downslope, stripping away earlier deposits and even carving deep canyons in the slopes. (One of these, Hudson Canyon, extends 150 miles beyond the mouth of that river.) Sediments lying on slopes can also "slump" downhill in much the same manner as a mud slide on land. And when the material in these "sea slides" comes to rest elsewhere on the bottom, it literally muddies the sedimentary record in that area.

But all these undersea changes, though they made the job of reconstructing accurate sedimentary records far more difficult, turned out to be a blessing in disguise. For it soon became obvious that the chances of getting a single core that would contain the entire record of the glacial epoch were almost prohibitively small. The reason was that even the improved coring devices which Ericson and his associates began pressing into service could thrust just so far into the compacted mud of the ocean bottom. The longest cores were only about 90 feet long; most were a great deal shorter than that. It was becoming clear that a complete sedimentary record of the Glacial Epoch would almost certainly be much more than 90 feet thick. Thus it was precisely the troublesome slumping which, by removing later sediments, made it possible for the corers, with luck, to sample deposits from the early part of the epoch and before.

With luck. For there was, of course, no way of telling, when the corer was lowered to the bottom, which sediments, if any, had been swept away at that point and which remained. Thus Ericson and his associates were forced to examine literally hundreds of cores, piecing the record together bit by bit in much the same way as tree-ring and varve chronologists had done before them.

Meantime, however, another oceanographer had been at work—Cesare Emiliani, of the University of Miami, whom we have already met as a partisan of the Milankovich theory of the Ice Ages. Emiliani, like Ericson, got into oceanography through petroleum geology; otherwise, however, the two are as different as *Asti spumante* and Medford rum. Temperamentally, Emiliani is no sea captain but a Renaissance condottiere, dashing forward impetuously into battle, vocally delighted with every victory, real or fancied, over his scientific opponents, but always ready, once circumstances change, to conduct a fighting retreat. He is also a good deal of a showman; he once published an article with six fictitious co-authors ("to make the mystical number seven") urging that scientific journals be replaced by wandering troupes of scientific minstrels, who would chant accounts of their discoveries at night-long banquets of their colleagues.

Emiliani had worked in Chicago with Harold Urey when the latter was developing his method of measuring ancient ocean temperatures, and went on to become perhaps the best-known expert in this area of oceanography. In the early 1950s, he began applying Urey's methods to a number of cores—some of them the same cores that Ericson had been working on. He and his associates examined the forams in dozens of core sections, separated out the species that lived near the surface (since only the surface waters could be expected to provide a clear picture of temperature changes), and analyzed the isotope ratios in the shells. The result was a series of graphs showing the presumed ups and downs in ocean surface temperatures during the past 300,000 years. Moreover, the general pattern of the graphs showed a strong consistency from one part of the oceans to another, indicating that the temperature changes were not local events. Finally, the "averaged" graph of temperature changes showed a marked resemblance to the van Woerkem–Milankovich graph of solar radiation changes.

Having found five major troughs in the ocean temperature record, Emiliani assigned them to the four "classical" ice ages (the last two troughs, separated by an interval of cool but not icy climate, he consolidated into the last ice age). The Glacial Epoch, he announced, had lasted only 300,000 years. This was half the length of the Penck-Brückner estimate (though other geologists had speculated that 300,000 years might indeed be closer to the truth). Emiliani then applied this chronology to human evolution, concluding, among other things, that Australopithecus, the first man-ape, generally thought to have appeared somewhat before the first Ice Age, had flourished a mere 400,000 years or so ago.

This was in 1958. A few years later, Ericson, who had

been plugging away on his own core work, came out with his chronology. It was based not on isotopic temperature measurements of forams, but on changes in the foram population—primarily, on the relative abundance of the species *Globorotalia menardii,* which present-day studies have shown to be abundant in warm waters but not in cool ones. Thus large numbers of this species in a core section would mean an interglacial period; few or none, an ice age. Ericson and his associates underwent appalling difficulties in working out ways of correlating one core with another. Indeed, their search for a complete Glacial Epoch record ended in a real Perils of Pauline finish. The core that finally closed the gap in the sediments was literally the last one in the house; had it proven unsuitable, they would have had to wait for another expedition to bring back more cores. With the gap closed, Ericson and company published their own glacial chronology. The first ice age, they declared, had begun, not 300,000 years ago, as Emiliani had said, or 600,000 years ago, as Penck and Brückner had estimated, but no less than 1,500,000 years ago!

This, one hardly need say, set off an argument.

Ericson criticized Emiliani's findings on the ground that the forams the Italian had used for his isotopic measurements were not, in fact, purely surface organisms; many of them spent part of their life cycle well down in the depths, where temperature changes might well be quite different than at the surface. Emiliani riposted that Ericson's species changes did not really reflect temperature changes; actually, he said, the same "ice age" core sections in which *Globorotalia menardii* was absent contained other species known to live only in tropical waters.

Meantime, other evidence had been coming in, much of it favorable to at least part of Ericson's theory. Some "hard" potassium-argon dates on Australopithecus showed that, whatever that creature's relationship to the first Ice Age, it had unquestionably flourished nearly two million years ago —a far cry from the 400,000 years that Emiliani had proposed. Cores from the southern Indian Ocean revealed a marked and significant change around 2.5 million years ago: the appearance in the sediments of angular, polished particles of rock, evidently produced by glacial action. These, it seems, must have been formed by ice sheets in Antarctica, which carried them to the coast and, breaking into icebergs, on out into the ocean, where they dropped the particles as they melted—even as they do today. Volcanic deposits from Antarctica provided potassium-argon dates showing that in at least one part of the coast, glaciation at some time prior to 2.7 million years ago was more extensive than it is now; similar evidence showed that glaciation had begun on that continent at least ten million years ago and perhaps even earlier. Finally, potassium-argon dates from Iceland gave the age of the earliest glaciation there at no less than three million years. All this pointed to a "long" rather than a "short" glacial chronology.

In the face of overwhelming forces, Emiliani conducted a strategic retreat; the minstrel changed his tune. He now concedes that the Glacial Epoch as a whole lasted much longer than 300,000 years. But he sticks to his chronology of the ice ages, which he still believes occurred at intervals of roughly 40,000 years. This means, of course, that there must have been far more than four ice ages; Emiliani himself now speaks of "scores" of them. Asked how this squares with the classical four-ice-age scheme, he declares that the geologists were wrong; it is, he says, almost impossible to assign most glacial deposits with certainty to one ice age or another, let alone to one "stage" of a particular ice age.

Meanwhile, Ericson and his friends have been revising their own chronology. Their original dates—like those of Emiliani—were derived from radiocarbon dating back to about 50,000 B.P., on protoactinium dating back to around 300,000 B.P., and on estimates, based on rates of sedimentation, before that. Now a new technique has evolved which can link potassium-argon dates—which, remember, take us back several million years, but are almost never obtainable from marine deposits—with ocean cores. This is based on the discovery of magnetic reversal.

In Chapter 12 we spoke of paleomagnetism—the "remnant" magnetism in ancient rocks dating from the time they were formed. Some ten years ago, paleomagnetic studies of fairly recent volcanic deposits revealed that the earth's magnetic poles periodically reverse themselves, north becoming south, and south, north. Why this happens is anyone's guess, but that it does happen seems now almost beyond doubt. And, providentially, the same volcanic deposits which provide a record of the reversals can also be dated quite accurately by the potassium-argon method. As a result, we now have a clear picture of magnetic reversals for several million years: normal (i.e., like the present) back to about 700,000 years; reversed, with two relatively brief interruptions, up to 2.4 million years back; then normal, and so on.

Subsequently, and importantly, it turned out that magnetic reversals could also be detected in ocean cores, meaning that the potassium-argon dates for the reversals on land could also provide some good dates for the older cores—and also help to fit cores into their proper chronological sequence.

Ericson and company, applying these new dates and correlations, discovered with some chagrin that many of their earlier cores had been misplaced in the sequence. Emiliani, of course, was delighted. But their revised conclusions were hardly such as to give him much pleasure. Reexamining the cores—including three that now turned out, quite unexpectedly, to provide precisely what they had originally been seeking and had almost given up hope of finding, i.e., a complete sedimentary record of the Glacial Epoch—they concluded that the "first" (Günz/Nebraskan) ice age had begun, not 1.5 million but a full two million years ago, with at least one (possibly less severe) ice age even earlier.

At this writing, then, we have two quite different pictures of the Glacial Epoch, differing sharply both from one another and from the traditional chronology. In one important respect, they agree—more or less. Ericson and his friends say the epoch began two million years ago, and Emiliani, though he has not committed himself to any specific figure, does not contradict this date. But on what happened thereafter, the gulf is wide. Ericson's school says four ice ages; Emiliani's school, though again unwilling to commit itself to hard figures, sees the total as much more like forty than four.

Considered in terms of the classical theory, both are equally implausible. By Ericson's revised chronology, the "second" (Mindel/Kansan) ice age lasted something like 500,000 years—which is nearly ten times the Penck-Brückner estimate. Granting that Penck and Brückner could have been—and probably were—wrong, could they really have been *that* wrong? But the same applies to Emiliani. He sees something like forty ice ages, as against the classical four, though these have been described by geologists in North America as well as in Europe and, with rather less certainty, in Africa as well. Granting all he says about the difficulty of distinguishing between one glaciation and another, the geologists may well have been wrong in singling out only four—but could they all have been *that* wrong, and in three different continents to boot?

The real difficulty is that we still lack reliable dates from where the action was—the glaciated regions of temperate North America and Europe. A glacial advance in Antarctica or in Iceland some three million years ago suggests but does not necessarily prove ice sheets in the temperate zones. There are, indeed, a few potassium-argon dates from these latter regions, but what they mean depends on who has discovered them. A German geologist has dated a glacial deposit along the Rhine, which he ascribes to the Günz ice age, at about 350,000 years—which has delighted Emiliani. But an American geologist has dated a deposit in the Sierra Nevada, allegedly formed by the succeeding (Kansan) ice age, at between one and three million years ago—which has tickled Ericson.

Finally, there is one more bit of evidence on Emiliani's

side. Two English geologists have dug into a prehistoric lake bottom supposed to date from the second interglacial. There are glacial deposits above and below the lake deposits, meaning, presumably, that the lake bottom includes a complete record of the interglacial period. And the bottom deposits are largely varved. Partly by counting varves and partly by estimating (the upper part of the deposits had been somewhat disturbed) they have obtained, not the date of the interglacial, but an estimate of how long it lasted. Their figure is about 35,000 years—very close to what Emiliani's temperature curves would suggest, as against Ericson's estimate of something like 500,000 years. If, that is, they are right in assuming that the lake *does* include a complete record of that particular interglacial and not some other glacial episode.

We should note also that Emiliani and Ericson agree pretty closely on the climatic history of the last 120,000 years or so—which is rather fortunate, since some very important events in human evolution occurred during that period.

And there, for the present, the controversy rests—and with it the dispute over the causes of the ice ages. If Emiliani is right, Milankovich can be said to have been vindicated. (Just to make things interesting, however, Wallace Broeker, a colleague of Ericson's at the Lamont Geophysical Laboratory, has come up with a different climatic chronology of the past 150,000 years or so, using still another climatologic method. He finds his chronology consistent with his own version of the Milankovich theory—which he has of course obtained by applying another set of bugger factors.) If Ericson is right, then Milankovich was certainly wrong, and the whole question of what causes an ice age is up for grabs again. Ericson himself is rather drawn to Sir George Simpson's theory in which an increase in solar radiation produces an advance of the ice—though most climatologists, for reasons already stated, find this unconvincing. But if the causes are still up in the air, the ice ages themselves are still very much on the ground. Let me conclude this chapter, then, by giving my own "working chronology" of this very critical period in man's story.

First, it is quite certain that a general and gradual cooling of most parts of the earth began at least thirty-five million years ago.

Second, it is almost as certain that no later than ten million years ago this cooling had progressed to the point where glaciers (but not necessarily ice sheets) had formed in Antartica, and perhaps in northern Greenland also.

Third, by something like three million years ago there must have been periods when the climate was more severe than at present, though probably separated by (longer?) periods milder than we now experience.

Fourth, at least two million years ago there was certainly a period of much more severe climate—an ice age, in fact.

From then up to about 100,000 years ago it is necessary to be arbitrary. I am going to assume that Ericson and the other partisans of four ice ages are right—not because I find their four any more or less plausible than Emiliani's forty or so, but simply because four is manageable. All the findings on human evolution have been organized around a framework of four ice ages, and short of completely reworking that framework—which I am by no means competent to do—there seems no alternative but to stick with it until the Emiliani-Ericson controversy is resolved—or until some other scientist with other bugger factors proves both of them wrong.

16. MAN'S BIRTHPLACE

African Climate, Now and Then

After all that has been written here about the two-million-year Glacial Epoch as the special time of man's evolution, it may seem ironic that for fully half that period mankind knew nothing of ice sheets and little or nothing of glaciers of any sort. The reason is that man evolved in tropical Africa, a region whose climatic peculiarities include a total absence of either snow or ice anywhere except on the highest mountains. This is true not only today but, for all practical purposes, even in the chilliest Ice Age times.

These climatic peculiarities go far to explain why man evolved in Africa and not somewhere else. Part of the explanation lies in the climatic and especially the dietary preferences of the primates, the class of mammals which includes man, apes, monkeys, and some more primitive cousins such as the lemurs. The primates are, almost without exception, tropical animals. Few species are found even in the subtropics and only two—man and the Japanese macaque, a monkey akin to the baboon—in any cooler regions.

The tropical primates are adapted, if one can call it that, to living where the living is easy. Their relatively unspecialized teeth and digestive systems restrict them largely to a diet of succulent fruits, shoots, and buds, obtainable only in a climate moist and warm enough to support abundant and tender vegetation at all seasons. They are creatures of warm forests and wooded savannas; in dry country, or temperate forest, where vegetation goes into a seasonal sleep, they would starve to death. Significantly, the only primates ranging beyond these warm moist habitats are man and the baboon-macaque group of monkeys—all of whom have more or less adapted to life on the ground, in the process adopting a considerably less limited menu.

The need for a tropical climate immediately rules out North America, Europe, and most of Asia as potential birthplaces for man. Fossil primates—including apes—have, indeed, been found in parts of Eurasia that are now distinctly cool, but the latest of them dates from a period some twenty million years back when these regions were still, at worst, subtropical; subsequently, we must infer, the apes were driven south or exterminated by the advancing cold and changes in vegetation.

South America is almost as uniformly tropical as Africa, and is endowed with a rich population of monkeys. How is it that none of these evolved into apes? The answer—one all too frequent in evolution—is that they just didn't.

The only remaining possibilities outside Africa are Indonesia and the southern fringe of Asia. Indonesia is warm and moist—it is, as already noted, one of the three main areas of Equatorial climate—and is populated by numerous monkeys and one ape, the orang-utan. The trouble seems to have been that most of it was *always* warm and moist, even during the ice ages. Thus its primate population had little incentive to abandon their *dolce vita* in the trees for a more down-to-earth existence. We are left, then, with southern Asia—meaning, chiefly, the Indian peninsula. Offhand, I can think of no climatic reason why that region might not have been man's birthplace; in fact, a species of ape thought to be closely akin to our ancestors lived there some fifteen million years ago and perhaps even later. Possibly India was—subtracting the areas that are either too chilly or too dry for ape habitation—too small an area for a wide-ranging animal like man and his immediate ancestors. Or perhaps, as with South America, the answer is that man didn't evolve there because he didn't.

In any case, we know that Africa almost certainly *was* man's Garden of Eden, and with the 20-20 vision that hindsight always brings we can see why. First, it is big—next to Asia, the biggest continent—meaning that it supplies both space for expansion of animal species and a large enough range of habitats to stimulate variety. Second, its climate, though it shows much variation from one region to another, is, over wide areas (in the past, much wider areas), warm and moist enough to support abundant plant life at all seasons —meaning, of course, that animal life can be correspondingly abundant. Finally, the changes in its climate over the past thirty million years or so, though their precise nature and timing are in some respects still debated, were certainly neither so trivial as to provide little climatic stimulus for evolution nor so severe as to threaten the survival of the rather rare species that man's ancestors almost certainly were. Adopting our usual procedure of taking the present as the key to the past, let us take a quick look at the kinds of climate found in Africa today.

The great African continent extends almost equally far north and south of the Equator; as a result, its climatic zones possess a rather nice symmetry on either side of that imaginary line. Astride the Equator, along the northern coast of the Gulf of Guinea and stretching east more than half way from the Atlantic to the Indian Ocean, lies a wedge of Equatorial climate—the perpetually hot, moist rain forest.

Rather surprisingly to some people, the rain forest is not particularly "typical" of Africa. Much more typical, in the sense of being more extensive, is the broad Savanna belt that surrounds the Equatorial belt to north, east, and south (the Atlantic is on the west). Here rainfall becomes less copious and also more or less seasonal—how much more or less depends primarily on how far one has moved from the Equatorial belt. And as the climate grows dryer, the trees, which as a group exceed all other land plants in their need for moisture, begin to thin out. At first, the forest becomes broken with grassy glades; then, as one moves into dryer areas, the glades expand and coalesce until at last one sees a sea of grass, broken by clumps of trees and, along river banks, by the meandering line of a "gallery" forest.

The Savanna belt stretches to the Indian Ocean over much of southeastern Africa; elsewhere, it in turn is surrounded by the Steppe belt, in which the rainy season is shorter and the rains themselves sparser. Here the vegetation is almost entirely grass, sometimes dotted with a few bushes or trees, such as the acacias, specially adapted to a dry climate. To the north and southwest again the Steppe withers away into true desert—in the north, the enormous and forbidding expanse of the Sahara, and on the southwest (where the continent is of course much narrower) the much smaller Kalahari, though much of what is labeled Kalahari on maps is actually steppe rather than desert. The deserts complete our sketch of African climates, except for a fringe of Mediterranean along the northwest coast and a corresponding snippet at the extreme southwest tip.

So much for the present. What about the past?

The climatic history of Africa is not as well known as we would like. Nonetheless, we can be certain that thirty million years ago—which is the point at which I have somewhat arbitrarily chosen to begin my account of man's evolution—it was overall a great deal moister than it now is. As you may remember from Chapter 11, that was a time of larger oceans and smaller continents than appear on present maps; in Africa, much of today's Mediterranean coast and, on the west, the valley of the Niger, were under water. Even more important, on the northeast a broad warm sea flowed where the arid Arabian peninsula and Iranian plateau now stand. Thus Africa in those days was entirely surrounded by warm seas from which the sun could draw copious supplies of moisture; the cold ocean currents to the northwest and southwest that today help create the Sahara and Kalahari were missing.

What the atmospheric circulation was like in those days is almost pure conjecture—apart, that is, from the thermodynamically necessary fact that it must have been less vigorous than now because of the lessened heat differential between the tropics and the poles. Fortunately, we need not bother much with conjecture, since more direct evidence indicates pretty clearly that Africa was forested almost from top to bottom. Thirty million years back, Aegyptopithecus —a probable ancestor of both man and apes which we shall meet more formally in the next chapter—inhabited a rich forest not far from the future site of Cairo, smack in the middle of today's Desert Zone.

Subsequently, the general line of climatic change is obvious. As we move closer to the present day, the climate must certainly have grown more like present-day climate. By two to three million years ago, climatic patterns in Africa must have been pretty close to those of today. But around that time, according to our Glacial Epoch chronology, the first Ice Age began. And just what happened at that point is the subject of the Great African Pluvial Wrangle.

Quite soon after the general pattern of the Glacial Epoch was laid down by Penck and Brückner, geologists and climatologists began wondering about Ice Age climates in the regions that were *not* glaciated. In fairly nearby areas, the picture seemed pretty clear. Immediately south of the ice sheets was a belt of cold and—except for the water melted from the glaciers—quite dry terrain, not unlike northern Canada or Siberia today. Both the cold and the drought made for sparse vegetation. And the intense winds powered by the speeded-up atmospheric circulation, working on the dry thinly-covered soil, blew up dust storms that would make the American "dusters" of the 1930s look like a sunny day. All across the central part of the United States and much of Europe, the moraines left by the great ice sheets are bordered to the south by a "loess belt," named for the rich, silty soil deposited by those dust-laden winds.

South of the loess belt, Ice Age conditions are equally clear. The Stormy Zone, forced well south of its present position by the outflow of cold air from the ice, gave such regions as the Mediterranean, Near East, and American Southwest and Great Basin a good deal more rain than they now enjoy. These same regions, moreover, must have been cloudier and, partly for that reason, cooler than they are now. Thus even as precipitation went up, evaporation went down. The results have been traced in detail in many areas. We have already mentioned, for example, Lake Bonneville, a great Ice Age body of fresh water that covered thousands of square miles of what is now steppe and desert; similar changes have been studied in North Africa, Palestine, and elsewhere. Thus the principle was established that an Ice Age to the north meant a rainy—pluvial—age to the south.

When scientists began probing into the climatic history of tropical Africa, they quickly found evidence that that region too had experienced pluvials. Raised beaches around lakes, raised terraces along river banks, indicated a climate in those areas considerably wetter than at present. It was perhaps natural—but rather superficial—to conclude, then, that the African pluvials too had been contemporary with the ice ages. Much earlier, it was noted that the dangers of inferring climatic change from one region to another increases with the distance separating the regions, or perhaps even as the square of the distance. The African pluvials are certainly a case in point, for recent research indicates that the original conclusion may have been the exact opposite of the truth. We are talking, remember, about *tropical* Africa. North Africa, including much of the Sahara, was certainly somewhat moister during the ice ages, and so, presumably, was the southernmost (Mediterranean) tip of the continent—though since there were, as we know, no large ice sheets in the southern hemisphere outside Antarctica, the changes there were probably less marked.

Now there is evidence that tropical Africa has not only undergone wetter periods during the Glacial Epoch but also dryer ones. We have already mentioned the "fossil" sand dunes of western and north-central Africa; these show that the Kalahari and Sahara at various times have thrust as much as 600 miles toward the moist Equatorial regions. And in coastal regions of west Africa, these former dunes can be traced out on to the ocean bottom of the continental shelf; Rhodes Fairbridge, who is something of a specialist in this subject, has made similar observations in northwest Australia, now a savanna but once a desert. The obvious conclusion is that the times of expanded deserts (and, obviously, contracted savannas and rain forest) were times of low sea level—i.e., times in which much of the world's water was locked up in the ice sheets. This hardly sounds like an ice-age pluvial.

Fairbridge has turned up other evidence pointing to the same conclusion, from examination of the middle course of the Nile. The Nile is a rather peculiar river. Its entire lower course of more than 1,000 miles passes through the Sahara, and hence has no tributaries. Its waters originate far to the southward; the Blue Nile flows from the well-watered Ethiopian highlands, the White Nile from central Africa. Both regions are chiefly moist savannas, with the heavy summer rains and dryish winters characteristic of that climatic type. Clearly then, the amount of water flowing down the Nile through Egypt or the Sudan measures the amount of precipitation in a large chunk of *tropical* Africa far to the south.

The Nile is also notable for carrying large quantities of silt during its flooding periods (which of course occur soon after the summer rains arrive farther south). For thousands of years, that silt has been deposited along the river's banks in the narrow Nile Valley and, farther north, on the fertile plains of the Nile delta—which, in fact, is built entirely from such sediments. Herodotus, the famous Greek traveler and historian, guessed this some 2,500 years ago and in a classic phrase described Egypt (i.e., its fertile, habitable regions) as "the gift of the Nile."

But the Nile was not always so generous to Egypt. Studying the river's banks near Wadi Halfa, on the Egyptian-Sudanese border, Fairbridge has found evidence of enormous silt deposits, rising in places to well over 100 feet above the river's present flood level. They begin some 300 miles upstream from Wadi Halfa and die out a similar distance downstream. The deposits, Fairbridge believes, point to a period when the Nile simply lacked the energy to carry its load of silt all the way to the Delta; its flow, he says, "must have been cut at times to the merest trickle." As a result, the relatively deep channel of the middle Nile became filled with silt, "lifting" the shrunken waters of the river which then deposited additional silt in the surrounding plain.

Fairbridge, using carbon from mollusk shells buried in the silt, has dated the middle layers of the deposit at around 15,000 B.P.; he estimates that the earliest silts go back to 25,000 or 30,000 B.P.—that is, to just about the time that the glaciers were beginning their last great advance. The uppermost silt layers have similarly been dated at something over 11,000 B.P.—the time at which the ice sheets were going into a rapid retreat.

The conclusion seems inescapable. When the ice advanced in Europe, silt began building up along the middle Nile, meaning that the river's flow was slackening, meaning that rainfall in Central Africa was diminishing. When the ice retreated, rainfall picked up, and the quickening Nile began clearing its channel of the accumulated debris and carrying its silt downstream into Egypt. Always assuming, of course, that Fairbridge has reasoned correctly. Personally, I am convinced he has, because the whole business doesn't make much sense any other way. We know from many lines of evidence that the temperate zones have been growing cooler for at least some thirty million years, while Africa, during the same period, has grown dryer. If we accept the ice age—pluvial equation, we must then believe that suddenly, when the temperate zones became really cold, the whole process reversed in Africa, which grew not still dryer, but wetter! It just doesn't figure. Again, taking the relatively recent past, it is generally agreed that there was a pluvial of some sort in Africa around 7,500 B.P. But in Europe this period was quite definitely not cooler than the present but distinctly warmer! All of which points to the ice ages as being times not of pluvials in Africa but of "anti-pluvials"—periods *dryer* than the present. The pluvials, then, would represent the interglacials, at least some of which are known to have been warmer than the present, in which case, presumably, Africa would "regress" toward the moister conditions of perhaps ten million years ago.

To be sure, if one accepts the Simpson theory (higher solar radiation, more evaporation, more ice) of glaciation, the African pluvials can be jockeyed back into phase with the ice ages. But almost nobody does accept it nowadays. For example, those doughty antagonists Ericson and Emiliani, contentious as they are on most things, both agree that during the last glaciation the central and south Atlantic were distinctly cooler than at present. For the Indian Ocean, the other great source of African rainfall, the evidence is considerably sparser, but what there is tells the same story. And with cooler tropical oceans, evaporation has simply got to be lower, and rainfall in the adjacent land areas must drop off.*

Interestingly enough, as the pluvial-glacial equation has fallen into disrepute, some of the African evidence for it has turned out to be pretty shaky. The first of the pluvials, for instance, is named the Kageran, after the Kagera River in Uganda; the evidence for it is partly taken from elevated terraces along the river's banks. But it has now become fairly clear that at some point during the past couple of million years, earth movements completely reversed the river's course! The all-too-pervasive influence of these and other geologic changes in that area has led one expert to conclude rather sourly that "it is seldom possible to infer the type of climate prevailing at any particular period."

* Recently, additional evidence for this view has come in from Lake Victoria, Uganda, where Daniel Livingston, a Duke University zoologist, has been digging out mud cores from the lake bottom. Pollen from mud layers firmly dated to the last Ice Age shows that the surrounding territory, now moist and densely forested, was a grassy savanna. These same layers, moreover, are heavily impregnated with salts such as calcium carbonate, indicating that evaporation must have lowered the lake level and concentrated the mineral content of its water—further evidence of a dry Ice Age climate.

Matters become more complicated when we consider the climates of mountainous regions. As already noted, a few of the highest African peaks (notably Kilimanjaro and Mt. Kenya) have snow-covered tops even today, and there is evidence that in past times the snowline reached some 5,000 feet farther down—low enough to bring permanent snow, and even glaciers, to several peaks now uncovered. These lowerings of the snowline (of which there were evidently several) have also been ascribed to the pluvials. In fact, however, many authorities have pointed out that snow cover depends not on precipitation (even today, the tops of Kenya and Kilimanjaro receive relatively little rain) but on summer temperatures. That is, the times of lowered snowlines must have been cooler, and therefore necessarily coincided with Ice Age times to the north, but not—or at least not necessarily—with pluvial times in the lowlands. It is, however, likely that the extended snow cover would produce small aggregations of cold air that might well bring on local "mini-pluvials" on the surrounding lower slopes.

Thus, the most plausible present picture of Ice Age climates in tropical Africa is one of widespread lowland drying: expansion of the Desert and Steppe areas at the expense of the Savanna and Equatorial regions. The latter, indeed, must have been squeezed into a very thin wedge indeed. It could not have disappeared entirely, however; otherwise the animal species adapted to that climate would have been wiped out during the last Ice Age (some 20,000 years ago), leaving the Equatorial wedge largely barren of life today. In mountain regions, however, the ice ages may have meant somewhat wetter conditions and certainly—because of the lowered snowlines, plus the generally cooler temperatures—distinctly cooler ones. The interglacials, on the other hand, would have seen a reverse set of changes: expansion of the Equatorial and Savanna regions, contraction of the Desert and Steppe zones and, in the higher areas, distinctly warmer and perhaps also dryer conditions.

Even accepting the validity of this overall pattern, however, many details are still hard to make out. The uncertainties in dating the various (four? forty?) ice ages plus similar uncertainties in dating human or quasihuman relics makes it risky to draw a priori conclusions on climatic conditions at any given fossil or archeological site. For safety, then, we shall stick as closely as we can to climatic evidence specifically associated with the sites.

17. PROLOGUE TO MAN

Life in the Forest

Some 60 miles south of Cairo, somewhat to the west of the Nile valley, lies the Fayum depression. Its name comes from the ancient Egyptian word *pa-ym,* meaning lake, and some thousands of years ago the entire valley was indeed a lake, its waters regularly replenished by the Nile's annual flood. Later, despite the river's slackening flow, the pharaohs were able to keep the valley watered by digging a canal westward from the river.

Today the canal is long silted up, and the lake, fed only by a niggardly flow from underground, has dwindled under the desert sun to the point where its brackish surface lies 150 feet below sea level. Surrounding it are arid badlands; across them blow the desert winds, which not infrequently strip away the baked sand and soil to reveal ancient bones. Among these, paleontologists have identified several species of primates; one of them, Aegyptopithecus (the Egyptian ape) is the very first ape in the fossil record, and very probably the earliest creature that can with reasonable certainty be placed upon man's family tree.

From what has been said here about the habitat preferences of primates, it can be deduced that when Aegyptopithecus and his kin flourished some twenty-seven million years ago, the climate of the Fayum must have been very different from its present ovenlike heat. In fact, the bones of our far-off ancestor lie amid fossil tree trunks, some of them nearly 100 feet high. What species of trees they were is unclear; paleobotanists cannot usually identify trees simply from their wood, and neither pollen nor fossil leaves have yet turned up. We can be sure, however, that they were tropical or at the very least subtropical species. The Fayum in those days must have been a rain forest or wet savanna, akin to those we now find in or immediately around the Equatorial Zone far to the south.

Some geologists, to be sure, deny that any region within the present Sahara could have been this moist; the trees, they say, must have been swept down by the rivers which flowed through the region as the Nile flows today. Elwyn Simons, of Yale, who has done much of the excavation in the area, denies this. The same layers of sand and silt which contain the trees, he points out, also contain the delicate jawbones of small rodents, which would surely have been ground to fragments had they been washed down any distance. Moreover, if the area was not forested, one wonders what all the primates—a notoriously tree-loving group of mammals—were doing there. We can take it that our remote kinsman lived in a dense forest.

What sort of life did he lead? How did that life shape the kind of creature he was? And the kinds of creatures his descendents would become?

Most of us tend to think of tropical forests, abounding in moisture and warmth, as teeming with life, but in fact this is an exaggeration. The reason lies in the peculiar character of the forest, based largely on its climate. The perpetual warmth and moisture do, indeed, make for a lush growth of vegetation; acre for acre, a tropical forest produces more vegetable matter per year than any other type of climatic region. But the vegetation is almost entirely giant trees. The ground beneath them is heavily shaded; moreover, debris from above rots away too quickly to form much of a soil, and the copious rains leach away nutrients from what soil there is. For these reasons, there is little underbrush except along watercourses, where the forest canopy is broken, and in swampy areas, where the saturated ground prevents trees from growing. Thus plant life on the forest floor is relatively sparse—and, necessarily, animal life as well. The chief vegetable food available at ground level is wood, digestion of which requires a complicated internal economy that only the termites and a few other animals have managed to evolve. Thus animal life in the tropical forest is most abundant amid the treetops, and it is there that Aegyptopithecus must have made his home.

His anatomy shows that he already had made considerable progress in adapting to that mode of life, notably in his sensory equipment. Ground-dwelling animals rely heavily on scent to find food and otherwise acquaint themselves with their environment; watch a dog sniffing its way around a strange neighborhood. For good smelling, you need a long muzzle with extensive nasal passages. In the treetops, however, scent does not "lie" well, and we find that Aegyptopithecus' muzzle, though longer than that of any existing ape, or even monkey, was a good deal shorter than those of the usual ground-dweller.

Failing scent as a clue to the environment, the obvious alternative is to rely on sight—and we can be sure that Aegyptopithecus did. It is probable, indeed, that he already possessed the special optical equipment that all higher, and some lower, primates—and no other mammals—have today: the fovea, an area of the retina which produces an especially sharp image. The fovea is the reason that, though we can see "out of the corner of the eye," we see clearly only when we look directly at things, thereby focusing the image on this peculiarly sensitive spot. Aegyptopithecus had another visual refinement whose use we still enjoy: he could see in 3-D stereo. His eyes, unlike those of most animals, were located on the front, not the sides, of his head. Thus his two "visual fields" overlapped—and we can be reasonably sure that his brain had evolved to the point where it could fuse the two images into one stereoscopic picture. Otherwise, there wouldn't be much point to the overlapping fields. Stereo vision is of course an enormous advantage to a tree-dwelling animal, since it provides true and accurate distance perception. A ground-dweller that misgauges the distance of a rock will at worst trip over it. But a monkey or ape 50 feet up who underestimates the distance of the branch he is leaping to will almost certainly not survive to become anyone's ancestor. Cover one eye and reach for a coin or other small object and you will see how much we have gained in inheriting the 3-D vision of our apish forebears.

Tree life, and the consequent change in sensory equipment, changed the brain as well. The olfactory ("smelling") lobes became less extensive, the visual cortex, more so. Moreover, we can assume that the areas having to do with balance and muscular control also expanded. Similar changes of a crude sort have been observed in squirrels today. In tree squirrels (such as the common gray squirrel) the olfactory lobes are distinctly shorter than they are in ground squirrels. Moreover, the cerebellum, that portion of the brain that deals with balance and, roughly speaking, with what we call muscular coordination, is more complex in the tree squirrels, having some forty-five folds as against only twenty-five in the ground-dwelling species.

Thus we can reasonably visualize Aegyptopithecus as being endowed with good sight, not so good smell, and well-developed muscular coordination for moving about the branches. Precisely how did he move about? Here we are uncertain. We have only a few foot bones *probably* belonging to Aegyptopithecus, and no hand or limb bones. We know that today's lower primates, as well as nearly all monkeys, are true quadrupeds, walking along the branches on four feet and, when they leap from one branch to another, landing on four feet as well. Apes, on the other hand, are "brachiators," swinging hand over hand (or sometimes hand over foot) through the branches. The important difference, as far as man is concerned, is that with quadruped locomotion the (four) legs move primarily back and forth in the plane of the body; sidewise motion is very restricted. Brachiators, on the other hand, can, and in fact must, reach sideways with their arms—as we ourselves do. Try kicking sideways and then grabbing sideways and you will quickly grasp the difference.

How far had Aegyptopithecus advanced toward apelike locomotion? Probably not far. In many ways he was as much a monkey as an ape, notably in the size of his brain and his probable possession of some sort of tail (Simons, in fact, describes him as having the teeth of an ape in the head of a monkey). Moreover, some of his descendants several million years later had apparently still not fully mastered the technique of brachiation.

Finally, what did he eat? Here we are more certain. If you live in the trees, there are essentially three basic types of food available. One is, of course, leaves. But leaves, though they are plentiful enough, are also very low in nutritional value; you have to consume them in enormous quantities to stay alive. This, in turn, implies an oversized digestive tract to handle the bulk of a leaf diet. We find this very specialized adaptation in such monkeys as the langurs and the colubus, who live almost entirely on leaves; when a female colubus monkey has dined well, she looks indistinguishable from one in the last stages of pregnancy. The teeth are also modified in certain ways—and we find no trace of these modifications in Aegyptopithecus. Not leaves, then.

Another possibility is insects—the ants and termites which crawl about the branches and the grubs that can be dug out from under the bark. There is, indeed, good reason to think that Aegyptopithecus' more remote ancestors were insectivorous, but they were tiny creatures less than a foot long. For an animal the size of an ape, or even a monkey, to live on insects, it must evolve very specialized equipment for collecting the creatures wholesale—as the anteater and aardvark have in fact done. Otherwise it is in the position of a man trying to eat rice grain by grain with chopsticks; he will expend more energy in picking up the tiny bits of nourishment than he will gain from digesting them.

It is not unlikely that Aegyptopithecus from time to time enjoyed a tasty grub or termite as an appetiser. But his main course almost certainly consisted of the fruits, buds and tender shoots that most monkeys and apes live on today. This was to have important consequences later on.

Aegyptopithecus was evidently a tolerably successful species. By twenty million years or so B.P., apelike creatures, presumably descended from him, were scattered around the Old World. At least one of them, Pliopithecus ("more-of-an-ape") had the free-swinging arms of the true brachiator; in fact, he is thought by some to be an ancestor of those singularly skilled aerialists, the gibbons. Others are much closer to our own ancestral line, and some of them may lie on it. But to identify these creatures, we must involve ourselves in another scientific controversy, the dispute between the "splitters" and the "lumpers."

The root of the conflict lies in the problem of what makes a species. This biological category is generally defined as a group of animals (or plants) more similar to one another than to any other group, and which under natural conditions interbreed, producing fully fertile offspring. This definition is good enough when we are dealing with living organisms, since it is easy enough to determine whether or not they interbreed. But with fossils it is, of course, quite impossible to determine who could, or did, interbreed with whom. We are forced to rely on similarity alone—and this immediately raises the thorny question of how similar is similar.

If you glance around any city bus or subway car, you will quickly note that individuals of our species differ widely in size, coloring, skeletal proportions, and many other traits; we know that similar (though usually less extreme) differences are found in most other present mammalian species. Twenty monkeys of a given species may look alike to the eye of a layman (as Orientals are supposed to look alike to the Occidental view), but to an expert on monkeys they are as individual as twenty monkey experts. In dealing with fossil apes or monkeys, however, we do not have twenty, or even ten skeletons known to belong to a single species, from which we could determine just how much individuals varied one from another. In fact we are doing well if we have even *one* complete skeleton; most fossil primates are known only in fragments, some of them only from a bit of jawbone or a handful of teeth.

At this point, the scientist who has just dug up a new fossil confronts temptation. His bit of bone may, of course, be so similar to a known species that its scientific identity is beyond doubt. But if doubt exists—and it usually does—who wants to report that he has turned up just another specimen of a species already discovered by someone else? Much better for the ego, and the scientific reputation, to report that you have found a new species—and if possible, a new genus. (The problem of defining a genus is even worse than with species. The most you can say is that members of the same genus will be somewhat less similar than members of the same species. Now read again what I have written above on similarity—and hold your head.)

Alas, too few fossil-diggers have had the courage to say: Get thee behind me, Satan! They were tempted, and fell—coining new scientific names that split what might well be one extinct species into several, and several species into several genera. A notable offender has been the great Louis Leakey, whose immense contributions to human prehistory are his only excuse. To every action there is a reaction. The name-coining frenzy of the splitters begot the lumpers, who try to squeeze as many individual fossils as possible into the same species and as many species as possible into the same genus. The splitters, of course, cry that the lumpers are ignoring vital distinctions among specimens—to which the lumpers retort that the splitters are making nomenclatural mountains out of fossil molehills.

Personally, I am firmly on the side of the lumpers—not because I know, or care, much about the scientific details of the controversy, but simply in the interests of clarity. The more names, the more confusion. So when I discuss the next candidate for a place on man's family tree, I shall not call him Proconsul, with the splitters, but Dryopithecus, thus lumping him with a wide-ranging genus of apes inhabiting Europe, Asia, and Africa. The name, for no good reason I have been able to discover, means "oak-ape." Certainly the Dryopithecines we are interested in—those whose remains have been found in western Kenya—did not live among oak trees, though they were certainly tree-dwellers. (That part of Africa is now savanna, but twenty million years ago it must have been still moist enough to support a dense forest.) The Kenyan Dryopithecines, both lumpers and splitters agree, came in several species and sizes. One was little bigger than Aegyptopithecus, but the others were distinctly larger —one as big as a gorilla.

Why, in the course of evolution, do animals grow bigger, as they do more often than not? Biologically, sheer size is an advantage in several ways. For one thing, the bigger you are (up to a point), the quicker you can move—thereby escaping other animals hunting for a meal and increasing your own food-gathering capacities. For another, big animals tend to be biologically more efficient. The main reason for this is something called the square-cube law, which states that, other things being equal, the weight of an animal increases as the cube of its height, while its body surface increases only as the square of the height.

If we can imagine an eight-foot giant with the same body proportions as a four-foot pygmy, the giant will have twice the height, four times (two squared) the skin surface, and eight times (two cubed) the weight. Now the amount of internal heat an animal generates depends on its weight; the amount of heat it loses through its skin and lungs depends on its body surface. Clearly then, our giant will generate eight times as much heat as the pygmy (other things being equal) but will lose only four times as much. In other words, he will expend considerably less energy simply in staying warm. Smaller animals, in order to compensate for their greater heat loss, have to have higher metabolisms, and therefore a considerably greater food consumption in proportion to their weight.* The tiny shrew (to take a quite extreme case) spends nearly every waking moment in the pursuit of food; if it is deprived of food, it will literally starve to death in hours.

Most species of Dryopithecus, being larger than Aegyptopithecus, were cheaper to fuel, so to speak, and therefore at a biological advantage. But being still tree-dwellers, they became embroiled in another application of the square-cube law which was, in itself, distinctly disadvantageous. For if their efficiency went up, their weight went up faster. And if you feed on fruits and shoots, which commonly grow right at the ends of the branches, increasing weight means an increasing chance of breaking the branch, and your neck, on the way to dinner. *Unless.* Unless you begin to distribute your weight among several branches—feet on one, hands holding on to another, or even two. And if you can do this, chances are you will soon learn not to bother creeping all the way out to the cracking tip where the goodies grow, but creep part way out and then—reach out your hand the rest of the way and grab your meal.

This reaching out is to us the most natural thing in the world; yet it is something that no cat, or dog, or cow, or rhinoceros can do—in fact, with very few exceptions, no animals but the primates can do it, and not all of them. Consequently, it should be no surprise to learn that Dryopithecus was no longer the complete quadruped that Aegyptopithecus probably was. His arms had a great deal of sidewise mobility; moreover, his hands could grip a branch, or a fruit, between fingers and palm (not yet between fingers and thumb).

* Also, for reasons still uncertain, a considerably shorter life-span.

His brain was bigger too, as it had to be if he was to control his feet on one branch, one arm gripping another, the other arm holding a fruit, and his jaws shucking and eating it. Watch a dog eating, and then a young child. Both may be rather messy, but the child is clearly performing considerably more complicated motions than the dog—even without having to hold on to a branch.

One final point, and the most important. If you can reach for a branch, your mobility is obviously improved if you can stand up and reach for it. Whether, and to what extent, Dryopithecus had begun to evolve toward this capacity we do not know; any prehistorian would probably trade one of his younger children for a Dryopithecine pelvic bone, which would tell us whether the creature's hind limbs were still set at right angles to the body or were beginning to straighten out. But at any rate the biological incentive was there, and if Dryopithecus had not yet begun standing up on his hind legs, we can be sure that before very many million years his descendants did.

Whether any of the dryopithecine species now known was in fact a human ancestor is something that specialists have argued pro and con; the evidence is technical, partly conjectural, and for us not terribly important. Some dryopithecines almost certainly were ancestral to the gorilla and chimpanzee, and if they were not also ancestral to man, we can be sure that our true ancestor of that period could not have been very different from them.

We are on somewhat (but only somewhat) more certain ground when we come to our next candidate for man's ancestry. This creature has been found in Kenya, for which reason Leakey the splitter calls it Kenyapithecus; the lumpers lump it with Ramapithecus ("Rama's ape," after one of the Indian gods), fragments of which had been found earlier not far from New Delhi. Our knowledge is equally fragmentary, unfortunately. We know that both creatures had arched roofs to their mouths—which we, but not Dryopithecus or the modern apes, also have—and that their teeth were arranged in a distinctly human manner. And that is about all the bones tell us.

But there are some things we know, not because we've seen them, but because they've just got to be. Ramapithecus lived some five million years after Dryopithecus, and I have no hesitation in declaring that if and when more of his anatomy becomes known, he will prove to have advanced further along the road that his predecessor was already traversing. He will be a brachiator, like the chimp, though doubtless not as skilled a tree-swinger because his arms will be shorter in proportion to his body. His brain will be bigger than those of the dryopithecines; when you have begun using your hands for other things than locomotion, an improved brain becomes a marked asset because it allows you to do still more things with them. The hands themselves will be able to grip between fingers and thumb. And he will possess a pelvis and hind limbs which will enable him to stand almost, if not wholly, erect on a branch to reach a toothsome fruit above his head.

Can we guess further? There is one slender and indirect clue. The Kenya fossil beds that contain Ramapithecus also contain an extinct and diminutive species of giraffe. Now a giraffe is an interesting animal. It browses on the leaves of trees and high bushes—in the case of a mini-giraffe, more bushes than trees. This means that Ramapithecus could not, like his ancestors, have inhabited wholly forested country; the 100-foot trees among which Aegyptopithecus played would have stumped even a maxi-giraffe. The Kenya climate was growing dryer and the forest must have begun opening out into savanna—though a considerably moister savanna than it now is.

With grassy or bushy glades beginning to chop up his forest home, what did Ramapithecus do? Did he conservatively stick to the trees, swinging his way around breaks in the forest? Or did he climb down and, balanced on his hind legs, trot across the glades and into the trees on the other side? Was it he, or some descendant, who took what was literally the most momentous step in human evolution? When someone digs up the foot bones of Rama's ape, we may find out.

18. DOWN TO EARTH

Life on the Savanna

We now reach an infuriating point in the story of man's evolution. It is rather like reading a suspense novel from the public library and finding, just as you get to the most exciting part, that some vandal has ripped out a whole chapter. Ramapithecus brings our story up to about fourteen million years B.P., and fossils from India suggest that he may have survived there as late as six million years B.P., disappearing with no descendants for reasons I have speculated about in Chapter 16. But in Africa, the record is blank for some twelve million years. Before that great gap (which we might as well call by its proper scientific name, the Pliocene epoch) we have what were almost certainly tree-apes—though, as already suggested, they may have traveled on the ground from time to time and very likely sought food there occasionally, as many tree-dwelling monkeys and apes do today. After the Pliocene, we have several species that are unequivocally ground apes, and at least one of them can be fairly classified as not simply an ape but an ape-man. In between, however, we have not so much as a tooth to help us reconstruct how this great change came about. We can only speculate.*

However, thanks to our knowledge of African climatic changes, supplemented by studies of present-day primates, our speculation is rather more than guesswork. Though we

* Thanks to recent diggings in Ethiopia by F. Clark Howell of the University of Chicago, we now do have a few teeth, and a couple of jawbones as well, from the Pliocene. Unfortunately, these fossils are so fragmentary, and date from so near the end of the Pliocene (only about five million years B.P.), that they add little to our knowledge of this key era.

have no direct knowledge of what happened to the descendants of Ramapithecus during the Pliocene, we know the conditions they had to contend with and can therefore put together a reasonably convincing picture of how they must have lived.

At the beginning of the Pliocene, world climates were still notably less extreme than they are today; fourteen million years ago there were perhaps already glaciers in Antarctica (and maybe Greenland), but no ice sheets there or elsewhere. At the end of the epoch, both Antarctica and Greenland were quite as Polar as they are today, and the other climatic zones had an equally contemporary distribution. In between, there must have been important climatic fluctuations, both long and short, but the overall trend in tropical Africa is obvious: cooler and dryer, with more seasonal rains. There much of the rain forest gradually gave way to the broken forest of the wet savanna, to the tree-dotted plains of the dry savanna, to the arid grassland of the steppe and at last, in some areas, to true desert. Which is to say, simply, that there were a lot less trees.

The great majority of tree species require fairly copious supplies of water—among other reasons, because they give off enormous quantities of it through their leaf surfaces. A few species have developed mechanisms for surviving under steppe conditions. The acacias, for example, get water by means of enormous underground root networks, but this very mechanism ensures that acacias must be widely separated for survival. On the African veldt today there are clumps of acacias, but there are no acacia forests.

Grasses, however, are quite another story. Biologically speaking, they are small businessmen, operating with limited capital for quick profits and modest returns. They can, therefore, survive in areas of sparse and seasonal rainfall where the cumbersome "fixed plant" of the average tree would quickly go into bankruptcy. A tree must grow for years before it can reproduce itself; a grass plant, by contrast, can spring up from seed to maturity in a few weeks of moderately moist weather and then wither away, leaving only its root system to survive the dry season—or, if the drought is severe enough, simply dying altogether, leaving its seeds to carry on the species next year.

For our ancestors, the expansion of grasslands during the Pliocene was at first merely a minor inconvenience. As long as the grass merely formed islands within the forest, a conservative-minded ape could simply travel around them. But in some areas the forest ultimately shrank to the point where it became an island in a sea of grass—and this, for a tree-adapted animal, was serious. A band of apes "marooned" on such an island might well multiply to the point where food would become a problem, unless it possessed enough physical and psychological flexibility to abandon the trees and come down to earth. It might, of course, try to evade the problem by simply migrating across the grassland to another island, but there it would likely find another band which, judging from the behavior of most present-day apes, would probably have resented the newcomers. The fact of the matter was that outside the shrinking Equatorial Zone—where the ancestors of the chimps and the gorillas seem to have holed up—the Pliocene was not a good time for tree-dwelling primates; failing some change in their habits, there would soon have been more primates than trees.

In addition to the absolute shrinkage of forest land, there was another problem. Such trees as remained outside the Equatorial Zone were themselves forced to evolve into, or be replaced by, species that could live in a climate with seasonal rainfall, carrying on their growing and fruiting only during the wetter parts of the year. Meaning, of course, that the trees' ape tenants could no longer count on a year-round supply of the tender fruits and shoots that were their favorite (though by this time probably not their only) food.

On the ground, however, there were other and less seasonal foodstuffs—not so palatable at first, perhaps, but quite adequate to sustain life. The grass itself, apart from its seeds, was little help. To live on grass requires highly specialized teeth and digestive apparatus that no primate has ever developed. But there were seeds, roots and tubers, eggs of ground-nesting birds, perhaps fish stranded by the dry-season shrinkage of pools and lakes. There was also water. The true forest ape or monkey drinks little if at all; his succulent diet supplies him with the water he needs. But an ape forced to adopt a dryer menu would have to supplement its reduced moisture content at lakes, streams, and waterholes—all of which could be reached only by descending from the trees.

This whole group of interlinked changes, then, offered a powerful biological inducement (in terms of both food and *lebensraum*) to primates that were willing and able to spend increasing amounts of time on the ground—provided, of course, that they could make the adjustment to a less finicky diet. In fact, two groups of primates did just that. One was a family of monkeys which became the baboons and macaques, the other was a family of apes, the descendants of Ramapithecus.

The baboons (including the macaques) are by all odds the most successful group of monkeys. Because they are well adapted to ground life, they can live in savannas and even steppes where no other monkey could survive; because they are omnivorous (almost as much so as man) they can get along in subtropical and even temperate climates in addition to the tropics. But their success has been limited by one important fact: they were quadrupeds when they lived in the trees, and quadrupeds they remained when they came down to earth. A baboon, like other monkeys, can sit on its haunches, grasp a piece of food, and lift it to its mouth. But when it travels, it travels on four feet. This means, for example, that it can carry nothing with it, or not at any speed; food must be consumed on the spot or not at all.

Matters were otherwise with the ground apes. Having already attained an erect, or almost erect, posture in the trees, they retained it on the ground, leaving their hands free for other things. And evolution finds work for idle hands. Part of the work, undoubtedly, must have had to do with food-gathering, which the apes had already learned about in the trees. Digging up roots with the hands is not easy, but it is easier than nosing them out with one's muzzle. Breaking off a grass stem and pulling it through the teeth to strip off the seeds is far more efficient—for a primate, anyway—than simply grabbing it between the teeth, no hands.

Hands also opened up an enormous and, for the apes, unprecedented source of nourishment: other animals. Grassland is as rich in terrestrial animal life as the rain forest is poor; the lessened rainfall means less destruction of the soil and less leaching of its nutrients, while the grass and shrubbery itself is a copious and ever-renewed source of food for animals that have evolved the necessary digestive apparatus to cope with it. It is not surprising, then, that the ruminants—the grazers and browsers—first became really abundant during the epoch immediately preceding the Pliocene. And along with the ruminants—the ancestral antelope, cattle, sheep and so on—of course came predators equipped with the necessary teeth, claws, and fleetness of foot to catch and kill them—the various ancestral cats, dogs, and kindred creatures.

Compared with even the first rough models of these meat-eaters, the ground apes were in many respects ill equipped for predation. They had no claws like the cats, and their teeth, though doubtless able to deliver a nasty bite, were not in the same murderous class with those of the ancestral dogs. But they did have hands that could seize a fledgling bird, or hare, or baby antelope, and batter it to death against the ground. Even baboons, basically vegetarians and with less efficient hands, not infrequently do just this—though apparently only when they happen across some potential prey, since deliberate hunting seems to be beyond their mental capabilities. But the ground apes, with better brains than baboons, no doubt learned before long to emulate the other predators, observing the movements of their prey, sneaking up on it, and seizing it with a final leap. Predation provided further biological pressure for an improved brain; not all predators are equally intelligent, but with almost no exceptions they are brighter than the animals they prey on.

We need not, however, suppose that the ground apes' meat diet was all acquired by predation; some of it, unpleasant as the thought may be, was very likely carrion. Again, they were poorly equipped to compete with the primitive hyenas who were also going into the carrion business (a pack of hungry hyenas can harass even a lion to the point where it will abandon its kill). But again, hands gave an advantage; a ground ape could charge in on a carcass—perhaps screaming for moral effect—seize hold of a detached leg or head, and run off with it, leaving the other carrion eaters to quarrel over what was left.

Did the ground apes do all this bare-handed? At first, perhaps, they did; but it seems likely that before many million years had passed they had learned better. To see why, we must look at the problem of predation from another angle: that of the hunted rather than the hunter. For there can be no doubt that the apes, as they became more consistent ground-dwellers, were considered fair game by other predators. How did they survive?

So long as they remained fairly close to the trees, they could always seek refuge in their ancestral home. But as the trees continued to thin out, this shelter became increasingly unavailable. They could, of course, have done what the baboons in fact did: evolve long, canine teeth, almost tusks, for defense. The males of a baboon troop can gang up on and face down even a leopard; indeed, of all the predators, only the lion is capable of making them beat a prudent retreat. But the apes must have evolved a better way during the Pliocene, for none of the post-Pliocene ape-men show any trace of the necessary dental equipment.

We know from present-day observations that many species of monkeys and apes will throw down objects—branches, fruits, excrement—at intruders. They apparently do this with no particular purpose but simply as a display of hostility and irritation. Chimpanzees, however—whose intelligence is probably pretty close to that of the ground apes—have gone a step further: they have learned to throw rocks, and throw them purposefully. The naturalist Jane Goodall, who has studied chimps extensively in the wild, saw one of them throwing stones at a photographer who was blocking its access to some bananas and observed others using the same tactics to drive baboons away from food they wanted. (Interestingly, the chimps threw both overhand and underhand.)

If the chimps, with no real need for self-defense (since they can escape up a tree in an instant) have gone this far, there can be little doubt that the ground apes, with fewer accessible retreats, must soon have learned to do as well and better. Before long, they did not need to depend on sheer surprise and speed to seize carrion; they could drive off their animal competitors with rocks, seize their chosen—if gamey—joint, and make off, discouraging pursuit with more rocks. And from this it was not much of a step to the discovery that a rock could sometimes bring down a meal of living game—or that a heavy chunk of tree-branch could stun game more effectively than brute force. If they could do as well as the chimps in one respect, they could surely do as well as the baboons in another, developing the capacity to act in concert. A threatening predator, then, would find himself confronted, not simply with one rock-throwing ape, but with a shower of rocks from the whole troop.

Slowly, almost imperceptibly, the ground apes were taking the first steps toward humanity. They were beginning to depend, not simply on their own biologically evolved physique, as all animals do, but on external, inanimate objects—tools, in fact. And they were beginning to recognize in practice (not in theory—they could not theorize) that a cooperating group of animals is a lot more effective than a single one.

Yet even with these advantages, it would be a mistake to think of the ground apes as especially effective or successful animals. They were not very able predators, compared with a leopard, not very competent scavengers, compared with a hyena, not very efficient vegetarians, compared with an antelope or pig. The only thing they were really good at, in fact, was doing many things inefficiently. They were masters of no biological trade—but jacks of many. In the course of some millions of years on the African savanna, they had begun acquiring what is unquestionably the most fundamental trait of man: adaptability.

19. HABITS AND HABITATS

The Ape-Men

We pick up man's fossil record at just about the time that the Pliocene was giving way to the Glacial or Pleistocene,* epoch. What the fossils reveal is a creature—several creatures, in fact—unequivocally adapted to ground life. The long drying out of the Pliocene had done its work well. When we examine the skulls of these animals, we find that the foramen magnum—the hole through which the nerves of the spinal cord enter the skull—lies almost at the bottom, rather than toward the rear as in tree-dwelling apes, meaning that the creatures' heads were balanced on the tops of their spines, like our own, rather than hunched forward. Leg, foot, and pelvis bones tell the same story: the animals stood and moved erect, or nearly so, though their gait was apparently a trot or waddle rather than the striding walk of

* Pliocene means "more recent," i.e., more recent than the preceding Miocene ("rather recent") epoch. Pleistocene means "most recent." To complicate matters, the epoch following the Pleistocene—the one in which we are now living—is called, simply, Recent.

our own species. Their feet were no longer capable of gripping a tree-branch, as their ancestors' feet presumably had been; doubtless they could and did climb trees for food or safety, even as men do today, but they did so no more efficiently than we do.

I have said "animals"—but were these creatures merely animals? Here we must drag in that old cliché of academic dialogue: Define your terms. What is a mere animal? Or, putting it in a more meaningful way, what is a man?

We have already noted that the Pliocene ground apes almost certainly used tools. Chimpanzees do—and not simply as missiles. They have been seen to use sticks for digging and rocks for cracking nuts; they poke thin twigs into termite nests and then withdraw them covered with delicious termites; on occasion they even carry the tools around with them. All these things the Pliocene apes must have learned to do. But that did not make them men, any more than a chimp is a man. Anthropologists, long aware that the use of tools is not restricted to our own species, have defined humanity more narrowly. Until quite recently, they employed a definition first put forward by that singular genius Benjamin Franklin: man is a tool*making* animal. Man, that is, does not merely use natural objects to manipulate his environment but also modifies them to make them more useful.

Now, it seems that even that definition is too wide. Chimps have been seen to make tools, after a fashion. They will not merely use a dead twig as a termite-probe, but will strip the leaves from a twig to make such a probe. They will also crumple up a leaf to sop up water from a shallow puddle—not a very efficient way of getting water, perhaps, but (as scientists have proved by experiment) far more efficient than using an uncrumpled leaf or one's fingers. These observations force us to define man as an animal that makes tools *to a pattern*. It does not simply modify objects, but does so in a uniform, customary, one might almost say habitual, way.

Thinking about it, I'm not sure that even this revision

quite meets the case. When the chimp strips a twig, he is certainly creating a pattern, of a sort. Human patterns have one more element: a certain quality of arbitrariness. Man does not simply modify natural objects, like the chimp with his crumpled leaf, nor does he merely remove portions of an object, making use of whatever part of the pattern remains, like the chimp with the stripped twig. Rather, he *fashions* an object, producing a pattern *that was not there before*—and, what is more, one that is peculiar to his culture or tribe or nation.

Judged by this criterion, the apelike creatures that inhabited southern and eastern Africa in the early Pleistocene stood on the very threshold of humanity, for they fashioned tools out of stones and, in some cases, fashioned them to a pattern. The tools are terribly primitive: pebbles ranging from ping-pong to billiard ball size, with a few chips knocked off one end to make a crude cutting or chopping edge. Some of them, indeed, were apparently chipped by nature—e.g., being rolled about in the bed of a stream—and are identifiable as tools only through being dug up on "living floors" miles from where similar pebbles are found today. Others are so similar to the "nature-made" group as to recall the classic remark of a French prehistorian: "Man made one, God made ten thousand. God help the man who has to distinguish the one from the ten thousand!" Yet by carefully examining the appearance and angle of the chipping, we can be reasonably sure that man, not God, made them. Still others show the very beginning of arbitrary patterning, by which some pebble tools from one locality can be distinguished from those found elsewhere—even as we can today distinguish a Japanese from an American hammer by the shapes of the heads.

Yet, for all that, were these toolmakers men? Putting it another way, do patterned tools by themselves make the man? I myself think not. So far as it goes, the definition is a good one, but we must recognize that it is partly a definition of convenience. Man is defined by tools because the tools are there to be examined, while the toolmakers themselves are present only as fragments of bone. Yet if we cannot imagine man without tools, neither can we imagine him without language, which is itself the most versatile tool of all. But language, of course, is gone with its speakers. Or is it? Though we certainly cannot say definitely whether or not the first African toolmakers had or had not developed a language, are there not, perhaps, facts by which we can infer it? I believe there are. But to explain why, I must first explain what language is, or rather, what it does.

Language is often described simply as a means of communication—but it is much more than that. Many animals communicate—by noises, by gestures or postures, by facial expression (e.g., chimpanzees), even by scent. The question is not whether an animal can communicate but what it communicates about. And what all animals but man communicate about is the immediate present. Even the most intelligent animals (the chimps) cannot talk about even the most familiar object or situation unless it is physically present. They can say the equivalent of "Come here," or "Go away," or "I want to mate with you," or "Give me that food, or else!"—but they cannot say, "We will eat bananas tomorrow," or "There are termites across the river," or "Who was that lady chimp I seen you with last night?" Only by means of speech—a system of arbitrary symbols—can we talk about things that lie in the past, or the future, or on the other side of the mountain—talk about them, make joint plans concerning them, and think about them. For it is by silent ("subvocal") speech that man characteristically mulls over his past experience, considers present possibilities and prepares for future contingencies. Language and thought are themselves closely allied to toolmaking. We can see this by watching a chimp making a termite probe from a leafy twig. He can make the twig, carry it about with him, and use it skillfully—but the whole process begins only when he sees a termite nest.

Now there is a fair amount of evidence, albeit indirect,

that the African toolmakers were, so far as language and thought are concerned, little beyond the chimp level. For one thing, they made relatively enormous quantities of tools. Granted, a stone chopper is far more durable than a human skeleton, but when we find thousands of choppers and only a few dozen skeletal fragments (the entire total of human remains up to about 100,000 B.P. would fit comfortably into a steamer trunk) it is hard not to conclude that the toolmakers made most of their tools on the spot—that is, to meet an immediate need. (The tools were so generalized in form that they could have met dozens of different needs, ranging from stunning a baby antelope, to chopping it up, to smashing its skull to extract the brains.) Tools were, certainly, carried about, and for considerably longer distances than a chimp will bother to carry his "tools." But the normal fate of a tool seems to have been to be used once or twice and thrown away.

The second point has to do with language not as a vehicle for anticipating the future but simply as a medium for thought, and the evidence is powerful that whatever thinking the first toolmakers did must have been on an inconceivably low level. The patterns of their tools—the stone ones, at any rate, for there were doubtless bone and wooden tools that have perished—show virtually no improvement for a million years or more. For at least fifty thousand generations they went on living in the same inefficient way, often hungry, sometimes starving, and not infrequently, we can be sure, making a meal for predators equipped by nature with more efficient tools than theirs. If they could think, it is hard to imagine what they could have been thinking about! No, if we cannot classify these creatures as apes—as, on the evidence of their toolmaking, we certainly can't—we equally cannot call them men. They were ape-men, or man-apes.

And here we get into one of those hassles over terminology which reminds us again that science is not quite as rational as some of its practioners would have us believe.

When the first skull of one of these man-apes was discovered in South Africa, nobody, not even its discoverer, was expecting to find a manlike creature in that region. Accordingly it was christened merely *Australopithecus*—the "southern ape." Subsequently, other specimens were discovered with tools which made it clear that they were more than apes, and were christened *Paranthropus* ("very like man"), *Plesianthropus* ("nearly man") and *Zinjanthropus* ("man of Zinj" —an old word for East Africa). Obvously the splitters have been at work here. Indeed, they have pushed one group of fossils into still another genus, our own—*Homo,* meaning "man" without trimmings. This step Elwyn Simons, who might be called the lumper's lumper, calls "a conjecture derived from a hypothesis based on an assumption." Simons and the other lumpers have tried to bring some order out of this terminological chaos by reducing all the genera to one. But they were stuck with the original generic name coined in 1924, so that the earliest creature that was definitely not an ape is still known as the southern ape—Australopithecus.

Much of what we know about Australopithecus we owe to the famous diggings at Olduvai Gorge in Tanzania, East Africa. The gorge, a sort of miniature Grand Canyon, was cut out by a now-vanished river which exposed fossil deposits covering, with only one important gap, something like a million and a half years of prehistory. Moreover, by a singular bit of geological luck, the gorge lies close to two now-extinct volcanoes, Ngorongoro and Lemagrut, whose periodic eruptions during the Pleistocene laid down numerous layers of ash by which, using the potassium-argon technique, the Olduvai deposits can be dated. Due largely to the skillful and devoted labors of Louis Leakey and his wife Mary, Olduvai has yielded up a singularly detailed and dated record of the ground apes' emergence into humanity. Their findings have also helped to make chronological sense out of man-ape fossils from South Africa's Transvaal, which, though quite as abundant as those in Olduvai, did not have the luck to be laid down in a volcanic area.

Perhaps the most interesting fact about the Australopithecines of Olduvai and the Transvaal is that they came in at least two (Leakey would say three or four) species. One, Australopithecus africanus, was a slender four-footer, no bigger than a modern pygmy. Its cousin, Australopithecus robustus (Paranthropus, Zinjanthropus) was, as its name suggests, bigger—about five feet—and burlier. From all the evidence, the two species seem to have been more or less contemporary (at Olduvai, much more than less) during the earlier part of their careers on earth. The simultaneous existence of two closely related man-ape species is puzzling, because it is exceptional. There is today only one species of gorilla, one species of chimpanzee, of orang-utan, and, for that matter, of man. The reason seems to lie in a general principle of ecology: Where two closely related species occupy the same territory (and at Olduvai we are pretty certain they did), they must be adapted to different ways of life. If both occupy the same "ecological niche," either one species will displace the other in fairly short order or, much more likely, they will not evolve into two species in the first place.

One clue to the explanation of the two Australopithecine species comes from the bones themselves. I have on my living room table a plaster cast of a robustus skull which I have doctored with furniture stain so that it makes a fairly plausible-looking fossil. It does not, needless to say, look much like a human skull; perhaps the most striking difference emerges when I turn it over and examine the upper teeth (the lower jaw was missing in the specimen from which the cast was made). The biting teeth (incisors) are not very remarkable; they are, if anything, smaller than my own. But the molars are enormous—massive, and so broad that they look as if they had been set in the jaw sideways.

The top and back of the skull are also missing, but from descriptions of it I know that robustus had a bony ridge along the top of his cranium, similar to, but smaller than, that on the skull of a modern gorilla. Mechanically, the function of such a ridge is to give a secure anchorage to the massive cheek muscles which move the lower jaw. Thus the ridge, with its vanished muscular attachments, tells the same story as the massive molars: robustus lived on a bulky diet which required lots of chewing. It seems very likely, then, that like the gorilla he was a vegetarian, or close to it, living largely or entirely on the traditional primate diet of fruits and shoots during the rainy season and on roots and seeds during the dryer parts of the year. This conjecture is strengthened by miscroscopic studies of robustus teeth; these reveal scratches apparently made by sand grains, which presumably were carried into his mouth with the roots.

Australopithecus africanus, with neither skull ridge nor massive molars, seems to have adapted to a considerably more catholic diet. From bones found on his living floors he ate lizards, tortoises, various kinds of small mammals, and the young of some medium-sized species such as antelopes —along with, we can be sure, such vegetable foods as he could get.

This hypothesis (and I should stress that it is still only a hypothesis) explains a number of other apparent facts about the two species. From what I have said about their contrasting life patterns, it will be obvious that while both could survive in a fairly moist savanna (which the Olduvai area seems to have been some two million years ago), only africanus could have managed to make a living on a dry savanna or steppe, since the vegetable foods to which robustus' anatomy—and perhaps other things—adapted him would have been too sparse for survival.

The South African deposits seem to confirm this, notably in the sand found in the various layers. Where we find africanus fossils, we find sand whose grains are rounded— evidently by being blown along for considerable distances, perhaps from the Kalahari some hundreds of miles to the west. And wind-blown sand, obviously, means a dry climate. The robustus fossils, on the other hand, are found with sand having sharp, angular grains—evidence that the climate was wetter. Still further confirmation comes from another curious fact. Robustus did not evolve significantly between his first appearance in the fossil record, some two million years ago, and his final bow, perhaps as much as a million and a half years later. During the same period, however, africanus changed into quite a different creature—larger, brainier, and in other respects more human—to the point where he had become a full-fledged member of the genus *Homo*.

It seems at least plausible that the difference stemmed from diet. I have already noted the important mental differences between herbivores and predators. Vegetable food, provided it is tolerably plentiful, offers little mental challenge, since all you have to do is find it and eat it. Animals, however, must not only be found but also caught—which, unless you are as superbly equipped by nature as the cats and dogs, means living by your wits. Africanus, it seems, must have been smarter than robustus. It is quite possible, indeed, that of the two species only he was a toolmaker. At any rate, no fabricated tools have ever been found with robustus bones—unless africanus bones were also present. Africanus fossils, on the other hand, often occur without robustus remains—but with tools.

Thus it appears that not all the ground apes learned the lesson of the savanna equally well. Robustus, sticking to a vegetable diet and adapting himself to it anatomically, remained tied to a particular climate, one that was neither too dry nor too wet. Africanus, on the other hand, evolved in the direction of less anatomical specialization, not more. From an evolutionary standpoint, he cultivated his mind, and reaped his reward by ultimately becoming the one animal that can survive amid rain forest, desert, or polar ice. Robustus remained a manlike ape, ultimately driven to the evolutionary wall (and quite possibly eaten) by his more adaptable cousin. Africanus became man.

20. THE RED FLOWER

The Start of Climate Control

The drama of human evolution is rather like a pantomime performance presented in a pitch-black theater, where every few minutes the lights flash on for a moment or so to let us see what is happening on stage. Only toward the end of the last act do these flashes come with enough frequency to give us any coherent notion of the plot. For the rest, our brief glimpses of the actors show mainly that as the evening draws on they are standing a little straighter and walking more freely, with higher foreheads and less hair. The activities they are engaged in—though it is hard to discern precisely what these are—seem to grow a bit more complicated. And that is about all we can see.

In one respect, of course, the picture is not quite as dark as this analogy suggests. In the evolutionary drama we do know beforehand what the denouement will be, since we ourselves are it. But in another respect things are even worse, since in human prehistory we frequently cannot tell which brief tableau precedes or follows which. As already noted, the chronology of the first scenes, with the Australopithecines as the dramatis personae, has been fairly well worked out, thanks to volcanic deposits and potassium-argon dating. Subsequently, however, the dates become sparse and increasingly unreliable down to perhaps 100,000 years ago, when first climatic changes and then radiocarbon begin to supply an assured chronological framework. For example, a recent article touching on Peking Man places that important human forebear in "the Second Glacial, the Second Interglacial and the Third Glacial"—which would mean, if one accepts the increasingly plausible "long" chronology of the Pleistocene, the improbably lengthy interval from perhaps 1,000,000 to 400,000 B.P. A single uranium-thorium reading, on the other hand, dates the creature "somewhere" between more than 500,000 and 210,000 years B.P.

Any would-be coherent picture of man's relationship to climate following his first appearance on the African savanna is bound to be largely guesswork—a fragile fabric woven of little fact and much conjecture which next year's excavations somewhere in the world may rip apart. With that warning, let us proceed.

The first thing that happened to our savanna- and steppe-dwelling forebears is that very little happened. For perhaps a million years, the sparse testimony of the spade merely confirms what we could have guessed anyway: their brains became somewhat bigger, their tools a trifle less crude. These leisurely changes are probably related—as cause, effect, or probably both—to the fact that man's habitat preferences also changed little. He remained a creature of the tropics and subtropics, in areas where the rains were sufficiently seasonal to produce at least some open country—savanna or warm steppe.

So far as Australopithecus himself is concerned, there is no evidence, apart from an ambiguous fragment of bone found in Indonesia, that he ever moved out of Africa or, indeed, north of the Sahara. Some of his brainier and more venturesome descendants, however, evidently reached North Africa, for their crude tools and one skull have been found in Algeria. Unless they made their way along the Nile Valley, the crossing was presumably made during the First Interglacial, when a climate warmer and wetter than today's would have driven summer rains into the Sahara far north of where they reach today. We know from Saharan rock paintings of a much later date that during a recent warm spell the fauna—and therefore the climate—there differed little from that of the South African veldt-steppe; indeed, beasts such as the elephant, giraffe, and ostrich roamed north to the Mediterranean almost to historic times. North Africa itself, though its climate most resembles that of southern California, is also not unlike southern Africa, except that its rains come in winter rather than summer.

From North Africa the creatures pushed east, through the Middle East (again, perhaps, during a moister period) and India. We have no clear traces of their journey, but in the Malay Peninsula crude tools (though no bones) have been turned up dating from something over a million years ago, either at the end of the First Interglacial or the beginning of the Second Glacial. My own preference is for the latter. During an interglacial, Malaya would surely have had a climate like that of today: Equatorial. And the rain forest, as we have noted, being poor in accessible vegetation, is inhospitable to animal life, including man. Even today, the Congo and Amazon jungles are sparsely populated; early man, with his subprimitive technology, could hardly have survived in such an environment. Certainly none of his traces turn up in other Equatorial areas until much later, so we may well believe that his migration through Malaya was blocked until a new glaciation cooled and dried it sufficiently to produce at least a reasonable facsimile of his preferred savanna.

Certainly his next step *must* have waited on a glacial period. By about a million years ago he had reached the island of Java which during an interglacial, as today, would have meant crossing the Straits of Malacca and Sunda Strait—both of them narrow and shallow, but quite deep and wet enough to daunt any man who had not learned to build at least a raft, which Java Man had neither the tools nor the wit to do. During a glacial period, however, the lowered sea level would have let him walk on dry land to Java along what is now the shallow Sunda Shelf to the north and west. Even today, parts of Java are savanna, and during a glacial period there was probably more of it. Java Man, in fact, seems to have been something of a conservative, both anatomically and culturally. Some of his kin, however, were more venturesome; instead of striking southeast they pushed northeast into China, where they have left their bones in a few places—notably in the famous Choukoutien cave near the modern capital of Peking.

Now the climate of Peking today is by no stretch of the imagination tropical or even subtropical. It is, in fact, Continental temperate, rather like that of Chicago but, because of the immense Eurasian continent to the west, even more continental, with hotter summers, colder winters, and less rainfall. During a glacial period, conditions would have been even more trying, with bitter winters like those of Winnipeg or Kiev. And even during a warm interglacial, conditions in winter would have been distinctly chilly.

All of which gives point to the fact that Peking Man was the apparent author of perhaps the most momentous discovery in human history: the controlled use of fire.* Charcoal deposits buried beneath the cave floor show where his hearths were laid; charred bones make clear that he roasted his game—and on occasion, regrettably, his fellows.

There have been many speculations on how, and why, man tamed fire: a taste for cooked food, the need to scare off predators, the sheer fascination of a leaping flame which even today can hypnotize us before a fireplace. But these notions, it seems to me, overlook one central fact: to a wild animal, fire is an enemy. Certainly fire could have been no stranger even to Australopithecus. The spewings of the volcanos around Olduvai must on occasion have started grass fires which sent him scurrying for his life, and in almost any savanna area a lightning storm at the end of the dry season can set a tree ablaze. But such wildfire, to Australopithecus as to his animal contemporaries, must above all have been something to get away from.

Had man remained a creature of the tropics and subtropics, he could hardly have had any very pressing incentive to fight down this ingrained terror, and in fact we find no

* Or at least one author; a hearth recently excavated in France may be even earlier than the Peking deposits, though in neither case are the dates more than rough approximations.

traces of hearths in Africa or any other warm region until much later than the Peking deposits. The conclusion seems inescapable: man learned to use fire because he was cold. At some point, perhaps as the ice of the Second Glacial expanded, he must have run into the coincidence of a small, not *too* frightening grass or brush fire and a spell of cold weather. The impact of the warming rays on his shivering hide must have thrust into his dull mind the notion that this terrifying, potentially painful red flower might be good for something—and therefore worth a closer look. He looked. And learned.

The taming of fire was revolutionary in many ways, perhaps most importantly as a means of controlling climate. Neither Peking Man nor, for that matter, modern man, could or can control climate in the large. But as one scientist has put it, the climate most important to man is that of the 1/16-inch layer next to his skin. Once he had discovered a way of keeping that layer reasonably warm, he could—and on the fossil evidence did—survive in climates that were a far cry from the warm savannas of his birthplace. His potential living space had expanded to include millions of square miles in the Temperate Zone. Significantly, traces of fire also turn up at Vertesszölös in Hungary, where the climate must have been almost as severe as that of Peking. Though the site cannot be dated, the tools found there are similar enough to those at Choukoutien to suggest that the two groups of firemakers were roughly contemporary, coexisting perhaps half a million years ago. Dating from somewhat later is carbon and charcoal found at Torralba, Spain; plant pollen found in the deposits shows that those fires burned at a time when the climate was rather colder than today (the onset of the Third Glacial, perhaps?). Not, certainly, as cold as Peking—Maritime temperature rather than Continental temperate—but still chilly enough to make fire a blessing if not a physical necessity.

In taming fire man had become a species which, for the first time in evolutionary history, could draw on sources of energy outside those supplied by solar heat and its own metabolism. With his body temperature less dependent on its internal heat, he could get along on less calories. (Central heating is one reason most of us no longer consume the mammoth meals of our great-grandparents.) Moreover, if man began using the energy of fire to keep warm, he quickly learned to put it to other purposes—notably, as we know from the Choukoutien deposits, cooking. And cooking, whether one likes one's mammoth steak rare or well done, is much more than a gustatory refinement. Man's digestive system is still apelike, and therefore by no means as well adapted to a meat diet as that of, say, a lion, whose evolutionary line has been eating flesh for tens of millions of years. Cooking, by breaking down the complex molecules of proteins and fats, renders meat more digestible—which is to say, more nutritious, with more usable calories per pound. The same is true even of some vegetable foods, whose carbohydrates are made more accessible when the plant's tough cell walls are broken down by heat. I don't know whether mickeys roasted in the ashes would taste as delicious to me as they did at age eight, but I am certain they would be more palatable than a raw potato—and a great deal more nutritious. As a minor metabolic economy, fire also cut down the energy expended in chewing. It was the first, and is still the chief, meat tenderizer.

Along with using fire to alter the chemistry of food, man also learned to use it to alter the chemistry of wood. If you char the end of a stick gently, it can then be whittled into a distinctly harder and more penetrating point (try the experiment with a kitchen match). Among the very few wooden relics of early man which have turned up (except under extraordinary circumstances, wood rots away in a few centuries at best) is part of a spear that had been fire-hardened in just this manner. Further chemical tricks with fire had to wait a long time. Nonetheless, the charcoal-tinged earth in Choukoutien cave foreshadows the potter's kiln and the smith's furnace—and, later still, the singular devices through which we make the fires of coal and oil move our goods, drive our machines, and light our homes and factories. Even the most modern atomic power plant, indeed, is fundamentally just a fireplace—with attachments!

Once having learned to overcome his own animal fear of fire, man may well have learned to use it to put fear into other animals. The deeper deposits at Choukoutien—the total is no less than 160 feet—show layers of human habitation alternating with others marking the occupancy of sabertooth or giant hyena. In the higher, charcoal-tinged layers, however, these fearsome carnivores have been permanently evicted. Cause and effect? And F. Clark Howell, who excavated the Torralba site, believes that the men there may have set grass fires to drive a heard of elephants into a nearby swamp, where they could be butchered at leisure. Certainly the traces of charcoal at the site are widely scattered—and the remains of several elephants have been dug up within a radius of a dozen yards or so, in deposits that once were bog mud. Someone, somehow, must have maneuvered them into the bog; as Howell has pointed out, nobody kills elephants and *then* drags them together into one place.

Fire gave man more than living space, more than better food, greater security, and better hunting tools and techniques. It gave him time. Living as we do, in a civilization where the flick of a switch floods our houses with light, it is hard to imagine what night must have been like in those far-off times. On a clear night in summer, I will sometimes turn off the lights in my cottage on Cape Cod and walk out to look at the stars which, undimmed by sky-glow. blaze far brighter than any city or suburb dweller can ever see them. It is a magnificent sight and, moreover, supplies sufficient light so that I can stroll along the white sand road, listening to the night noises. But there is nothing to do but watch the stars and listen.

Before man had fire, all his nights—barring only those with a bright and unclouded moon—were like that, fit only for star-watching or sleep. Whatever he needed to do, whether hunting, toolmaking, root- or berry-gathering, he had to do between dawn and dusk. Fire gave him the night hours. If, like us, he could get along on eight hours sleep a night, his "working day" was immediately expanded by whatever remained of the night—on the average throughout the year, a gain of perhaps 30 percent in available time. He could not, to be sure, hunt at night, unless—which is conceivable—he learned to use torches to bedazzle and befuddle his prey, as some hunters still use searchlights illegally to "torch" deer. But he could chip his tools around the fire, and use his new leisure to devise better tools.

No less fascinating than the things fire did for man are the things its use tells us about him. You will remember that we viewed Australopithecus as weak in thinking, very weak in foresight, because, so far as we can tell, he lacked words—the vehicle by which man characteristically thinks about the past and plans for the future. Unless you use words—an arbitrary noise that "stands for" an object—you can only deal with the object present, never with the object not-present. One line of evidence leading us to this supposition was the fact that all of the man-apes' tools were—or in any event, could have been—improvised to meet a fairly immediate need. But man-the-fire-user is clearly a different creature. *You cannot improvise a fire;* even the most primitive techniques of fire-making require preparation. To strike a spark from two rocks, they must be the right kind of rocks; if it's a matter of rubbing the proverbial two sticks together, just any old sticks won't do. And in either case, you must have ready some tinder—shredded bark or dried moss—on which the precious spark can be caught and nursed into flame. Making fire, whether in a cave or a Boy Scout camp, forces you to Be Prepared.

We do not, actually, know whether Peking Man, or his descendants at Torralba, could *make* fire at all. But even if they couldn't, they still required foresight. They may have gotten their fire from a natural grass or brush-fire, but they then had to keep it alive, meaning that they needed wit enough to have fuel in readiness when the flames began to die. If you wait until your fire is almost burned out before going to pick up wood, you are likely to end up mighty cold in an Ice Age winter. The Keeper of the Flame is an ancient and pervasive concept in human mythology and religion; the Vestal Virgins of Rome guarded an ever-burning fire, and even today eternal flames burn at various memorials around the world.

The presumption of foresight (and its prerequisite, language) which we derive from man's use of fire confirms inferences from other facts. Peking Man, as we know from the bones at Choukoutien, dined chiefly on venison, but on occasion killed rhinoceros and even elephant—a job which certainly required the cooperation of many men. Even more must this have been true at Torralba; to drive a herd of elephants into a bog, by fire or any other means, a group of men must be in the proper place at the proper time, primed to do what is needed. Somebody, that is, must be able to *say* the equivalent of, "You hide there. When I shout, stand up and start yelling."

The evidence that our ancestors of half a million years ago possessed language—albeit, no doubt, far less expressive than any tongue we know or can imagine—seems to be clinching evidence that man had made a decisive step in his evolution. Not only was he brainier than Australopithecus—the mere bones tell us that—but brainier in a different way, a way embodying the beginnings of planning and foresight, of the language that organizes human cooperation and human thought. He well deserves the promotion to a higher grade than the taxonomists have given him: no longer is he Australopithecus, or even, in the old terms, Sinanthropus (Peking Man), Pithecanthropus (Java Man), and so on, but *Homo erectus*—upstanding *man.* The flickering flames of those far-off campfires signal the fact that the hairy creatures huddled around them had become, if not "sapient," as we believe we are, at least human.

21. PEOPLING THE EARTH

Climate Control and Migration

When we look at the chipped lumps of rock that Homo erectus left behind him, we note a striking difference from the "pebble tools" of Australopithecus. Those rudimentary artifacts are at best only crude approximations of tools; at worst, they can be identified as quasi-human handiwork only through the circumstantial evidence of where they were found. With erectus, by contrast, we are clearly dealing with objects which somebody has gone to some trouble in shaping to a useful pattern.

But even here, though the patterns are unambiguous, the uses are anything but. One of the most distinctive tools associated with some erectus cultures, a pear-shaped, flattened object usually called a hand-ax, may have been used as an ax—but was surely just as effective in grubbing up roots, sharpening a spear, bashing in the skull of a deer, or skinning a rhinoceros. (The old-time backwoodsman's ax was hardly less versatile; it was said that a good axman might even "edge her up a bit and shave with her on Sundays.") Thus when we examine a leaf-shaped flake of stone with a smooth, fairly sharp edge along one side, it is hard to say whether we are looking at a scraper—which is what such objects are usually called—or perhaps a crude knife. Still, considered in the light of the expanding human toolkit of a few hundred thousand years ago, scraping seems as good a bet as any. For heavy work, these men had hand axes, as well as tools which, from their size and shape, can only be described as cleavers. Whatever the smaller "scrapers" were used for, it must have been a rather light and delicate job

—quite likely, removing the fat and other tissue from the interior of a hide so that the skin, which once warmed its animal owner, could warm a hunter. For man's efforts at climate control—still referring, of course, to the climate next to his skin—were not limited to the use of fire. There are, after all, other ways of keeping warm.

As soon as man became crafty enough to kill fairly large and tough-skinned animals, which on the evidence of the fossils at Choukoutien and elsewhere was better than half a million years ago, he found himself compelled to cut away the hide before sitting down to dinner. We can imagine, then, a beetle-browed hunter working himself into a sweat one chilly day to hack and haul the hide off a deer or wild sheep and then, sitting back to await the broiling of his mutton or venison dinner, beginning to shiver as he cools. Almost without thinking, he pulls the still-bloody hide about him—and relaxes in comfort, having just invented the prototype of all subsequent blankets, robes, togas, Nehru suits, and miniskirts.

At first, these archetypal insulators must have been rather temporary affairs; only after many generations could man have worked out ways of treating hides so that, in a few days or weeks, they would not stiffen into worthless rigidity or, in a damp climate, decay into a slimy and evil-smelling mass. But wherever and however men learned about tanning, they must have discovered the need to scrape the hide clean as a first step. This other venture in climate control was quite as eloquent of man's growing foresight as was his use of fire. A fur robe, however crude, can no more be improvised than can a flame. Thus a scraper, as the writer George R. Stewart has said in an almost classic passage, means "not only a scraper but . . . a knowledge of how to scrape, and a desire for scraping, and enough leisure (beyond the struggle to get food) to allow time for scraping. All this means self-restraint and *thought for the future* . . ." (my emphasis).

Not surprisingly, what may be the first scrapers turn up amid evidences of a cool, or even cold, climate—among the tools of the Clactonian culture, named after a town in southern England. The Clactonians, by some accounts, lived in England and adjacent parts of Europe as early as the Second Glacial, but such accounts are difficult to accept—until somebody digs up a Clactonian camp site with evidence that they could use fire. The English climate is bad enough today, but with an ice sheet covering Scotland and the Midlands, it must have been uncomfortable even for men with furs and fires to supplement their body hair, such as it was.*

The Clactonians, in fact, are a puzzling people. If in fact they did live in western Europe during the Second Glacial, there is evidence that they abandoned it during the succeeding interglacial, being replaced by another culture, evidently moving up from Africa, which we shall call by the best known of its varied names, the Acheulian. There is nothing very odd about this; human cultures had disappeared before and would do so many times after. But as the comfortable climate of the Second Interglacial gave way to the renewed chill of the Third Glacial, the Acheulians themselves drew back toward their African homeland, to be replaced in turn by what, from their tools, seem to have been descendants of the original Clactonians. There is no evidence of where these folk had been in the meantime. Logically, we would expect them, habituated to a chilly climate and accustomed to consuming the characteristic animals and plants of that habitat, to have followed the ice northward, into Scotland, Finland, and perhaps Scandinavia. But there is, in fact, no

* When, where, and why man lost the coat of hair which he must have inherited from his ape ancestors is very obscure. Climate does not seem to have much to do with it. To be sure, the present natives of tropical Africa tend to possess less body hair than "white" populations, whose original home was in cooler climates, but against this is the fact that the tropic-dwelling great apes are without exception far hairier than the hairiest white. The most we can say is that in the tropics the loss of hair might have occurred at almost any time during man's evolution, but that in temperate climates it almost certainly could not have happened until man had developed fire, and probably fur robes, as replacements for his own vanished pelt.

evidence of Homo erectus, whether tools or bones, north of latitude 55. The Clactonians, for all the fragmentary evidence tells us, simply vanished and reappeared. Which is ridiculous.

The Clactonian puzzle is not the only one concerning early man's migrations into and out of various parts of the Temperate Zone. (There is no evidence that he ever abandoned the tropics and subtropics, though doubtless the expansion or contraction of the inhospitable Equatorial and Desert areas at times forced him to shift his ground.) It has generally been assumed that man did not make his way into Siberia or Central Asia until almost the dawn of recent times, and certainly no relics of Homo erectus have been found there. Not even during interglacials.

Now the winter climate of Siberia today is notoriously severe. Warm winds from the south are blocked off by the mighty ranges of the Himalayan system, while air from the frozen Arctic Ocean has free passage southward, to be cooled still further by the immense stretches of winter-chilled land. Moreover, large parts of this great land mass are either desert or dense forest of pine and spruce which, though not so poor in animals as the tropical rain forest, would hardly have been a happy hunting ground for so primitively equipped a creature at Homo erectus. But between the forest and desert lies a sizable belt of grassland and open forest which during an interglacial period must have supported a large population of horse, antelope, deer, bison, and aurochs (an ancestor of our domestic cattle), all of which, as we know from European findings, were consumed with relish by the early hunters. If Homo erectus could survive successfully in Europe of the Third Glacial, how is it that he does not show up in Siberian deposits of the preceding interglacial—at least parts of which were considerably warmer than at present? Even during the Third Interglacial, Neanderthal Man—a far brainier specimen than Homo erectus—does not seem to have gotten much past the southern tip of the Caspian Sea, and to the North, not even that far east.

The simplest explanation is that we simply have not found all the evidence.* Central Eurasia is a big place and only moderately populated even today; there may be a lot of chipped flints lying under its soil (yet traces of "modern" stone age man *have* turned up).

A second possibility emerges when we look at a map of winter temperatures in Eurasia today. Summers could not have been a problem even to Homo erectus; during the Second Interglacial, even as today, southern Russia and southern Siberia must have been, if anything, warmer than western Europe. But winters are—and were—another story. When we examine the winter isotherms (the curves marking zones of temperature) on the map, we see an upward bulge along the European coast, due to the maritime influence of the Atlantic, and a sharp downturn over Scandinavia and the Baltic lands, due to the Continental influences to the east. The result is that January temperatures around the north end of the Caspian, for example, are as severe as those at the uppermost tip of Norway, some hundreds of miles to the northwest.

It is quite possible, then, that winter temperatures were simply too severe for the climate-control techniques of our more primitive ancestors. Even fire and fur together weren't enough to keep them from freezing. Moreover, the Russian steppes are very poor in any kind of natural shelter from the howling Siberian winds. Even the modest protection provided by a thick forest would involve migrating north—into colder weather and poorer hunting. Nor are there

* Recently, in fact, a rather remarkable discovery was made in northern Russia, during excavations at a mammoth-hunters' camp dating from the end of the last Ice Age. About 15 feet below these remains, the Soviet diggers found 20 tools of "Mousteroid" type—which elsewhere are pretty consistently associated with the Neanderthals. The tools have not been firmly dated, but the best present guess seems to be the last interglacial. The sparse number of artifacts indicate that it was a temporary (presumably summer) camp, not a year-round home, and the climate during parts of the last interglacial was, as we know, warmer than at present. Even so, these findings, only about 100 miles south of the Arctic Circle, seem to be fairly impressive testimony to Neanderthal enterprise in the face of distinctly demanding climatic conditions.

many caves, such as served some Europeans well during glacial times. (A cave, or "rock shelter," is of course another, though crude, type of climate control. Its walls can intensify the warmth of a fire by reflecting back some of its heat, and at the same time keep off the worst of the wind.)

In theory, of course, our hypothetical interglacial Siberians could have migrated south each autumn. The glacial Europeans almost certainly did; we know that the reindeer, one of their chief food sources, migrates today and even taking the dimmest view of Homo erectus' intelligence and foresight, we must at least grant him the wit to follow the herds on which he depended. But in Siberia, you cannot go very far south before you reach desert, and starvation, or mountains, and dropping temperatures. Yet if interglacial Siberia was too climatically tough for early man, how could he have survived in glacial Europe? One intriguing possibility is that the glacial climate of Europe was not so severe as one might think.

On the face of it, the climate a few score miles south of a mile-high wall of ice should be almost intolerable to anyone but an Eskimo—and the Eskimos, during the Third Glacial, were still tens of thousands of years in the future. But in climate as elsewhere, things are not always what they seem. Conceivably, a glacial winter in Europe might have been no colder—even warmer—than one today!

Frankly, I am not sure that this could have happened in Europe. There is evidence, however, that it not only could but did happen in North America. Reid Bryson, a climatologist at the University of Wisconsin, has noted the presence of armadillo fossils in the central United States, far north of where the animals are found today (around the Gulf of Mexico) and dating from a *glacial* period. Bryson concludes that at that time winters must have been milder than they now are—and not in spite of the ice but because of it!

His explanation is reasonable enough, when you think about it. Bitter winter weather in the Mississippi valley today is produced by air flowing down from central Canada and (occasionally) even farther north. These masses of "continental polar" air can spread out amazingly (at times their leading edges will cross the Gulf of Mexico into Yucatan), but they are also very shallow—no more than a few thousand feet thick. Now during at least part of every ice age, the great ice sheets of Labrador and Keewaytin joined up with those of the Rockies to form a gigantic wall, one to two miles high, along (roughly) the present U.S.-Canadian border. Result: little or no polar air, with winters *milder* than at present. (Summers, of course, were much cooler.)

So far, so good. But what about western Europe?

Today, that region gets most of its really cold winter weather from two kinds of air. The most important is maritime polar air moving south from the berg-filled Greenland sea; less significant, though even colder, is continental polar air from Russia and Siberia, which despite the generally westerly winds in the regions manages to push as far as the Atlantic several times in most winters.

Now during the Third Glacial, the period under discussion, the British and Scandinavian ice sheets were joined in the middle of the present North Sea—it was dry land then. The result was a wall of ice much like that in North America, stretching from Ireland well into Siberia. Thus the coldest maritime polar air was blocked off from direct access to the lands south of the ice sheet; if it got there at all, it would have had to make an extensive detour over the Atlantic, which would necessarily have mellowed it a good deal.

So much for one source of bitter winters. What about the other—air from Siberia?

There was no real ice barrier separating western from eastern Europe. The northern ice reached no farther than central Germany, a couple of hundred miles short of the Alpine ice cap. Through the gusty corridor between them a certain amount of Siberian air doubtless reached western Europe, but because of the narrowness of the corridor, probably less of it than today.* Looking farther east, we find another par-

* Rhodes Fairbridge, in fact, believes that the corridor at times narrowed to only a few miles.

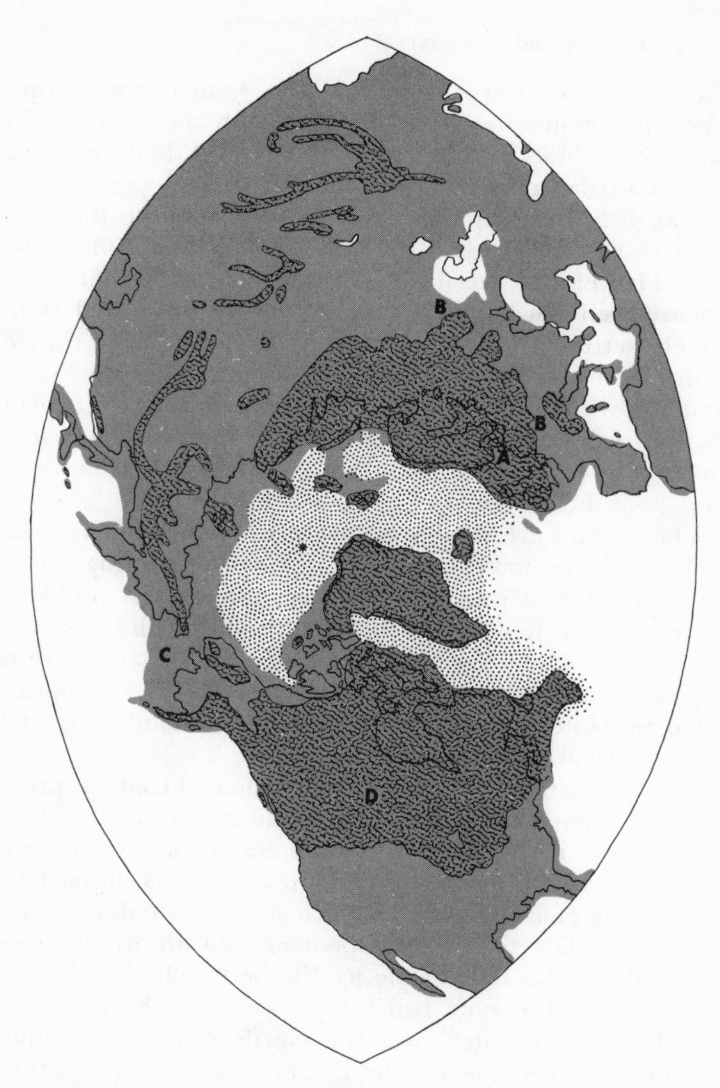

Ice sheets, probably caused in part by land emergence, further expanded land areas by lowering sea level. In Europe, ice barrier to north (A) and narrow corridors to east (B, B) may have limited extremes of winter climate; in North America, Bering Sea bridge (C) provided a migration route for men and animals, but ice barrier (D) blocked movement farther south.

tial barrier, the Carpathians and the Transylvanian Alps. These low mountains (low compared with the Alps) would hardly have blocked Siberian air, but would doubtless have hindered it far more than does the featureless topography of the north European plain, now its usual route but then ice-covered. Farther east still, we find another interesting situation. The Third Glacial ice sheet, at least part of the time, reached well down into the valleys of the Volga and Don, thereby setting up another, broader corridor between the ice to the north and the mountains of the Caucasus to the south. And across the eastern end of this corridor lay the Caspian sea—so swollen by runoff from the ice and reduced evaporation (cooler temperatures!) that it extended well north of its present shores.

Thus almost every cubic foot of Siberian air moving westward would be forced to cross the Caspian, picking up its water-stored heat on the way. True, the sea was probably fresher than it now is, and likely frozen for at least part of the winter, but even a frozen sea can warm the air above it appreciably. Coastal areas along the Arctic Ocean, though frigid in winter, are distinctly less so than inland areas well to the south of them.

There remains, of course, another source of cold air, probably more important than the preceding two put together: the ice sheet itself. There can be no doubt that over the ice in winter lay a permanent "high-pressure" area, formed by the chilling of air over the ice, such as we find today in Antarctica and Greenland. This pool of cold air would have flowed down the sides of the ice like water off the edges of an umbrella. But note two things: First, much of this air would have originated over the North Atlantic, bringing moisture to keep the ice sheet going. Losing its moisture over the ice, it would simultaneously gain heat. This is Conservation of Energy: if evaporation of water absorbs heat (ever sit around in a wet bathing suit?) then condensation of water—over an ice sheet or mountain range, for example—will inevitably liberate the same amount of heat. Second, in flowing "downhill" off the ice, the air would gain more heat, as descending air always does; dropping a mile or more, it might grow 15 or 20 degrees warmer. Of course this doesn't tell us much, since we don't know air temperatures over the ice, but enough to at least suggest that the winds off the ice sheets were by no means as frigid as might appear.

Adding it all up, I am going to offer a guess which some professional meteorologist or climatologist will probably have my hide for: During the Third Glacial, *average* midwinter temperatures in the ice-free parts of western Europe were rather colder than at present: perhaps several degrees below freezing rather than several degrees above. At the same time, however, *extreme* winter temperatures—when the bottom drops out of the thermometer—were higher. There would be few or no warm spells, but to balance them, fewer and milder cold spells. A long, cold winter—but a bearable one.

When we look at the opposite end of Eurasia, we find a much less tolerable situation. To the extent that winter Siberian air was blocked off from Europe, it had to go somewhere else. Not south—the Himalayan ranges formed a three- to four-mile-high barrier—but east, where most of it still goes today. The frigid Siberian "winter monsoon" must have poured into China even more violently and frigidly than at present. It makes me question, in fact, those stories of Peking Man occupying his caves during the Second Glacial. Unless, of course, he migrated south every winter, which he may have done. As for the lands north and east of China —forget it! Northeast Asia even today is about as inhospitable a place as exists on earth outside Antarctica; you are invited to use your imagination—preferably in front of a roaring fire—on what it was like during a glacial winter.

But there are still, of course, the interglacials, in which, perhaps, Peking Man or his descendants might have pushed north into Manchuria and the Soviet far east. On the archeological record thus far, they never did, nor did the much shrewder Neanderthals. Maybe even the interglacial climate was too tough a hurdle—but we need not speculate about that. For during the interglacials most of the area was covered, as it now is, by a dense "taiga" forest of spruce, pine and larch—no place for a primitive hunting people that cared about eating regularly. All of which explains why neither Homo erectus nor Homo neanderthalensis has ever turned up north of Peking, and, a fortiori, why neither of them ever made it from Eurasia to America. Between fire, animal skins, and caves they had achieved a fair degree of climate control, but not enough for roaming about eastern Siberia or Alaska. They could manage cold parts of the Temperate Zone, if the winters weren't too cold (or the rainfall too scant). But a true subpolar climate was evidently too much for them. The men who conquered the subpolar parts of the earth were men like ourselves. And there is reason to believe that they did it by means of two important inventions.

A curious fact about the tools of early man is that none, or next to none, of them were made of bone. The Neanderthalers, and perhaps some more primitive peoples, *used* bones as tools—notably, in the so-called "baton" technique in which a bone was used to tap small flakes from a flint "core," thereby producing either thinner flakes (if flakes were what they wanted) or a much more accurately shaped core (if they were making, say, a hand-ax). But the evidence that the bones were ever shaped to make them more useful is sparse and ambiguous.

Almost simultaneously with the appearance of modern man, some 40,000 years ago, we begin to find pieces of bone which have clearly been shaped. Some of them seem to be daggers; others, spear or javelin points. Still others are not simply old designs in a new medium but new kinds of tools. Looking, for instance, at a collection of tools found in southern Germany, one sees something called a split-base projectile point—a narrow bone rod some two inches long, pointed at one end with a deep cleft cut into the other. According to the experts, this point was set on the end of a stick that had been trimmed to fit the cleft, presumably lashed into place, and used as a projectile to kill game. Maybe. It would have been a tiny projectile, with little weight and, if thrown by hand, correspondingly little striking power. It would, indeed, have made a good arrowpoint, but there is no reason to think that the bow and arrow had been invented at that period. I can think of something else it could have been used for. If a thread of sinew or a strip of rawhide were jammed into the angle of the cleft, the tool would serve excellently to pull the thread through holes that had previously been punched along the edge of a hide. The "projectile point" may, in fact, be a bodkin—the prototype of the needle. Be that as it may, we know quite definitely that modern man soon began making true needles, eyes and all. If you have needles, obviously you have something else, and something of vital importance for life in a really cold climate: sewn clothing.

And the climate *was* cold—and getting colder. Fortunately, we are by now close enough to recent times to be quite clear on what happened and when. And, for a change, just about everybody agrees on it—Emiliani with his isotopes, Ericson with his foramanifera, another foram specialist who has studied different cores, with different species, from the Indian Ocean, and Rhodes Fairbridge who has worked out a climatic curve from changes in sea level (lower sea means more ice). What they agree on is this:

From about 100,000 to 70,000 years ago (give or take 5,000 years), there was a long warm period. Depending on whom you talk to, this was either the last interglacial or a warm phase of the Fourth (and last) Glacial. Whatever it was, it lasted about 30,000 years. At its peak (perhaps 80,000 B.P.) conditions in Europe were appreciably warmer than at present; a sizable proportion of subtropic animal fossils (forest elephant and rhinoceros) have been found in Central Europe, for example. It was during this period that the Neanderthals flourished. About 70,000 B.P., conditions started to deteriorate, but not suddenly or steadily. The European ice sheets reformed and advanced, but in a series of back-and-forth movements, like an old-time minuet; during one of these climatic oscillations, Europe may have been warmer than now. But the overall trend was toward cooler and dryer conditions. This reached its first maximum about 45,000 years ago with reindeer in central Europe, after which things warmed up a bit. During some 25,000 years of irregularly worsening climate, then, the Neanderthals apparently continued to flourish. But it was during the same period, so far as we know (and that isn't very far, yet) that modern man was acquiring his present form—probably somewhere in the temperate or subtropical Middle East.

From about 35,000 B.P., the temperature went down and down for some 17,000 years. The glaciers reached their maximum extent, musk-oxen appear in Europe, while most of even the Temperate Zone animals vanish. And just about at the beginning of that mammoth cold snap, the Neanderthals vanish; modern man takes over. The relative suddenness of that takeover makes one wonder whether better climate control techniques didn't have something to do with it.

The Neanderthals, remember, seem to have had only hide robes (though they had doubtless learned to belt these around them). Such clothing would at best have covered the shoulders, torso, and upper legs, with a gap over the chest where the skins came together. Many European (and American) men preserve putative traces of these inadequate garments in the distribution of their body hair (a small bit of evidence, by the way, in favor of the theory that at least some Europeans have Neanderthal DNA in their genes). A European male's head and neck are well covered by hair and beard—assuming these are allowed to grow freely in the modern, and presumed Neanderthal, fashion. On the chest, below the beard line, he will have at least traces of hair and often a thick mat. Finally, there will be fairly noticeable hair on the forearms and shins. (Hair in the armpits and around the pubic area probably had no protective significance, but rather served a sexual function. The body odors which we now seek to banish by baths and deodorants are thought, in their time, to have done in reality what today's advertisements delicately suggest that perfumes will do.)

Male readers can now imagine themselves outdoors in winter, dressed in a heavy blanket that covers *only* the hairless parts of their bodies—and perceive how the Neanderthals, tough though they must have been, were handicapped by the increasing cold of the last glaciation. Their successors, however, with bone needles, stone punches, and sinew thread, could sew themselves jackets, pants or high leggings, moccasins and boots. Like the Eskimos of today, whom culturally they somewhat resembled, they could venture abroad in all but the very bitterest weather.

And for really frigid days—as well as to provide a shelter for their women and children—they achieved another invention: houses. In Czechoslovakia and, significantly, southern Russia, archeologists have turned up circles of mammoth bones and other evidence showing that in these cave-poor areas man had begun to build huts of some sort. The shelters, some of them dug into the ground, were roofed either with earth-covered branches or with hides over a sapling framework; the circles of bones evidently served to anchor the hides around the circumference. Heated by fires, they established tiny islands of temperate climate in the glacial steppe country—appallingly smelly by our standards, and dreadfully stuffy by any standards, but for all that the prototypes of all man's subsequent log cabins, castles, temples, split-levels, and skyscrapers.*

How effective these two new techniques of climate control

* Recently, French archeologists working along the Mediterranean have uncovered what are said to be traces of fairly elaborate windbreaks—circles or ellipses of closely-set stakes or branches. And these are alleged to date from around 300,000 B.P.! If their interpretation is correct (I have not examined the evidence in detail), this discovery will write a brand new chapter in the history of human habitation—and also, I would say, force us to considerably upgrade our notions about the technological sophistication of the men of those times (Homo erectus, presumably). My present feeling, however, is that one should be very cautious in interpretation; to move back the first constructed human dwellings 250,000 years at one jump is a risky leap without due and careful consideration.

were can be seen from the fact that their possessors migrated not only west, toward the Atlantic, but east, into the frigid heart of Eurasia. Modern man's relics—and his only—have been found in central Siberia; soon after the glaciers had reached their maximum extent he had reached southeastern Siberia and the Pacific. Sewn clothing and houses had opened up to man another few million square miles of living space.

Nor did he stop there. In fact he *couldn't* stop there; a hunting people must follow the game, and if the game wanders east, the people—unless the climate is intolerable—must follow. We know the game did wander east; early in the last glacial period the reindeer and woolly mammoth wandered from Eurasia to America, even as the horse had earlier wandered in the opposite direction. They wandered, of course, across the present Bering Sea, transformed by lowered sea levels into a broad isthmus which the geologists call Beringia.

Which brings us to an old and fascinating scientific controversy: when did man discover America? We are not talking, of course, about the rival claims of Columbus, the Vikings and so on. An American Indian, listening to one of these arguments, is said to have remarked, "Discover America? Hell, *we* knew it was here all the time!" Anyway, for a lot longer than Europeans have known it. The question is, how long?

During the later nineteenth century, when scientists had finally accepted the fact that man must be a great deal older than the 6,000 years or so the Bible allowed him, all sorts of wild theories were hatched on this point. An enthusiastic Argentinian claimed to have found human bones in Argentine strata some fifteen million years old—which if true, would have been quite a trick. Argentina, he claimed, was the center from which man had spread to populate the globe. This was an extreme case, but other, more reputable scientists made guesses almost as dubious.

Nonsense of this sort inevitably begot a reaction, and for

decades the notion that man had been in the New World for more than a few thousand years was virtually taboo. Credit for this ban goes largely to the great Aleš Hrdlicka (Herd-*lich*ka, not *Hard*likker), for more than a generation the dean, not to say dictator, of American anthropology. In theory, Hrdlicka was willing to concede that man might have arrived in America as early as 15,000 years ago; in practice, however, he knocked down evidence of it with a polemical and vituperative skill which made any scientist concerned with his professional future reluctant to risk a run-in with The Master. The situation somewhat resembled the later dictatorship over Soviet biology by the unlamented Trofim D. Lysenko. To be sure, Hrdlicka, unlike Lysenko, never exiled anybody—but some anthropologists may well have thought wistfully about emigration after the Grand Old Man had savaged their papers ("wishful thinking, imagination, opinionated amateurism and desire for self-manifestation" was one of his better efforts). Hrdlicka's death in 1943, followed a few years later by the discovery of radiocarbon dating, opened up the subject to really scientific inquiry. And the hard evidence now available—a good deal of it climatic—adds up to a pretty consistent picture, though not all anthropologists are willing to accept it.

In the first place, we can assume fairly safely that man must have made the trip from Siberia to Alaska during the last Ice Age, at a time of low sea level. We arrive at this assumption by elimination. In a warm period, man could have made the journey by boat—as the Eskimos did much later. But boats argue a culture oriented toward the sea, living off fish, seal, and the like. And the earliest human tools found in either Siberia or America point, without exception, to their makers being hunters, not fishermen. A hunting tribe, arrived at the ocean, is not suddenly going to sit down and build boats—particularly when it does not know there is anything on the other side of the water.

Alternatively (and still in a warm period) man could have crossed the strait in winter, on the ice. I am told that a few years ago, when the Cold War was somewhat hotter, the U.S. Army actually conducted tests to determine whether tanks could be driven from Alaska to Siberia, or vice versa.* They concluded that it was possible—and where a tank can go, so can a mammoth, let alone a mammoth-hunter.

Against this are the habits of the larger arctic mammals in winter. We do not know what habits mammoths had, but the reindeer, or caribou, sticks pretty close to the forest edges during the winter for the excellent reason that there is where he is most likely to find food in the shape of twigs and such. I find it hard to believe that any reindeer, or mammoth, herd would take off across a treeless, featureless, 50-mile stretch of ice in mid-winter, and unless the herds went, there seems no reason for their hunters to have gone.

The need for a land bridge limits the time of the possible migration to from 50,000 to 40,000 years ago, during the first stage of the last glaciation, and from 28,000 to 10,000 years ago, during the second, more severe stage. The first period, being somewhat before the time that modern man appeared (and some 8,000 miles from Bering Strait at that), we can disregard. But there was another climatic barrier which enables us to narrow things down even further. Having gotten to Alaska, man still had to make his way into the heart of America. And for much of the Ice Age, the way was blocked by the wall of ice mentioned earlier. This lasted, with trivial interruptions (if any), from about 23,000 to 13,000 years ago, or later. Thus man could have made his way into America either between 28,000 (opening of the land bridge) and 23,000 (closing of the ice barrier) B.P., or between 13,000 (opening of ice barrier) and 10,000 (closing of land bridge) B.P.

It may be sheer coincidence, but every one of the generally accepted carbon dates for early man in the New World falls within the latter span. The so-called Llano culture of New Mexico, best known for its typical "Clovis (projectile)

* The notion that either nation could mount an effective invasion along that route is, of course, too fantastic for anyone but a general to believe.

points" dates from about 12,000 B.P.; a site at Onion Portage, Alaska *may* date from as early as 13,000. And, just to make the picture complete, the earliest human traces found in Siberia date from perhaps 16,000 B.P., while the first indications of sewn clothing (in Europe) may be as early as 30,000 B.P. Still, hope springs eternal in the anthropological breast. The great Louis S. B. Leakey, of Olduvai Gorge fame, has dug up some very crude tools in southern California which he claims may be as much as 50,000 years old. But his dating is less than watertight, and the tools are so crude that some specialists are unwilling to concede they are tools at all. Indeed, I suspect that had anyone but Leakey made such a suggestion, the laughter would have echoed from California to Kenya. Again, a fragment of what is probably a tool has turned up in a layer of Indiana loess (wind-blown silt) supposed to date from 35,000 to 40,000 years B.P. But this dating too is indirect and the object itself dubious enough to lack conviction.

The shortage of evidence does not keep scientists from speculating—nor, of course, should it. Unfortunately, not all of them draw a clear line between speculation and fact. A Swiss anthropologist, Hansjürgen Müller-Beck, in summarizing one of his recent papers, declares that he has "reconstructed" the expansion of hunting cultures across Eurasia "and their crossing of the Bering land bridge . . . about 28,000 to 26,000 years ago." Alas, Müller-Beck is drawing the long bow—or the long projectile. When you read the text of his paper—a very well written and valuable one, by the way—you discover that what he has "reconstructed" is not what happened but merely what he thinks must have happened. He explains in detail (and, I must say, fairly convincingly) how man *could* have made his way to America at that early date. But as regards direct evidence that man did, his paper—like the Siberian archeological record before 16,000 B.P.—is a total blank.

Siberia is a big place, and there are unquestionably many more discoveries to be made there. So, for that matter, is

America. But, subject as always to change without notice if further evidence turns up, climate and carbon together date the discovery of America at about 13,000 years ago.*

Of course those early hunters didn't *know* they had discovered a new continent. But then, neither did Columbus!

22. CLIMATE'S UGLY LEGACY

The Problem of Race

I wouldn't like to swear that more nonsense has been written about race than about any other topic; men have been writing nonsense for a long time and on many subjects. Where race is outstanding, I think, is in its capacity to elicit nonsense from scientists. For example, the great Louis Agassiz, who discovered the glacial epoch, in later life managed to "discover" that Negroes have smaller brains than whites. From this nonexistent "fact" he then drew a non-sequitur conclusion, that Negroes are less intelligent than whites.

Of course that was some time back. But only a year or two ago, Dr. William Shockley, a Nobel Laureate no less (he invented the transistor), succeeded in topping the great glaciologist. Shockley suspects that Negroes suffer from "evolu-

* I should not drop this matter, however, without mentioning some stone tools in Chile that may date from as early as 15,000 B.P., which would put the initial discovery of America at least a thousand or two years earlier. Even more remarkable is a stone artifact dug up in Canada from beneath deposits which, Rhodes Fairbridge is convinced, were laid down by the last glaciation. So the question is still very much open. We can, however, be certain that if man did reach America at the early date these findings imply, he must have done so in extraordinarily sparse numbers—perhaps only a single small band of hunters.

tionary adolescence," whatever that is, and are therefore dumber than whites. He wants the government, or somebody, to set up research projects to prove his theory "objectively." Thus far, the worst Shockley could be accused of is silliness. His stated reason for urging the research, however, is that proof of Negro inferiority *would make Negroes feel better*—it would be, he says, "the greatest relief to the frustrated agony of black Americans."

A man who can believe that can believe anything.

Admittedly, being an expert on transistors doesn't make one an expert on people. But even some anthropologists and psychologists, whose scientific business is precisely with people, are by no means immune to the bug of racial nonsense.* It can be argued, indeed (and other anthropologists have done so), that the very concept of race is somewhat nonsensical. A race, if it means anything, means a sizable group of people, descended from the same ancestors, who are more similar to one another than to any other group. Well, remember the trouble we got into with "similarity" in defining a species? Races are worse.

Take, for instance, the "race" variously called Black, Negro, Negroid, or Congoid. It includes some of the tallest people on earth (the Dinka and Watutsi) and some of the shortest (the Congo pygmies). It includes people with broad flattened noses, medium straight noses, and narrow arched noses. Some Negroids are broad and chunky, some tall and willowy, some betwixt and between. They are characterized as a "race" because almost all of them have quite dark (though seldom "black") skin, rather protruding (prognathous) jaws, rather small brow ridges, and kinky or woolly hair. Yet in parts of India we find people with equally dark skins, tightly curled hair, and occasionally with prognathous jaws who are classed as "Caucasoids" or even "Whites." In Australia, the brown-skinned aborigines are markedly

* The latest victim is Prof. Arthur Jensen, of the University of California, though his nonsense is far more sophisticated than Shockley's.

prognathous but are classified as "Australoid," as are their now-extinct relatives in Tasmania, who had woolly hair to boot. Finally, in parts of the southwest Pacific, such as the Solomon Islands, we find peoples who by any anatomical criterion ought to be Negroids and were once classified as such—but are now lumped with the Australoids because nobody could explain convincingly how they could be genetically related to the African Negroes some 10,000 miles away.

With all this confusion over what makes a race, it will come as no surprise to learn that the number of races recognized by various anthropologists ranges from three to over two hundred. And it will come as hardly more of a shock to learn that I do not propose to discuss "races" at any length in this chapter. What we shall talk about instead is physical differences between various groups of people—or at any rate, the differences that seem to be somehow related to climate. Whether the differences in any given case add up to a "race" is something we can let the professionals wrangle about. Meanwhile, however, we can note that whether or not "races" are a physical reality, they are certainly a social reality, meaning that too many people not only believe in races but believe all sorts of nonsense about them. Thus to the extent that climate is responsible for physical differences among populations—apparently, a pretty considerable extent—it is also responsible, indirectly, for one of the nastiest problems that plagues human civilization today.

In trying to relate human physical differences to climatic differences, we immediately run into a major snag. Human populations have been migrating from one climatic region to another for a long, long time, meaning that a population with presumed physical adaptations to one climate will now be found in several. And often we have no very certain means of knowing which was really the group's "original" home. Often, indeed, we find ourselves arguing in a circle; because a people is thought to possess physical adaptations to a particular climate, it is assumed to have originated in that climate; and because it "originated" in that climate, its physical characteristics must have climatic significance.

Having erected our usual warning signs, then, let us first consider peoples that live in (or presumably originated in) hot climates. The most obvious fact (and also the most socially explosive fact) about such people is that they tend to have considerably darker skins than residents of colder climes. Very nearly all the dark-skinned peoples are found in Equatorial, Savanna or Tropical Steppe climates, as in most of Africa, southern India, New Guinea and its neighboring islands, and northern and central Australia. The only important exceptions are the aborigines of southeast Australia and their massacred relatives that once lived in Tasmania—both regions with climates of a warmish Maritime type. But there is no doubt that these peoples, whenever they reached their temperate homes (the earliest radiocarbon date from Tasmania is about 8,000 years ago) must have made their way there from the north via Indonesia and southeast Asia, in which tropical regions they must have spent 10,000 years or more. Conversely, when we find relatively light-skinned peoples in tropical areas (as in present-day southeast Asia, Indonesia, and Polynesia), there are strong archeological reasons for believing that they are fairly recent immigrants. (In the case of Europeans now living in tropical climes, we of course *know* that they are immigrants.)

All this applies only to the Old World. In the New World, though the hot-climate Indians tend to be rather darker on the average, the differences are much less marked—thus giving further support to the belief that man is a relative newcomer to the New World. In settling it, he presumably migrated through so many climatic zones that he adapted very little to any of them.

We find the very lightest-skinned peoples apparently originating in the cool Maritime climate of northwest Europe—the blond, pink-cheeked group sometimes called Nordics. (It is perhaps significant that the lightest-skinned American Indians, sometimes called "white Indians" by explorers, were found in the Maritime climate of the U.S.-Canadian northwest coast.) This association between ultralight skin and Maritime (rather than merely cool or cold) climate suggests that the climatic factor most relevant to skin color is perhaps not temperature per se but sunlight—in particular, the ultraviolet component of sunlight which, on striking the skin, manufactures the vitamin D that our bodies require for normal bone formation. According to this theory, people living beneath the often-cloudy skies of a Maritime temperate climate, with weak sunlight in winter even during clear weather (and most of their skin necessarily covered by clothing to boot), could only survive with the aid of ultralight skins which, by allowing nearly all the ultraviolet rays to pass through, would manufacture a maximum of vitamin from a minimum of sunlight. Physiologists have estimated that "Nordics"—but not darker-skinned peoples—could obtain their "minimum daily requirement" of vitamin D from only a few hours of sunlight on their faces each day.

Farther south, in the Mediterranean zone, with fewer clouds and longer winter days, vitamin D production would no longer be a problem; sunburn, and skin cancer, however—to which the Nordic skin is particularly prone—would be. Hence the need for a swarthier skin, and one that under summer skies could tan dark enough to screen out the burn- and cancer-producing ultraviolet rays. Since complexions of precisely this sort predominate around the Mediterranean lands, so far so good. Unfortunately when we move still farther south, into the lands of dark brown or even black skins, the theories begin to break down. There is, in fact, no really convincing reason yet discovered why anybody should have black skin—except that several hundred million people do.

Protection against sunburn (or skin cancer) doesn't seem to be the answer, in the sense that black skin seems to give no better protection than a good dark tan. I myself am rather light-skinned (though not quite Nordic), yet after a couple of weeks of cautious July tanning, I can spend several hours a day on the beach with no unpleasant effects. The swarthier North Africans and Levantines, who tan to a coffee color, can spend all day half naked in the summer sun.

Some anthropologists have suggested that just as "Nordic" skin protects against vitamin-D deficiency, so black skin protects against vitamin-D excess. Now there is no doubt that chronic overdoses of vitamin D can produce a number of unpleasant results, including kidney stones and arteriosclerosis. But it has been calculated that to manufacture the vitamin in this amount, even a light-skinned man would have to spend the entire day in direct sunlight, never seeking the shade, and expose every square inch of his skin to the sun. Since at least half our skin is necessarily in shadow at all times, this would take a bit of doing. The problem is further complicated by the fact that dark skin demonstrably absorbs more sunlight than light skin, meaning the "heat load" on the body should be increased—hardly an advantage in the tropics. (This, however, would be mitigated by a simple fact of physics which none of the anthropologists I have read seem to be aware of: though black absorbs more light than white, it also radiates more heat, meaning that if dark skins gain heat more readily, they also lose it more readily.)

To this problematical fact we can add a further one: many, and perhaps all, of the dark-skinned peoples seem to have originated not on the blazing tropical steppe or the sunny grassland of the savanna, but in the moister, shadier tree-dotted savanna, or even in the dim Equatorial forests. This has given birth to the most eccentric theory of all, held among others by the anthropologist Carlton Coon, who has written several books on "races." According to this notion, black skin is not a protection against heat but against cold (!) because of its greater capacity to absorb heat. I judge from this that Dr. Coon and the others have never lived in the tropics; I have, and I do not recall suffering from the cold at any time, even without a black skin to help me out. Moreover, in the moist tropics, in which there is of course no "winter," even moderately cool temperatures occur only at night—when even the most sunlight-absorbent black skin would be of no practical advantage.

If we assume what is by no means certain—that the dark-skinned peoples took on their present tints in an Equatorial or near-Equatorial forest environment—the reason may have been the one first put forward (so far as I know) by Rudyard Kipling in the *Just-So Stories:* camouflage. There is no doubt, certainly, that a black man is less visible than a white man in a dim light or forest shade, and it may well be that black hunters in the tropical forest were more successful than white ones because they scared the game less. The same theory might also explain why the Bushmen of South Africa, who now live in sunny steppes and deserts and so far as we know have never been forest dwellers, are not black or even chocolate-colored but a yellowish desert-brown.

But there are objections to the camouflage theory too. When it comes to black or dark-brown skins, one is tempted to echo the cry of the anthropology student who, having reviewed the inconclusive evidence for all the theories, exclaimed, "For Chrissake—skin has to be *some* color!"

It does, of course; the pity is that color has been made the basis of an elaborate mythology in which all sorts of totally unrelated traits are linked up with it. Or even with the mere hint of color; there are American "Negroes," 7/8 or 15/16 "white," who are nonetheless subject to all the social disabilities based on their presumed genetic mental and moral inferiority. If 1/16 African ancestry makes a man a "Negro" genetically, there must be something remarkably potent about African genes—though I don't suppose your average racist thinks of that.

There are a number of other physical traits that seem to represent adaptation to hot climates. Some tropical peoples are rather sparsely endowed with body hair, though the geographical inconsistencies in the quantity of body hair (hairy people in the tropics, relatively hairless ones in the temperate zones) suggest that this is not a very important climatic adaptation. Rather less ambiguous is the form of hair. Woolly or kinky hair is altogether limited to people living in, or originating in, the tropics. Now when hair of this sort is allowed to grow naturally (as in the "Afro" hairstyles currently popular among American—but few African—blacks), it generally forms a thick, bushy mass covering the top and sides of the head. On the one hand, this bush serves as a fine insulator for shielding the brain against the direct rays of the sun; on the other, hair is kept away from the neck and shoulders, allowing these parts of the body to sweat and cool freely. The neck is a specially important cooling area because it contains so many major blood vessels so close to the skin.

We also find specialized types of body build among various tropical peoples. The Nilotic black tribesmen along the torrid upper Nile, such as the Dinka and Shilluk, are notably slender and long-limbed—both traits which increase the ratio of skin surface (i.e., cooling surface) to body volume. Indeed, the majority of tropical peoples tend to be rather slenderly built (e.g., the Indonesians, Vietnamese, etc.), though it is not always easy to say whether this represents adaptation to climate or simply the result of inadequate diet. However, the slight stature of the various groups of pygmies is unquestionably of genetic, not nutritional, origin. These tiny peoples are invariably found in Equatorial or near-Equatorial climates—the Congo, southeast Asia, the Philippines, and New Guinea—and one is tempted to see their miniature bodies as heat-elimination devices (because of the square-cube law, a pygmy has a notably higher ratio of skin surface to body weight than a similarly proportioned individual of normal size). But there are other theories; some have suggested that the small build of the pygmies enables them to move more easily through the thick forest. On this, I would want to see more data; the true Equatorial forest, because of its dense shade, has little underbrush and presents few barriers to individuals of any size. More convincing is the suggestion that the pygmies' small stature evolved as an adaptation to an environment of limited food resources. As noted earlier, the Equatorial climate is poor in animal life and in accessible plant life as well. A food-gathering people living in such a climate would have to make do with a pretty sparse diet—and one especially deficient in the protein necessary for growth.

Still another hot-climate adaptation may be represented by what is called "steatopygy," which is Greek for fat behind. Among the Bushmen and the related Hottentots of southern Africa, many women and some men possess enormous, shelf-like buttocks, consisting largely of fatty deposits. Now women of all populations possess more body fat than men; this, whatever biological functions it may have (a food reserve, certainly), helps provide the rounded feminine contours that most men find attractive. On the other hand, fat is also a good insulator (hence the thick blubber of the seals and whales of Polar and Subpolar seas). In a hot climate, it may perhaps represent some sort of advantage to a woman to concentrate her reserves in one place, leaving the rest of the body uncovered by its insulating blanket. The camel's hump represents just such an adaptation, and so may the similar hump of the zebu or Indian cattle. Still, it isn't always safe to jump to conclusions about the reasons for the shapes of ladies' behinds. A Hottentot man, asked why he thought the women of his tribe possessed such massive buttocks, replied with a shrug, "We like our women that way!"

Perhaps the subtlest climatic adaptation is found in certain tropical and subtropical peoples. It is an abnormality of the blood, first discovered some sixty years ago, called "sickle-cell anemia," a disease which, as its name suggests, is marked by peculiarly shaped red-blood cells which, moreover, are easily destroyed, leading to a troublesome and sometimes fatal anemia and other unpleasant symptoms. "Sickling" was originally described among American Negroes, and for a long time was considered a Negro "racial trait." So dogmatic did white physicians become on the subject that on the few occasions when the disease turned up among whites, it was thought to indicate "Negro blood" in the family. A further puzzle arose when sickling was found to exist only if the condition was inherited from both parents; "carriers" of the trait, with only one sickling gene in their chromosomes, showed few and generally trivial symptoms. Yet if two carriers mated, they would have one chance in four of producing a "sickle" child—who was estimated to have only one fifth the normal chance of surviving to adulthood. On the face of it, then, the sickle trait was a biological disadvantage. The geneticists calculated, in fact, that early deaths among anemic children would remove something like one sixth of the sickling genes in the population every generation. Nonetheless, in some African tribes something like 40 percent of the population gave evidence of possessing the gene, and its incidence showed no signs of diminishing.

A clue to the mystery turned up when it was noticed that the sickling gene in Africa seemed limited to the warm and reasonably moist areas; it was, in fact, almost unknown outside the Equatorial and Savanna belts. And one characteristic of these belts was their high incidence of malaria, which, being carried by mosquitoes, does not "thrive" outside a reasonably warm, moist climate (the insects breed in stagnant water). In fact, as it turned out, the incidence of the sickle gene pretty well paralleled the incidence of one kind of malaria—the especially deadly "malignant tertian" form. This, we now know, is because the sickle gene, though markedly disadvantageous in a "double dose" (i.e., from both parents), is actually protective in the single dose. In some unknown manner it protects its possessor against some of the effects of the malaria parasite. Thus a tribe in a high-malaria region would, biologically speaking, be trading off the excess deaths from sickle-cell anemia against the larger number of deaths from malignant malaria.

The association between the sickle trait and malaria also explained its presence—without "Negro" blood—in some white people. With few exceptions, these individuals turned out to hail from parts of Italy, Sicily, Greece, or Turkey where malaria was, or had been, common. Similar malarial conditions had begotten similar effects; the "racial" trait was apparently an independent climatic adaptation. Since then, similar blood abnormalities have been discovered in malarial regions of Asia, though it has not yet been determined whether they too provide protection against disease.

When it comes to temperate climates, man's physiological adaptations are considerably less marked. This is not very surprising if we recall that men in these climates have been adapting artificially to cold—i.e., through fire and some sort of clothing—for several hundred thousand years. Temperate Zone peoples are uniformly of fairly light color and possess hair which is either straight or wavy, but which in either case will hang down sufficiently to form a sort of scarf around the neck and shoulders. Otherwise, apart from the "Nordic" complexion, which, as already noted, seems to be an adaptation to the cool, cloudy Maritime climate, the Temperate Zone people show few marked climatic adaptations. Carlton Coon, among others, believes that a narrow, high-bridged nose represents an adaptation to a distinctly cold and/or dry climate, on the theory that it helps warm and/or moisten inhaled air. But there are far too many exceptions, both ways, to take much stock in this. In fact, when we examine the people who seem to be habituated to the dry and very cold Subpolar to Polar climate, we find their noses are flattened, not arched. These are the people of northeast Siberia, such as the Tungus and Chukchi, and, of course, the Eskimos, who are closely akin to them both anatomically and linguistically. In fact if we had to come up with a particular "race" specialized for a given type of climate, these far-northern people would be the closest thing that Homo sapiens has yet produced.

The cold-climate physical traits of these people are, as a matter of fact, more or less scattered through all the so-called Mongoloid peoples, whose present habitat extends from the Arctic Ocean on the north to Equatorial Indonesia on the south. This, and other evidence, suggests that the ancestors of these peoples, or some of them, must have migrated from the far north, after having adapted to the severe climate there during the last Ice Age. But the most notable concentration of cold-adaptation traits is found among the northernmost groups. First, their body build is stocky, with relatively short limbs, thereby cutting surface area, and heat loss, to a minimum. Second, their faces are flattened, notably as regards the nose, which barely projects beyond the overall line of the face (the Chinese, in fact, often refer to Europeans as "Big Noses," and the beaky white imperialist is quite as much a stereotype in the Chinese press as is the slant-eyed toothy Oriental in our own). The utility of this flattened profile will be clear to any "white" reader who has developed the characteristic red nose of cold weather.

Body hair is sparse, at first glance rather unexpectedly. In fact, however, a man who can survive at all in a Polar or near-Polar climate can do so only with the aid of special clothing—sometimes several layers of it. Body hair under those circumstances is of no added value, and facial hair is a positive disadvantage. As "Caucasoid" polar explorers have found to their irritation, a beard and mustache collect moisture from the breath which, freezing, covers the hair with a layer of ice. For hair the true Polar-adapted population "substitutes" fat. The most exposed portions of the face—the eyelids and cheekbones—are covered with fatty layers, which give special protection to the sinuses and incidentally give the eyes their characteristic "slantiness" (actually, more narrowing than slanting).

The remaining physical differences between peoples have, so far as we can tell, no climatic or other biological significance. They are, it would appear, the result of random variations, of what is called "genetic drift," operating in various small and isolated bands of hunting and food-gathering people. It is in precisely such populations that the effects of genetic drift will be most marked—and of course it is in precisely this way that Homo sapiens has lived for most of the 40,000 years or so he has been on earth. So long as a given trait had no significance for survival, pro or con, it could appear and proliferate within a tribe for no reason other than sheer happenstance or, perhaps, because the men of the tribe "liked their women that way"—or vice versa.

Given the unquestionable existence of hereditary *physical* differences among peoples, climatic or merely by chance, the question is sometimes asked, "Why shouldn't there be mental differences as well?" Insofar as temperament, emotional makeup, is concerned, there probably is no reason why not—though temperament is so bound up with tradition and culture that its genetic component is, and will remain, almost impossible to establish. The "reserved" Englishman or "phlegmatic" German are neither of them, after all, much different in "racial" makeup from the "moody" Spaniard or "demonstrative" Italian.

Of course, the people who generally ask this question—usually quite disingenuously—are for the most part thinking not about temperament but about intelligence. And here there is in fact plenty of reason "why not"—in particular, the fact that in any culture you choose to name (except, possibly, our own highly-civilized one), intelligence is of clear and positive value in survival, while stupidity is the reverse. Recalling that man has been a hunter and gatherer for most of his existence, ask yourself: Is there any such society, known or imaginable, in which the clumsy toolmaker, the inept hunter, the dumb broad who has trouble telling edible roots from inedible or poisonous ones, will be *more* likely to survive and leave descendants? Is there any natural environment, any climate, which will not yield up more of its rewards, whatever they are, to the clever man rather than the stupid one? Unless one can conceive of such a society, it is difficult to explain why any population should be more, or less, intelligent than any other.

Even more dubious is the attempt of some ignorant people to draw inferences about "racial" intelligence from the physical traits of various groups—in particular, their alleged "apelike" appearance. Just for the hell of it, I recently got up a check-list of "apelike" characteristics (sloping forehead, profuse body hair, heavy brow ridges, thin lips, and the like) and scored various "races" on how they rated. Well, guess who came to dinner as the *least* apelike guest of honor!

Beyond all this, the question of whether particular groups are or are not more intelligent than others is a foolish one because even an "objective" answer to it, such as some racists are hunting for, would tell us nothing of value. Differences in "racial" intelligence, even if we assume they exist, are differences in group *averages*. And in dealing with people, as in dealing with climate, averages are good for very little.

The average annual temperature of St. Louis is 56 degrees—topcoat weather. Well, any St. Louisian who was fool enough to go about in a topcoat in July (average 80°) or January (average 32°) would deserve what he got—which would probably be heatstroke or pneumonia. Just so, an employer who hired an executive or engineer simply because the man came from a social or "racial" group of high average intelligence would deserve what he got—bankruptcy. The average American family has, I believe, something like 3½ children. I have yet to hear of any parent troubling his head over how to feed, clothe, and educate half a child.

PART THREE

Climate and Civilization

23. AFTER THE GLACIERS

The Remarkable Recent Era

From the standpoint of the natural sciences there is nothing about the Recent era—roughly, the past 12,000 years—that would set it off from the preceding Pleistocene. The earth's climate has, of course, changed markedly since the last Ice Age, but no more than it had changed half a dozen times already; the planet's animal and plant species differ little from those of 20,000 years ago; and man, from a strictly anatomical standpoint, has changed hardly at all. Yet from a sociological and ecological standpoint, this unremarkable era is very remarkable indeed. During these few thousand years, amounting to perhaps 1/200 of the time since the Australopithecines battered out the first crude pebble tools, man's culture, and with it his relationship to the rest of nature, has changed immensely.

At the end of the Pleistocene, man was still a hunter, a gatherer, and therefore, necessarily, a wanderer on the face of the earth, pushed to and fro by the seasonal migrations of game and the waxing and waning of vegetation. In a mere 7,000 years thereafter, he learned to manipulate plants, animals, and occasionally climate to a degree that permitted him at least a semi-sedentary life over wide areas and, in a few favored spots, large permanent settlements—cities. From a rare, though widespread, animal he became the dominant species on earth, able to survive alike in Equatorial rain forest or Polar tundra—everywhere, in fact, except the very coldest parts of the Polar Zone and the very dryest parts of the Desert. Man's climatic tolerance far exceeds that of any other species—apart, of course, from parasites such as the louse, the tapeworm, and the bacillus, for which man's own body supplies heat and moisture.

During the Recent era, man's interaction with climate, and with climatic change, became increasingly a regional and at times even a local matter—and in subsequent chapters will be discussed as such. First, however, we need to take an overall view of climatic history in the Recent era on a world scale, along with some of the theories that explain it.

The first point to keep in mind is that we are examining climate from much closer up, and therefore on a smaller scale, than we have done up to now. An Ice Age or an interglacial is measured by tens and perhaps hundreds of thousands of years; the Recent climatic shifts, once the glaciers had vanished, lasted for a few centuries or at most a few millennia. The scale of the changes is equally small; instead of the spectacular contrast between glacial and interglacial, we see only a few degrees' rise or fall in average annual temperatures, a few inches' increase or decrease in rainfall.

The methods by which climatologists have reconstructed these mini-changes differ little from those used to plot the history of the ice ages: pollen, tree-rings, "fossil" lakes, and, especially, the "fossil" beaches and other evidence that point to a rise or fall in sea level. The oceans respond rapidly to an increase or decrease in the "continental" glaciers of Greenland and Antarctica as well as the mountain glaciers found even at the Equator; a further increase during warm periods occurs because water expands as it grows warmer.

The overall picture developed from these various lines of evidence looks about like this: In 10,000 B.C. (note the change in dating) the glaciers were still with us, though they had been receding for some 6,000 years. By around 7000 B.C. the British ice sheet was gone and the Scandinavian sheet survived only as a remnant in the mountains. In North America, the ice retreated more sluggishly, essentially because there was so much more of it, persisting for a thousand or two years longer.

By 6000 B.C., then, world climates were much like what they are now. But the warming process did not stop; rather quickly the world passed into what is sometimes called the Climatic Optimum, with annual temperatures in the Temperate Zones several degrees higher than at present. This reached a peak between 5000 and 4000 B.C. Around the first of these dates, however, there seems to have been a markedly wet period in parts of northern Europe, with peat bogs expanding to cover areas that once were forest. Things then cooled off slightly; by about 2500 B.C. climates were again much like those of the present; but again things did not stop there. The cooling continued until around the beginning of the Christian era—unfavorable times for northern Europe but optimal for much of the Mediterranean, since rainfall was higher and apparently better distributed throughout the year. According to British climatologist H. H. Lamb, there was no really dry season in the Mediterranean even as late as the second century A.D.—in contrast to today, when the climate, as he observes, "would be improved by some summer rain."

Beginning around 200 A.D., or perhaps earlier, things warmed up again, though not to the extent they had during the climatic optimum. This "Little Climatic Optimum" peaked between 800 and 1200 A.D., and produced some remarkable historical effects in northern Europe which we will talk about in Chapter 35. Once again, world temperatures began swinging back to the cold side, with a return to near-present climates between 1200 and 1400, a slight warming for a century or so thereafter, and a renewed cold spell between about 1600 and 1850. During this "Little Ice Age"—a name fraught with exaggeration—European glaciers, at least, seem to have expanded farther than they had at any time since the final retreat of the ice sheets.

Since 1850, things have warmed up appreciably (it seems to be quite true that "we don't have cold winters like we used to"), though the warming trend has stopped, temporarily at least, since 1950. And here we reach the point at which climate begins to overlap with weather. A century warmer or colder than the present unequivocally represents a change in climate; half a dozen unusually cold winters, or dry summers, can hardly be described as anything but a run of unusual weather.

The theories which purport to explain these Recent climatic shifts are not much more satisfactory than those cited in Part Two, which attempt to explain the ice ages. As one would expect, most of them involve changes in either the sun's radiation or the earth's atmosphere; one, however, explains climate shifts by changes in the tides.

An "atmospheric" explanation that has received a good deal of attention is the volcano theory, associated particularly with the distinguished American meteorologist Harry Wexler, who, prior to his death a few years ago, headed the U.S. Weather Bureau. Volcanos, Wexler pointed out, can eject enormous quantities of fine dust into the upper atmosphere. Thus the catastrophic explosion of Krakatoa in the East Indies in 1883, which directly and indirectly killed some 30,000 people, threw enough fine particles of ash into the stratosphere to produce unusual and lurid sunsets over much of the world for several years thereafter. And dust, of course, blocks out part of the incoming sunlight. Indeed the Krakatoa eruption seems to have been followed by a series of unusually cold winters in Europe and North America; the famous Blizzard of '88 may have represented the volcano's last climatic kick. Wexler saw, or thought he saw, a similar relationship between other major eruptions, or "runs" of minor ones, and other series of cold years,* and it is cer-

* Notably, the second decade of the nineteenth century, which was markedly colder than average over most of the northern hemisphere. In the United States, this miserable period culminated in the catastrophic year 1816, known in our folklore as "the year of no summer" and "eighteen hundred and starve-to-death." For once, tradition does not exaggerate. Spring crops froze in the ground all over the Northeast and in June some lowland areas of New York State, where temperatures for that month now average around 65°, had three feet of snow.

tainly possible that a century or so of high volcanic activity could lead not to a few cold years but a Little Ice Age. Unfortunately, nobody has managed to prove it. Proof would require a fairly comprehensive census of volcanic eruptions over several thousand years—and historical records in most parts of the world (including such volcano centers as the East Indies) don't go back that far.

There is another problem with Wexler's theory. Volcanos produce more than dust; they also liberate tons of carbon dioxide, which affects climate in precisely the opposite sense. Carbon dioxide, along with water vapor (though to a lesser degree), is responsible for the atmosphere's "greenhouse effect" which traps outgoing infrared (heat) radiation. And if it is difficult to estimate the amount of ash thrown out by an eruption even a couple of centuries ago, it is well-nigh impossible to do more than guess at the amount of carbon dioxide produced, or to calculate the extent to which the one would offset the climatic effects of the other. So as far as the major Recent climatic fluctuations are concerned, the best we can say for the volcano theory is "unproven"—and probably unprovable.

Carbon dioxide, and perhaps dust as well—from sources other than volcanos—are suspected of playing a considerable role in the climate of the past century or so. But this, since it represents a rather special problem, can be deferred to the final chapter.

The tidal theory of climate was devised by the great Swedish oceanographer Otto Pettersson, much of whose long and productive life was devoted to studying "internal waves" in the ocean. These disturbances are not unlike ordinary waves, except that they are bigger (sometimes more than 200 feet high) and slower. As their name indicates, however, they surge well below the surface, at the boundary between two layers of water, one of which is distinctly colder or saltier (i.e., heavier) than the other. Though they are of course invisible, they can be plotted by recording the minute-to-minute changes in temperature or salinity at a particular depth, as the lower layer of water surges upward and then drops. In measuring these waves, Pettersson found that some of the biggest seemed allied to the tides, showing variations in height at the regular 12½-hour interval between one high tide and the next. Moreover, they underwent further variation in strength from one part of the lunar month to another, precisely as the tides do.

So far, so good. It then occurred to the oceanographer that the tides do, and the internal waves should, show long-term variations in strength because of slight changes in the astronomical distances between earth, moon and sun; these reach a peak of maximum tides (and, presumably, maximum waves) about every 1,700 years. Periods of peak waves, Pettersson suggested, should also be periods of cold stormy climate in Europe. He reasoned that the waves would carry large quantities of relatively warm Atlantic water into the depths of the Arctic Ocean. The gradual warming of the polar waters would partially melt the pack ice that covers them, thereby sending many more icebergs south into the North Atlantic. The melting ice, in turn, would weaken the climatic effects of the Gulf Stream (which as we know does much to make western Europe habitable); the result would be a cold spell in northwest Europe several centuries long.

So far, so possible, at least. Some climatologists, at any rate, find Pettersson's reasoning plausible, though I am inclined to suspect that they may be influenced as much by his quite deserved scientific prestige as by his scientific logic. For the fact is that whatever the plausibility of his reasoning, the results it yields simply don't square with the facts. Astronomical calculations show that the last "tidal maximum" would have occurred in 1433, while the preceding minimum—according to Pettersson's theory, marking a warm spell—would have come around 600 A.D. But in fact, as noted, the Little Ice Age occurred roughly between 1650 and 1850—about 300 years later than Pettersson's theory would indicate. Similarly, the Little Climate Optimum peaked between 800 and 1000; again about 300 years later than the theory predicts. And the dating of these climatic episodes, let me emphasize, rests on evidence far more diversified—and far less speculative—than the tidal theory. It is conceivable that Pettersson's internal tides may indeed have something to do with short-term climatic variations of a few years or few decades. But for long-term shifts, they just don't work.

When we consider the theories involving changes in solar radiation, there is first of all our old friend Milankovich. The last Milankovich maximum was about 8000 B.C. This was some 4,000 years before the peak of the climatic optimum, but the lag is not unreasonable when you consider that in 8000 B.C. there was still a fair amount of ice to be melted and the enormous mass of the oceans to be warmed up from their glacial temperatures. But alas, when we look at the Milankovich curve *since* 8000 B.C., we find that though it has been slowly dropping, it has been higher than its present level until only a few decades ago. Yet, as we have seen, temperatures have by no means been higher than at present in all the centuries since the climatic optimum; for fairly extended periods, in fact, they have been lower.

This would seem to be another bit of evidence in favor of Sir George Simpson's contention that the Milankovich changes are simply not big enough to have much effect on climate directly. They may, indeed, have affected it indirectly by triggering ice ages, as Emiliani and Geiss have suggested, but so far as Recent climatic changes are concerned, they can be ruled out.

This leaves us with the final, and by all odds the most plausible, theory: sunspots.

During my school days, sunspots were undergoing a vogue rather like that of astrology today. Anything that could not easily be explained on sensible grounds, from the 1929 Wall Street crash to the migrations of lemmings, was chalked up to the waxing and waning of sunspots. Since then they have fallen out of fashion with almost everyone except the climatologists—and even these experts are not certain about the relationship between these solar disturbances and climate. But the evidence of such a relationship is, if not conclusive, at least persuasive.

Astronomers have been looking at sunspots ever since Galileo clapped a bit of smoked glass across the aperture of his crude telescope and discovered blemishes on the sun's face. Nobody yet knows what causes them, or why there are more at some times than at others. However, sunspot "counts" over several hundred years have established that their numbers do vary markedly. In particular, they reach a peak at intervals (very roughly) of eleven years; some astronomers believe that longer cycles also exist, but this is controversial. Though actual sunspot records are not very old, their numbers can be estimated indirectly even for pre-telescope centuries from the activity of the aurora borealis (northern lights). Sunspots are known to disturb the earth's magnetic field, and one symptom of these disturbances is more frequent and extensive auroral display. Now it happens that Chinese astronomers noted the appearance of auroras for many centuries before Galileo, and from their records it is possible to estimate the numbers of sunspots almost back to the beginning of the Christian era.

We also have an independent check on these estimates in the shape of tree-rings and carbon-14 dating. You will remember from Chapter 7 that tree stumps and other chunks of wood can, for certain periods and places, be dated literally to the year. For this reason, wood samples of this sort were used to calibrate and check the carbon-14 dating technique; if a wood sample is known to date from around 1400 A.D. but the carbon readings give 1600, either the technique is faulty or the sample has somehow been contaminated with "younger" (i.e., more radioactive) carbon. Over the years, it became apparent that carbon dates were subject to certain systematic errors. Not very big errors—in fact no bigger than the normal error to which all such dates are subject. But these errors were consistent. During certain periods, that is, all or nearly all the carbon readings would come out later than the known age of the wood; in other periods, earlier. It seemed, in fact, as if the nuclear reactions that form carbon-14 high in the stratosphere must be somewhat more active at some times than at others.

Now the motive power of these reactions are "cosmic rays"—high speed particles moving in from space which, striking stable atoms of atmospheric nitrogen, turn them into radioactive atoms of carbon. And the quantity of particles reaching the earth is known to be influenced by the earth's magnetic field, which, as just noted, is itself influenced by sunspots. In fact, when we compare a graph showing sunspot numbers averaged over a period of a century with another showing the systematic carbon-14 errors, they are a pretty good inverse match. Periods averaging high in sunspots, that is, were low in carbon-14, and vice versa.

Rhodes Fairbridge and others have carried matters a step further and shown that graphs (and other types of data) reflecting climatic changes—such as, for example, rising and falling sea level—match the sunspot and, inversely, the carbon-14 curves. A long period high in sunspots, or low in carbon-14, means warmer-than-average climates; few sunspots means a cool or cold spell.

Nobody is certain why this should be so. It has been suggested that the intense bursts of ultraviolet radiation known to be produced by sunspots step up the production of ozone in the upper atmosphere, and ozone is still another participant in the creation of the greenhouse effect. But not enough is yet known about changes in ozone concentrations at great heights to be sure that ozone is responsible. The relationship between sunspots and climatic cycles has at any rate been traced back to around 1000 A.D., and with less certainty for a thousand years before then. Like most natural (or human) relationships, it is not perfect, but it is certainly striking.

Enter now the villain: David Shaw of Lamont Geological Laboratory, who has tackled the problem from another angle. He has put together monthly average temperatures from a few places where records have been kept for long periods—New York City (back to 1822), the Netherlands (back to 1735), and central England (back to 1698). He then ran all these figures through a computer programmed to perform what is called "power spectrum analysis." Only a mathematician could explain what that is! What it is supposed to do, however, is reveal whether the temperature readings show peaks, or troughs, at particular intervals. Shaw's raw temperature figures showed a pronounced and sharp peak at intervals of twelve months—which was expected, since it reflects the yearly temperature cycle from winter to summer. When the figures were averaged to eliminate this cycle (Shaw, borrowing a term from the detergent industry, called this "pre-whitening"), the graphs showed no pronounced peak of any sort. In particular, they showed no peak corresponding to the eleven-year sunspot cycle—though a power spectrum analysis of sunspot numbers showed such a peak quite clearly. His conclusion was that "no demonstrable relationship exists between midlatitude temperatures and sunspots." (He also found, rather surprisingly, that there seems to be no relation between winter and summer temperatures; that is, a bitter winter is just as likely to be followed by a blistering summer as by a cool one.)

On the face of it, unless somebody manages to pick apart Shaw's math, this would appear to dispose of the question. It doesn't. For Shaw was dealing with cycles—that is, *regular* changes in temperatures and sunspots. But the Recent climatic changes we have been talking about are anything but regular. Moreover, all of them are a great deal longer than eleven years. What Shaw has apparently proved is that sunspots have no immediate, or short-term, cyclical effect on temperature—which is something very much worth knowing. But he has not proved that long-term irregular changes in temperature could not be produced by similar changes in the *average* number of sunspots. And in fact, there is good reason to believe that if such an effect exists at all, it would manifest itself *only* in the form of long-term averages. To suppose that sunspot changes could significantly change temperatures over any brief period would involve, it seems to me, neglecting the "inertia" of the earth's existing temperature—in particular, that of the oceans.

Since water, as we know, has an enormous "specific heat," the oceans act as a climatic "flywheel," smoothing out short-term temperature changes even as, in the oceanic Maritime climatic zones, they minimize temperature shifts from winter to summer. So that even if sunspots can't speed or slow the flywheel quickly—which Shaw seems to have proved—there seems no reason why they could not do so gradually. The sunspot explanation is a long way from being conclusively established, but for the present, at least, it is the best explanation we've got.

24. THE ROAD TO CIVILIZATION

Some Definitions—and Myths

Before going into the story of how man, with some assistance from and nudging by climate, made his way to civilization, it might be good to define what we are talking about.

Civilization, in the sociocultural sense, is the third main stage of man's social evolution. In the first stage, savagery, he lived on what he could get, hunting or trapping animals and collecting plant foods. Typically, he operated in small wandering bands. In the second stage, barbarism, he lived chiefly on what he could produce, raising plant foods and keeping domestic animals (though often supplementing their meat with that of wild game). Typically he lived in semipermanent villages of a few hundred souls.

These two types of society, despite their marked differences, are alike in being very simple structures. Every family pursued much the same occupations as every other; different kinds of work, to the extent they existed, were parceled out according to age or sex. Barring an occasional priest-medicine man, there were no full-time specialists or craftsmen.

With civilization we get a very different picture. First, though many people still live in villages, a sizable proportion live in cities with populations in the tens (later, hundreds) of thousands. Moreover, in both village and city, notably the latter, we find hosts of specialists: potters, smiths, carpenters, masons, weavers, sailors and ship captains, merchants and clerks, soldiers and tax collectors, priests and acolytes—individuals who eat food that others produce. And of the latter groups—this is an important point—at least some are scribes, keeping written records of their commercial, governmental, or religious activities.

Because of the popular meanings attached to "savagery," "barbarism," and "civilization," I must emphasize that as applied to human culture they carry no implications whatever about the manners or morals of the peoples labeled by them; they define basic types of societies and that is all. The Bushmen of South Africa are culturally savages, but are otherwise a great deal less "savage" than almost any "civilized" nation; the general quality of their life is accurately summed up by their own name for themselves: The Harmless People. Of all the millions who over the ages have been murdered, massacred, speared, shot, bombed, or otherwise violently cut off, at least nine tenths—at a conservative guess—were victims of civilization, not savagery.

From what I have said up to now, it will be obvious that the journey from savagery to civilization must have been made in two stages—first the transition to barbarism, and then, in a few favored places, the transition to a complex urban society. The first step was taken independently in two regions certainly—the Middle East and Middle America (southern Mexico and northern Guatemala)—and probably in at least two more—China and Peru. The second step, which required further special climatic and geographic conditions, may have occurred independently only twice—in Mesopotamia (the Tigris-Euphrates valley) and in Middle America. It is now generally agreed that two other Old World civilizations—Egypt and the Indus Valley—were initially set moving by stimuli from Mesopotamia, though both quickly acquired their own very distinctive qualities. What role outside stimuli played in pushing China from barbarism to civilization is still uncertain, and the same goes for Peru (vis-à-vis Middle America, that is). Peru, indeed, is rather a special case, perhaps not quite a true civilization; despite its high culture, its cities were relatively small and it never developed writing.

I have gone through these geographical listings partly because they delimit the regions in which we will shortly be making climatic "closeups," but partly also because of some myths currently being circulated. For example, some of today's Black Nationalists have begun referring to Africa as "the cradle of civilization." With all due respect, this just isn't so. Africa was, as we have seen, the cradle of Man (a not inconsiderable distinction!); it was not the cradle of civilization. Egypt, to be sure, was *a* cradle of civilization, and one that subsequently exerted some influence on Europe to the north and much more on Black Africa to the south. But though Egypt is geographically part of Africa, it is culturally part of the Middle Eastern and Mediterranean world. The African ancestors of today's Black Americans had about as much to do with founding civilization as my own (white) north European ancestors—which is to say, nothing whatever. Some Black Nationalists have gone a step further and claimed that the Sumerians (who were unquestionably the first civilized people) were "black"; occasionally even the Israelites become "black" (though their modern descendants, the Jews, often end up as "white" oppressors of their close kin, the "black" Arabs).

All of which merely tells us that Black nationalism is just

as ridden with myth, self-deception, and plain nonsense as the various varieties of white nationalism. This, while hardly surprising, is a bit depressing. One had hoped they would do better than we did.

Just to keep the record straight: of the pioneering civilizations cited above, two (Mesopotamia and Egypt) were established by swarthy, brunette "whites," meaning that their skins ranged from cream to coffee-with-cream (depending chiefly on how much of a tan they had). They looked, in fact, much like the inhabitants of these lands today. The Indus Valley people were very probably "black" (dark brown), but with straight-to-wavy hair like the Mesopotamians and Egyptians; the odds are that they closely resembled the present inhabitants of southern India. Chinese civilization, needless to say, was founded by yellow-skinned Chinese; and the American civilizations, of course, by American "redskins" (actually, reddish-yellowish-brownish skins). None of the early civilizations, that is, were founded by blue-eyed blonds, redheads or other northern European types; none were founded by kinky-haired blacks. I don't suppose this fact will have the slightest effect on partisans of White Anglo-Saxon or Black African superiority or inferiority, but it is a fact nonetheless.

Having established the where and the who of civilization, let us look back for a moment at the climatic and human developments that set the stage for it at the very end of the Pleistocene—meaning, essentially, during the last Ice Age.

In reading about conditions in Europe during the coldest parts of the Würm glaciation—tundra or cold-grassland vegetation populated by reindeer, muskox, and the like—it is easy to picture a climate like that of today's tundra regions in the extreme north of Canada and Siberia. But this bleak picture is a bit too bleak. Ice Age or not, Europe was still located nearly 1,500 miles south of the present-day tundra, with a corresponding increase in incoming solar heat. The summer sun was higher and stronger, though this was partly offset by the fact that summer days were shorter, with no midnight sun. And there was no four-months-long winter night; even at the winter solstice there were some eight hours of weak sunlight every day.

Winters, no doubt, must have been taxing times—but not, as we saw in Chapter 21, so taxing as all that. The "corridor" between Europe and Russia may, indeed, have been broader than in some preceding glaciations, leaving more room for incursions of bitter Siberian air. Nonetheless, there seems little doubt that even in winter the ice-free parts of Europe supported a rich to very rich animal life. And man, having at last developed a technology really adequate to cope with an Ice Age climate (sewn clothing, huts where necessary) seems to have lived fairly high on the hog—or the reindeer.

Certainly his culture was far richer than any previous one. The most eloquent evidence of this are his achievements in art, for it was during the onset and height of the last glaciation that man invented both painting and sculpture. His cave pictures and bone carvings are considered, by knowledgeable critics, as aesthetically equal to anything produced by later artists. Whether the invention of art implies a qualitative leap in human intelligence between the Neanderthals (who produced no *objets d'art* of any kind) and their modern successors is arguable. (I happen to think it does.) What is not arguable is that art implies leisure. The finest of the cave paintings represent weeks of labor—not merely in producing the painting but in preparing the pigments, securing torches for light in the dark depths where the artists often worked, and training hand and eye through still visible "practice sketches." Clearly the artists were either able to take considerable time off from hunting or, more likely, managed to persuade their fellows to hand over a share of the game, presumably in exchange for a share in the potent hunting magic which the cave paintings are thought to have represented.

Ironically, what seems to have destroyed this rich culture was an "improvement" in the climate. As the ice sheets retreated, grassland and tundra were replaced by forest—

and less game. Deer, aurochs, and wild pig were certainly present, but the enormous herds of reindeer, mammoth, and wild horse followed the ice northward (if, indeed, they had not been decimated by a prehistoric overkill). Some of the hunting peoples followed them—to become, perhaps, the ancestors of today's Lapps and Finns. Those that did not seem to have fallen, for a while at least, on relatively hard times.

But man's ingenuity proved itself equal to the challenge of the receding ice, devising means of exploiting new food resources to replace the vanished herds. The harpoon, originally used to hunt reindeer, was pressed into service as a fish spear. The dart or javelin was miniaturized and combined with a new invention, the bow, to shoot wildfowl and small game. And man turned once more to collecting fruits and nuts, roots and berries—and to aid in this purpose, invented basketry.

It is at this point that the curtain rises on the advance to civilization.

25. UPLAND CLIMATES

Settling into the Middle East

As noted in the last chapter, the transition from savagery to barbarism, which is to say the invention of agriculture, is known to have occurred independently in only two places: the Middle East and Middle America. Far Eastern agriculture was quite probably an independent, or partly independent, development, but not enough has yet been discovered about it to be sure of when, where, or how it occurred.*

* Recent excavations in northern Thailand point to that region as the probably home of Far Eastern agriculture.

Peruvian agriculture may have been an independent development at the beginning, but was certainly influenced quite early on by Middle America, notably in its acquisition of domesticated corn. (A neighboring South American culture in coastal Ecuador is now thought to have acquired the important notion of pottery from, of all places, Japan; whether it received the idea of agriculture at the same time has not yet been determined.)

It may be a coincidence that both the Middle Eastern and Middle American heartlands of agriculture are mountainous uplands. (So, for that matter, is much of the apparent Peruvian heartland.) But coincidence or not, it is a fact—and a fact which, climatically speaking, appears to have had much to do with why agriculture first developed in those places and not elsewhere.

The most obvious climatic fact about uplands is that they are cooler than the adjacent lowlands. The higher the altitude, the lower the atmospheric pressure (this is why the cabins of jetliners must be pressurized); the lower the pressure the more the air can expand, and expanding air grows cooler (the expansion, and therefore cooling, of previously compressed gas is the basic mechanism of most household refrigerators). As a rule of thumb, temperatures drop from three to five degrees for every thousand feet of altitude, so that a shirtsleeve 80 degrees at the base of a 15,000-foot peak may drop to below freezing at the top.

The effect of this temperature fall is to compress a whole series of climatic zones into a very small compass. A mountain in the Equatorial Zone, for instance, might have rain forest at its base; farther up one might find deciduous trees comparable to those in the Stormy (temperate) zone. Still farther would come a Subpolar coniferous forest, followed by an alpine meadow that is a fair imitation of the Polar tundra, with an "ice cap" of perpetual snow (and perhaps even a glacier or two) over all. Thus in a climb of perhaps a dozen miles, one can experience climates which at sea level would stretch over some 4,000 miles.

This somewhat idealized climatic picture is further complicated by the other climatic variable, rainfall. As mentioned many times already, cooler air is more humid and therefore likelier to produce precipitation. Thus almost without exception, uplands are moister—sometimes much moister—than the adjacent lowlands. The wettest spot on earth is not in any Equatorial lowland but in the Temperate Zone—some 4,300 feet up. Cherrapunji, in northeast India, gets an average of 37 *feet* of rain per year; in one incredible twelve-month period (1860–61) it was deluged with 86 feet. (For comparison, New York averages under four feet a year and London considerably less than that.) During a good, brisk, Cherrapunji downpour, the raindrops are said to reach the size of baseballs. Going to the other extreme, even in the bone-dry central Sahara the Tibesti plateau will sometimes catch some summer rain by "lifting" moist air moving up from the Gulf of Guinea hundreds of miles away.

Unlike temperature, there is no simple rule relating rainfall to height. Not infrequently the lower slopes of a mountain range will get most of the rain; by the time the air reaches the top it will have already lost nearly all its moisture. As it begins to descend (and warm) on the other side, we have the rain-shadow effect, so that the windward side of a range can be lush forest with the leeward side near desert* (though still not as dry as the lowlands farther on). Add in the additional variables supplied by mountain topography —from rolling foothills to high valley to scree slope to rocky cliff—and you have a dozen or more distinctive habitat "niches," each one differing from the others in climate and/or topography and supporting its own characteristic community of plants and animals, all in an area of a few score square miles.

With these considerations in mind, let us take a closeup look at the climate of the Middle Eastern homeland of agriculture, the first place on earth where hunting and gathering

* Contrasts of this sort are especially common where wind directions are relatively constant—as in Hawaii, located in the northeast trade wind belt.

savages learned to be farming and stock-raising barbarians.* The region we are talking about is bounded on the west by the Mediterranean, on the north by the Anatolian plateau, on the south by the Sinai and Arabian deserts, and on the east by the Zagros Mountains; on today's maps it includes southern Turkey, Syria, Lebanon, Israel, Jordan, Iraq, and western Iran. Climatically, the region can be considered an extension of the Mediterranean Zone; it obtains nearly all its moisture from that body of water. (Moisture from the Black and Caspian Seas is blocked by mountains, and from the Persian Gulf, to the southeast, by the prevailing north-to-west winds.) The western coastal regions have a typically Mediterranean climate. In the summer they are almost outposts of the Arabian Desert to the south; in winter, storms from the west (many of them originating over northern Italy or the waters west of it) move eastward along the axis of the Mediterranean, bringing rain or, in the higher places, snow. Inland, however, the climate rapidly grows dryer, since much of the incoming moisture is grabbed by low mountains along or near the coast, which form a partial rain shadow over the Syrian desert and the only somewhat moister lands north of it. As the land drops again, to the low, flat valley of the Euphrates and Tigris, the air grows dryer still. At last, forced up over the high ranges of the Zagros, it deposits nearly all its remaining moisture there before descending again to the Iranian plateau.

Thus the climate of this area ranges from moderately dry along the Mediterranean coast and in the mountains to extremely dry in the desert areas. At one time it was thought that this was a recent development, and that in fact a postglacial shift from moist to dry climate, by reducing game, stimulated the development of stockraising and agriculture. But this theory neglected one important fact. During the ice ages, the southward shift in the Stormy Zone undoubtedly

* The first reports on the Thailand excavations now suggest that agriculture may have appeared there a bit earlier than in the Middle East. It is doubtful that either event influenced the other.

brought more rain, especially in the summer, to coastal and mountain regions. But while the storm tracks could and did shift, the mountains, and their rain-shadow effects, did not. And in fact, such evidence as we now have indicates that the interior sections at least were if anything dryer than they now are.

As noted before, action often begets reaction, and several archeologists have recently argued not merely that the climate didn't change much at the end of the Ice Age but that it didn't change at all. They based their argument on animal bones, which indicated that the fauna, chiefly mammals, had been about the same during the last Ice Age as at present (allowing, of course, for the effects of recent human depredations). Somebody wasn't thinking very sharply here. In the first place, most mammals are reasonably flexible in their climatic preferences, and secondly, the radical change from glacial to interglacial *must* have altered the climate, notably by reducing average temperatures. In fact when geologists began to examine the Zagros ranges carefully, they found moraines and other evidence of glaciers now extinct; they estimate that the snow line, now above 10,000 feet, may have dropped well below 5,000, which argues persuasively for a markedly cooler climate there and, necessarily, in the adjacent lowlands.

Exactly where in this region agriculture was first invented, if indeed it was invented in any single area, is still uncertain. Man has been living in the Middle East for at least 50,000 years; Neanderthal bones have been found in both Palestine and the Zagros and, as mentioned in Chapter 21, the region is also suspected of being the "birthplace" of modern man. But if one had to pick a particular part of the area for intensive examination, by all odds the best bet would be the hill country between Mesopotamia on the west and the crest of the Zagros on the east. Botanists and zoologists who have studied the wild ancestors of the plants and animals that man first domesticated—wheat and barley, sheep, goats, pigs, and cattle—have found them widely distributed in varying patterns around the whole Middle East. But only in the Zagros foothills, it seems, did all these species occur together. In the same area, moreover, have been found remains of very early, perhaps the earliest, agricultural settlements.

Prior to about 9000 B.C., the climate in this area was, as we have said, colder than today and apparently dryer as well. Pollen from lakebeds shows few or no trees; the predominant woody plant seems to have been the sagebrushlike *Artemesia,* now common in parts of the dry Iranian plateau on the other, rain-shadowed, side of the Zagros. Within a few thousand years, however, the climate—or, more accurately, climates—had shifted to something very like the present pattern, with sizable groves of pistachio and oak.

Now it was during this same period that agriculture began. Cause and effect? Did the encroachment of trees on grassland thin out the game sufficiently to nudge man toward discovering more reliable food sources? The experts disagree, and there seems no present way of resolving their dispute. What seems almost certain, however, is that the new climatic pattern of the region helped set up certain human living patterns which set the stage for agriculture and stockraising. At the very least, the change in environment must have forced men to change their cultures. So long as man's ways of surviving work reasonably well, he is, not unreasonably, prone to stick with the patterns he knows. But once circumstances force him to exercise his ingenuity, there is always a fair chance that the new pattern he devises will be better than the old.

If, around 9000 B.C., a prehistoric Mesopotamian had begun walking northeast from the banks of the Tigris toward the Zagros, he would have passed through a series of climatic zones much like those we would find on a similar trip today. First he would encounter the flat arid plains, whose annual rainfall of less than 10 inches reduces both plant and animal life to a minimum. The only exceptions are the riverside swamps, which doubtless supported a sizable population of small mammals and a few large ones. Whether they also supported human beings is unknown;

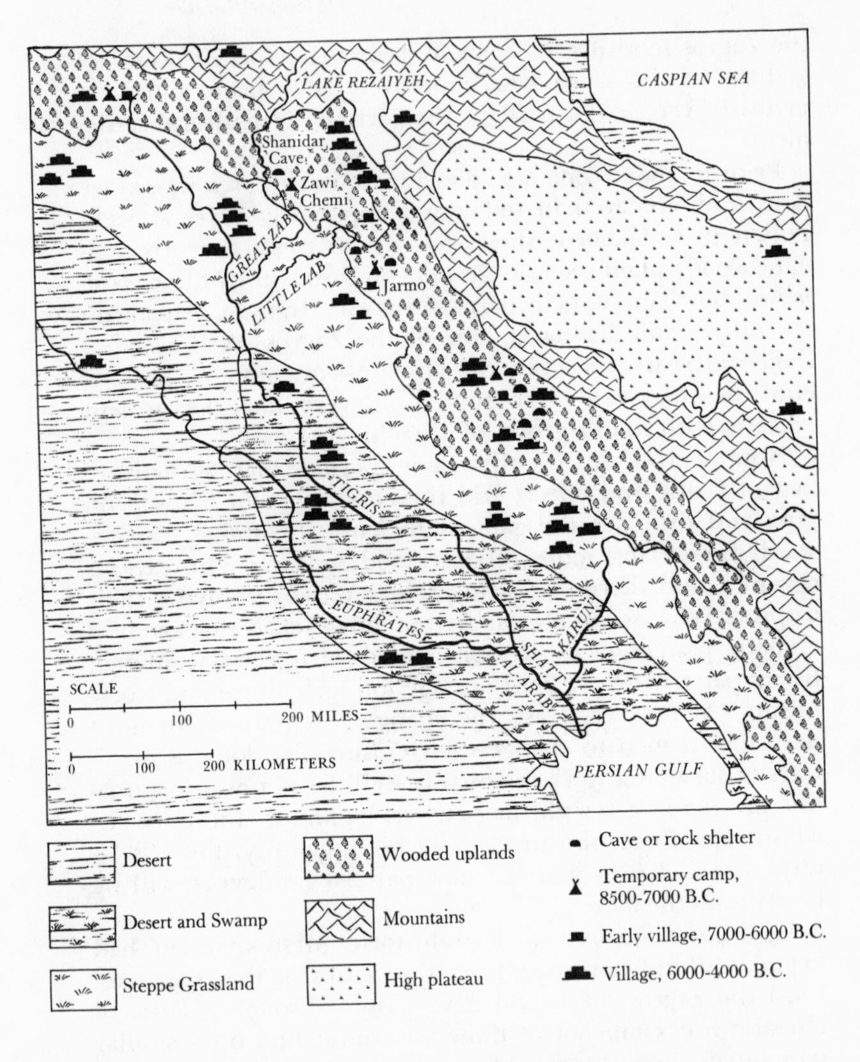

Diverse habitats of the Mesopotamian region helped stimulate the beginnings of civilization there. The driest areas were settled late—i.e., only after the invention of irrigation.

shifting river channels and the silt left by the annual flood waters have almost certainly obliterated whatever evidence there once was. As the land rose, he would reach the Assyrian steppe, in summer almost as hot and dry as the desert behind him, but transformed by 10 to 15 inches of winter rain into lush grassland. At that season our prehistoric traveler would have found herds of gazelle, wild ass, and wild cattle, and in the rivers, carp and catfish. If he knew where to look, he would also have found something else: natural asphalt —a very handy substance if you want to cement a flint arrowhead to its shaft. Still farther on, he would reach the true foothills, and then the mountains themselves. Here rainfall (depending on altitude and topography) ranges from 10 to a respectable 40 inches a year. Moreover, in many places the predominantly winter precipitation is soaked up by the porous mountain soil and released in springs which keep flowing even during the dryer months. Trees begin to break the monotony of the grassland—groves and an occasional forest of oak and pistachio, with poplars along the stream banks. At the higher altitudes, winters are distinctly cold, but in summer, both plants and animals flourish as they could not on the steppe below. Here, in those far off times, were wild sheep in the grassy areas, wild pigs champing acorns beneath the oaks and, in the ruggeder and dryer areas, herds of goats. And along the moister slopes were fields of hard-seeded grasses, emmer wheat and wild barley. If he chose to press on farther, he would reach the really high ranges of the Zagros—but these, being both cold and rugged, had and have little attraction for man or beast.

This entire journey would be a matter of perhaps 150 miles—not much more than a week of steady walking. And even with plenty of allowance for the slower pace of his women and children, and for hunting and seed-gathering along the way, our primitive Mesopotamian could have covered it comfortably in a month to six weeks. There is every reason to believe that at least some Mesopotamians of around 9000 B.C. did just that, following the game from winter to summer pasture and en route taking advantage of the different growing seasons for plants at different altitudes. Two tiny-seeded relatives of the lentil, *Astralagus* and *Trigonella,* for example, are found from the steppe to the mountains, but ripen in different months in different locations. Our migrants could have harvested them on the low steppe in March, moved up to the foothills at around 2,000 feet for another harvest in April or May, gotten a change of vegetable diet in June through the wheat and barley belt between 2,000 and 4,000 feet, and in July harvested still a third crop of *Astralagus* around 5,000 feet in the cool upland pastures, which were the local equivalent of a summer resort.

Mesopotamian man was still a wanderer, but, thanks to the compressed climatic zones of the uplands, much less of a wanderer than man in earlier periods. Consider, for example, the Ice Age reindeer hunters of Europe, whom we can assume followed the migrating herds which were their walking larders. Today's barren-ground caribou (another name for reindeer) of northern Canada annually cover up to 1,600 miles airline distance, and considerably more than that on the ground. Their Ice Age relatives, living in a Europe with considerably narrower climatic zones, doubtless migrated in a narrower compass, but we will still not be far off if we imagine the reindeer hunters as covering a thousand miles a year—meaning that they must have spent something like half the year on the move.

By contrast, the Mesopotamian hunters and gatherers, following sheep or goats or wild cattle from winter to summer pasture and back again, would have traveled perhaps a fifth of this distance. Moving more slowly, they could carry more equipment with them, meaning they had an incentive to make more complex tools. The first sickles—half a dozen carefully chipped "microblades" of flint, cemented with asphalt into a grooved holder of wood or bone—date from this period. And nobody is likely to engage in such elaborate toolmaking if the result must be jettisoned after a season's use to save weight. Thus the aborigines who inhabit today's

Australian desert, forced by their impoverished environment to cover hundreds of miles a year in foraging, must travel light—so light that they still improvise most of their stone tools on the spot as Australopithecus did.* (To avoid misunderstandings, Australopithecus improvised because he was stupid; the Australians do so because they are intelligent enough to have adapted to a very inhospitable land and climate—one that has claimed the lives of several score white explorers.)

Even more important, perhaps, is the gain in detailed local knowledge. Traveling over a migration path of hundreds of miles, and one that may well have varied somewhat from year to year, man simply could not hold in mind all the small details of terrain, vegetation, and animal life, a knowledge of which can, in a pinch, make the difference between bare subsistence and a modest if primitive prosperity.

Exploring the few square miles around my summer home in Cape Cod, I have in just a few years amassed a store of information which would be invaluable to food-gatherers (which I am) or hunters (which I am not). I know where the best blueberry patches are and the approximate order in which they will ripen; I know the little tidal stream where one can gather mussels at low water,** and the landmarks on North Truro beach that guide me to the sandy shallows where I can dig a bucketful of chowder clams in five minutes. I know the grassy hillsides where bobwhite nest, the meadows where the deer bed down at night, the tidal flats where plover and yellowlegs pause on their migrations south from arctic breeding grounds. If I were a fisherman, I would also know the best beaches for flounder and stripers, and which ponds are richest in bass and sunfish. Without realizing it, I have gone through much the same process as little bands of Mesopotamians must have gone through in the centuries around 9000 B.C.: what has been called "settling in" to a habitat of relatively narrow compass. As I have learned,

* The tools, of course, are better made than those of the ape-man.

** or could, until overgathering destroyed the mussel bed.

they too must have learned—the meadows where the wild wheat grew and quail foraged, the groves that yielded nuts and acorns, the mountain pools where fish lurked, the craggy goat trails where hunters could lay an ambush.

In the words of Keith Flannery of the Smithsonian Institution, "an incredibly varied fare was available to the hunter-collector who knew which plants and animals were available in each season in each environmental [i.e., climatic] zone: which niche or 'microenvironment' the species was concentrated in, such as hillside, cliff or stream plain, which species . . . was most practical to hunt or collect." Nor does local knowledge stop there. When you are following a herd of 5,000 reindeer (not large as such herds go) you will doubtless learn much about reindeer herds, but little if anything about individual reindeer. But when you have hunted a particular flock of sheep or herd of wild cattle for a few years, you will probably have learned a good deal about each of the dozen or two individuals that compose it: the skittish ewe, too wary to catch, the lowering bull that is likely to charge and gore you.

One important result of this sort of knowledge emerges from diggings at two sites in and around Shanidar Valley in the northern Zagros. In the upper levels of Shanidar Cave (the lowest levels go back to Neanderthal times) and in the valley-bottom site of Zawi Chemi, we find relics of a people who in the centuries around 9000 B.C. hunted goats and, less often, sheep. Soon after that date, however, we see a remarkable change. The sheep bones jump from a minority to more than 90 percent of the total. Moreover, the proportion of young animals, hitherto about 25 percent (about what one would expect in a wild herd), shoots up to 60 percent. The conclusion is inescapable. The Shanidarians no longer hunted sheep, but herded them. They had, moreover, learned the wisdom of slaughtering primarily the lambs and yearlings, leaving most of the adults for breeding stock.*

Just how this stockraising came about we may never know.

* Lamb, moreover, is tenderer than mutton.

One theory, as plausible as any, ascribes it to the very human impulse to acquire pets. It may well be that some young lamb, its mother killed by huntsmen, might have been carried back to camp alive and there, at the insistence of the children, kept alive. We know that many young animals undergo a process known as imprinting; very early in life, they become "attached" to almost any moving object with which they are constantly in contact—normally, the mother, but under abnormal circumstances a human being or even a mechanical dummy. Thus our prehistoric lamb might have formed the habit of tagging along after its captors; two such animals, a male and female, could have become the nucleus of a domesticated herd. Certainly it would have taken no great stroke of genius to note the utility of having one's meat hanging about the camp instead of running away on the open range.

The Zawi Chemi people seemingly had domesticated only the one species of animal, and no plants. But there is evidence that they collected plant foods intensively; the diggings have turned up not only fragments of sickles but also bits of stone mortars (try carrying one of *those* on a thousand-mile trek!) for grinding seeds and acorns, and remains of woven twigs which may well have been the first baskets. Deep pits apparently served to store seeds—the first granaries. And if they stored grain, it is no wild leap to assume that they also carried it with them in baskets or skin bags, as a food reserve on their autumn migration to the winter pastures of the lowland steppe.

Now wild wheat and barley, though their natural range is tolerably restricted, will flourish under considerably wider environmental conditions provided only that the soil has been disturbed. They will grow, notes Flannery, "on the back dirt pile from an archeological excavation and they probably did equally well on the midden outside a prehistoric camp." Thus we can visualize our prehistoric herders in autumn carrying their baskets or bags of grain down to the steppe and, by sheer accident at mealtime, dropping a handful of grains on the garbage heap, or perhaps on a patch of loose earth where the children had been idly digging. By spring, when it was almost time to move upcountry again, the seeds had sprouted and grown into wheat, which some thrifty housewife (or hutwife) gathered to supplement the family rations on the march. Before many generations had passed, they would begin digging the ground and dropping the seeds deliberately, perhaps even leaving a small plot at each halting place for gathering on the return journey.

And here we reach a key point in our hill-country saga: instead of merely moving *himself* from one narrow climate zone to another, man has begun to move his food resources. The time is not far off when part of the tribe will settle down at a permanent camp—in fact, a village—in some moist part of the lower foothills. The women, children, and old men will stay put, cultivating wheat, barley, and gradually other crops. The boys and younger men will each spring drive the tribal herds of sheep (and by this time goats) up to the summer pastures, returning in the fall to the milder lowland climate and a festive tribal reunion.

The effects of moving food plants do not stop here. It is likely that even before the dawn of deliberate digging and sowing, the mere fact of moving the plants had begun to improve them. Wild wheat and barley differ markedly from the cultivated grains we now know, notably in that the rows of seeds are affixed to a stem, or "rachis," which, as the plant matures and dries, becomes very brittle. When the head of such a plant is struck by a gust of wind, the rachis falls apart and the seeds are scattered far and wide. This mechanism serves excellently to disperse the seeds—which is, of course, its biological function—but would represent a real nuisance to prehistoric reapers, who would lose much of the grain merely in the process of harvesting it. In every field, however, there are a few mutants whose rachises are much tougher. Under natural conditions they do not spread, because their seed-dispersal mechanism is inefficient. But for precisely the same reason, these tough-rachis plants will be over-represented in the baskets of grain collected by industrious primitives. By the same token, they will also be over-represented in the little patches of grain sown, by accident or design, some dozens of miles away from the nearest stand of wild grain. It would not take many "biased" transfers of this sort to arrive at a strain from which the wild, brittle-rachis type had been largely eliminated. Quite fortuitously and unconsciously, man has made his first venture into "improving the breed," obtaining a variety of grain which, though a poor risk for survival under wild conditions, will give a distinctly better yield under cultivation.

This is an important point, because many of us tend to think of the invention of agriculture as a mere matter of learning to dig the ground and plant seeds in it. In fact, the process was much longer and more complicated. The domesticated plants and animals which, some thousands of years later, were to feed and clothe the first true civilizations were a far cry from their wild ancestors. They were as much a product of human activity and human ingenuity—for man soon must have learned to do by design what he had first done by accident—as were the sickles that harvested them and the axes that slaughtered them.

Another notable example of man's "invention" of his domesticated species is the sheep. When these animals were first herded (as at Zawi Chemi), we have every reason to believe that they were not the fat thick-fleeced animals with which we are familiar today. They could hardly have differed significantly from the wild sheep still found in a few places around the Mediterranean, such as the Barbary sheep seen in some zoos—or, for that matter, the bighorn of our own Rockies. These animals are lithe, active, and above all hairy rather than woolly; their "wool" is merely a short undercoat lying beneath the hair. Again we must assume a mutation, or mutations. In some early flock of these domesticated hairy beasts, one or two animals must have developed an undercoat long enough to lap over the outer hairs. The owner, when the time came to slaughter and skin the animal, would have noted that a coat made from the hide was distinctly warmer than the usual hairy jacket. (The sheepskin coat, often elaborately embroidered, is still a common garment in the colder climates in the Middle East—notably, for example, in mountainous Afghanistan, whence thousands of them are currently being imported to meet the demands of fashion.) At some later date—probably a good deal later—somebody grasped the fact that one need not skin the sheep but could cut the fleece away from the skin, leaving the animal to grow a new crop of wool during the following year.

The transition from savagery to barbarism—from hunting to herding, from seed-gathering to cultivation—was long ago termed the Neolithic Revolution by the late, great prehistorian V. Gordon Childe. During the 1940s and '50s many of his colleagues rejected the term, declaring that the transition had not happened in a "revolutionary" way—i.e., in a few years or generations, as had such revolutions as the Industrial, the American, or the Russian. But their rejection of Childe's terminology was influenced, I suspect, as much by political expediency as by scientific precision. During those Cold War years, "revolution" was a dangerous word to throw about, and, moreover, Childe himself was an outspoken Marxist (though not a communist). It is certainly true that the transition from savagery to barbarism was nothing like as sudden as the overthrow of Louis XVI or Nicholas II, nor do I believe for a moment that Childe was fool enough to suppose it was. But just what is "sudden"? The transition occupied something less than 2,000 years—from 9000 B.C., when man was still a food gatherer (albeit a highly sophisticated one) to 7000 B.C., when we find agricultural villages not only in the Mesopotamian hills but also in Palestine, southern Turkey, and perhaps even Greece. And this after modern man had been around for at least 30,000 years—a time ratio of 30:2. If a country which had stuck with one form of government for 300 years were to radically change it in 20 years, I doubt that anyone would claim this was not a revolution.

But quite beyond time scales and ratios, the Neolithic Revolution was revolutionary because it brought about a profound, radical change in man's relationship to his environment. Instead of having to take what nature offered, he was now compelling nature to offer what he wanted to take. Since the days of Australopithecus he had been Man the Collector, though the tools and intelligence he brought to the task had grown ever more sophisticated and efficient; now he was Man the Producer. And if that isn't a revolution, we'd better find a new definition for the word.

26. VARIATION ON A THEME

The First Mexican Revolution

Though my acquaintance with archeologists is not extensive, I would suspect that as a group they tend to be somewhat offbeat. Their digging expeditions are likely to take them to out-of-the-way places, where they must engage in often difficult dealings with people of strange languages and exotic customs. The scientist who lacks the flexibility and imagination to cope with the odd and unexpected is not likely to last long in archeology—even assuming he gets into it in the first place.

If there is anything to this generalization, then an excellent illustrative case is Richard ("Scotty") McNeish, of the University of Alberta at Calgary. More than any other individual, McNeish has helped elucidate the way in which agriculture came to Middle America, and his notable scientific contributions are not unconnected with his own colorful personality. By temperament and experience, McNeish is cosmopolitan, a New Yorker transplanted to the Midwest and later to Canada. Small and wiry, he has a fund of nervous energy which can keep him bustling about in temperatures ranging up to 120 degrees in the almost nonexistent Mexican shade. Though his work has drawn admiration from his Mexican scientific colleagues, in the field he associates by preference with the local natives rather than academic (or political) bigwigs. Professional convenience marches with personal preference; the natives, he points out, know the ground. If the diggings turn up a fragment of dried vegetation, for instance, the locals can tell him immediately at what season it was gathered 5,000 years ago.

McNeish takes an impish pleasure in the fact that he originally learned Spanish at digs in Mexico's "bandit country" just below the Rio Grande—and as a result sounds like the Mexican equivalent of Humphrey Bogart. This too has practical advantages: though his hoodlum diction scandalizes some Mexican servants, it has proved useful on occasion in preserving order among the diggers. His driving is as uninhibited as his speech, and his habit of gesticulating wildly with both hands while jockeying a jeep along a dirt road at forty miles an hour has shaken the nerves of several eminent scholars. There is also the hilarious, but alas unrepeatable, story of his dealings with the team of big-city American moviemakers who came to film his back-country diggings. . . .

The hunt for the origins of Middle American civilization, to which McNeish has made such major contributions, began around the beginning of this century, but became sharply focused only about thirty years ago, when it became, in effect, a hunt for the origins of corn. Cultivated corn was the staff of Middle American life long before Columbus, as indeed it still is. But the most assiduous search had failed to turn up any variety of wild corn (in contrast with the Middle East, where wild wheat and barley still grow). Some botanists even speculated that wild corn had never existed. The cultivated grain, they suggested, must have originated as a hybrid between certain species of wild grass. This particular speculation was finally laid to rest in 1953, when investigators drilled into the silt of a prehistoric lake bed that now lies beneath Mexico City. There, some 70 feet down, they found cores of compacted mud containing corn pollen, dating from perhaps 80,000 B.C., or long before man had arrived on the American scene.

Even before this finding, the hunt for corn had been quickened by several discoveries in New Mexico and northern Mexico. There, McNeish and other researchers had unearthed corn cobs—tiny but evidently cultivated—dating from about 3600 B.C. In 1959, with evidence of wild corn under Mexico City and only cultivated corn north of it, McNeish turned south, but diggings in Guatemala and Honduras were barren. Suspecting that he had overshot the mark, he began hunting out likely sites in southern Mexico—dry, highland valleys where vegetable remains would have some chance of surviving the long millennia. Finding nothing in his first valley, he moved on to Tehuacan Valley in the state of Puebla. There, he and his local guides scrambled in and out of thirty-eight caves. In the thirty-ninth they struck pay dirt—tiny corn cobs which, by carbon dating, proved to be several centuries older than any yet discovered. Partly by hunch, McNeish decided that the area was worth a major archeological effort. Having found backing from two private foundations and the U.S. government, he got the diggings underway in 1961.

The succeeding four years saw what is perhaps the most intensive archeological study ever undertaken of a limited area. The professional field staff alone numbered more than a dozen, plus scores of local diggers (some of whom may well have been unearthing relics of their own remote ancestors). The diggers, all told, unearthed nearly a million remains of human activity: delicate arrowheads chipped out of glassy obsidian, grinding stones laboriously pounded out of boulders, a bit of bark cord, still looped into a prehistoric figure-eight knot, and a two-inch bit of hollow cane charred at one end, in which some ancient Indian had smoked tobacco. Something like half the total was made up of potsherds, those fragments whose incredible durability has made them the archeologist's staff of life. Plant remains may rot and bones may crumble, but pottery goes on virtually forever.

To classify and interpret this massive hoard, the Tehuacan expedition drew on the expertise of some thirty scholars scattered across North America. The specimens of corn alone required the services of four men from Harvard and the Rockefeller Foundation; beans and squash, textiles and pottery, each had their respective specialists. A University of Chicago student classified animal bones; a Frenchman and a Mexican contributed specialized knowledge of geology and land forms; a professor in British Columbia studied shells. Perhaps the most recondite speciality was that of a Scotch-Canadian, who examined dried human feces for clues to the diets of the long-dead producers.

The story all these experts have pieced together begins, of course, with the natural setting of the human drama. Tehuacan Valley is something over seventy miles long by twenty wide; it is set among mountains whose rain shadows block off most of the incoming moisture. But its sparse rains (less than 24 inches a year in even the moister sections) are concentrated during the summer, which is the growing season. Moreover, springs at several points provide a permanent if not copious supply of water.

Considered in more detail, the climatic picture is not unlike what we have already seen in Mesopotamia: a series of narrow zones depending on altitude, each with its characteristic population of more or less nutritious plants and animals. The lowest, and therefore dryest, area is the alluvial plain along the Rio Salado; cottontails, jackrabbits, gopher, and quail live there all year round, and in the rainy season mesquite pods provide a vegetable course. Farther up the slopes we find a forest of cactus and thorn. Here the diet is rather more varied: cactus fruits in the spring, plenty of deer and peccary in the autumn, and cottontail, skunk, and doves the year around. On the highest slopes, as well as in some moisture-catching canyons, are oak groves interspersed with maguey. These, too, yield deer in the autumn, and also acorns; wild avocados ripen in the rainy summer, and maguey all year.

Dryer, and therefore less productive than the Mesopotamian hill country, Tehuacan is also less attractive to man. When the human story opens, however—about 10,000 B.C.—it was a bit less forbidding. With glaciers still covering much of Canada far to the north, the climate was distinctly cooler and (because of lowered evaporation) rather moister than at present. Jackrabbits were more abundant, and antelope and horses (of species now extinct) made a sizable addition to the available meat resources. But the really large economy-size packages of meat—bison and mammoth—which roamed the Great Plains farther north were missing. The first Tehuacanos, McNeish observes dryly, "probably found one mammoth in a lifetime and never got over talking about it—like some archeologists."

The size and nature of the occupation sites from this early date makes clear that the population was, as one would expect, migratory, wandering season by season from one temporary camp to another much like today's South African Bushmen. The initial population was probably not much more than twenty. By around 6000 B.C. the population had risen somewhat, but still was little more than a hundred. With the North American ice sheets all but gone, the climate had grown warmer and dryer. Perhaps as a result, the antelope had moved away to the north; the horses had disappeared completely, as they had from all the Americas—for reasons still debated.

These events forced the Tehuacanos to rely more heavily on plant foods, to prepare which they hammered out a variety of stone scrapers, choppers, and grinders; they also wove baskets and nets. Their diet included a kind of squash, a seed-bearing plant called amaranth, tiny avocados with pea-sized pits, and chili peppers—all of which were subsequently cultivated in Mexico, as indeed they still are. Significantly, the squash was gathered not for its bitter flesh but for its seeds. As in Mesopotamia, we can assume that the seeds were on occasion carried about in baskets and occasionally dropped in some propitious patch of earth, perhaps near a spring, a logical camping place. At any rate, well before 5000 B.C. the Tehuacanos had drawn conclusions and were soon cultivating not only squash but also chilies, gourds, and beans. At just about this time corn makes its first appearance in the archeological record—wild corn, with cobs no bigger than half a cigarette. Like the wild wheat of Mesopotamia, it had loosely-attached seeds (also, an open husk); like wild wheat, it underwent a process of selection, at first fortuitous and later, with cultivation, deliberate, which gradually converted it into something close to the rich-yielding corn of today.

But here the story takes a rather different turn. For the Tehuacanos, despite their increasing reliance on agriculture, took much longer to reach a fully sedentary life than did the Mesopotamians. Even as late as 1500 B.C., they may have been forced to quit their little hut villages during part of the year to gather wild plants and game. In 4,000 years they had achieved less than the Mesopotamians had accomplished in 2,000. In part the reason is one common to all of America: lack of domestic animals. For the American farmers, remarkably adept at exploiting plants (in addition to corn, squash, and beans, they have given the world peanuts, tomatoes, chocolate, guavas, avocados, and both sweet and "Irish" potatoes), were extraordinarily backward in dealing with the animal kingdom. Apart from the dog (a source of food as well as a hunting companion) their list of domesticated animals includes only the turkey, the guinea pig and, in the Andes, the load-carrying llama and the woolly alpaca, both also used for meat. (Significantly, the latter two animals were derived from a single wild species, the guanaco, a distant relative of the camel.) Thus the men of Tehuacan, unlike the Mesopotamians, had no milk, butter, or cheese (also no manure for fertilizer), and for meat were still largely dependent on hunting.

Much of the agricultural lag at Tehuacan, however, must be traced to the marginal climate. Given the brief rainy season, a full reliance on agriculture apparently had to wait until at least the beginnings of irrigation, which probably occurred around 1000 B.C. Precisely where Middle Americans first achieved permanent settlements is uncertain. It may have been at some more propitious, still undiscovered, spot in the highlands, less rain-shadowed than Tehuacan. American archeologist Michael Coe, on the other hand, plumps for the coastal lowlands of both the Pacific and the Gulf of Mexico, regions of tropical scrub forest and quasi-savanna with more rainfall and a much longer rainy season. Permanent agricultural settlements may have existed in these less hostile climates well before 1500 B.C., but it is difficult to be sure. In these environments, so much damper than Tehuacan, plant remains, whether wild or cultivated, only survive under very exceptional circumstances.

It is ironic that the semiarid climate of Tehuacan (and similar Mexican uplands), which has enabled science to reconstruct the first Mexican Revolution in far more detail than any other Neolithic Revolution anywhere in the world, also ensured that the revolution would proceed more slowly in the New World than the Old. The difference in climate between Middle America and the Middle East is probably as responsible as any single factor for the eventual conquest of the former by the cultural heirs of the latter.

27. BLOSSOMING AS THE ROSE

The Waters of Civilization

As we have seen, man's Pleistocene development was marked by two major achievements in climate control. Fire enabled Homo erectus to move out of the subtropics into the Temperate Zone and even allowed him, or his descendants, to survive the rigors of Europe during the third glaciation. Sewn clothing and hut-building took Homo sapiens into the Subpolar and even Polar zones, and thereby allowed him to take over the virgin hunting ground of the New World.

The Neolithic Revolution, for all its importance as a step toward civilization, involved no comparable advance in climate control, but rather a growing ability to take advantage of climatic variations in time (i.e., the seasons) and space (i.e., hilly terrain). As we move toward the culmination of the "civilizing" process, however, climate control comes back into the limelight. It is, moreover, climate control of a wholly new kind and on an unprecedented scale. It involves not heat but the other climatic variable, moisture; and instead of warming the "1/16 inch next to his skin," or the few cubic yards of a hut or cave, we find man irrigating land by acres and square miles. Under his increasingly skillful hands, guided by an ever more sophisticated mind, the desert bloomed as the rose—and civilization blossomed along with it.

Actually, the Biblical phrase is literally true only of two, or perhaps three, of the pioneering civilizations. Lowland Mesopotamia was a desert then as now, and Egypt, a desert now, was little if at all better then. The Indus Valley is largely desert at present, but it may then have been somewhat less forbidding, as we shall see later.

China is a rather special case. The valley of the Yellow River, where the first Chinese civilization developed, lies well to the north of the three regions just cited; in theory it "ought" to have a moist Continental climate like that of New York or Washington, which lie in about the same latitude. But the north China plain lies to the east of the enormous Eurasian land mass over which, during the colder months, there forms a tremendous pool of frigid air. This constantly accumulating air mass has to break out somewhere. But (as pointed out in Chapter 21) the "obvious" southern route is blocked by high mountains, while escape to the west (i.e., across Europe) is seldom possible because the prevailing winds follow an opposite course. The bulk of it must therefore pour southeast, across Mongolia and northern China and on out into the Pacific where, turning south and then southwest, it may reach as far as southeast Asia. Thus for something like half the year, north China is swept almost continuously by this intensely cold and intensely dry airstream, the Siberian monsoon. Annual rainfall is something like 20 inches, nearly all of which falls between May and September.* Intensive agriculture in the region must have required irrigation for the spring sowing—and, not infrequently, flood control during the rainy summer months.

The relationship of irrigation to Middle American civilization is still an open question. Archeologists disagree on whether the first true civilization there, that of the Olmecs, originated in the relatively dry highlands or in the hot moist lowlands along the Gulf of Mexico. It was in the latter area, certainly, that the Olmecs' successors, the Mayas, raised their remarkable temples and devised their sophisticated calendar. But there is no question at all that the climax of Middle American civilization—its most powerful states, its mighti-

* In effect, the climate could be described as abnormally warm Subpolar.

est cities—was set in the highlands, and that it was based at least in part on irrigation agriculture. (The same is true, incidentally, of Peru's proto-civilization.)

This consistent, or almost consistent, relationship between irrigation and civilization naturally suggests that civilization was somehow caused by irrigation. Some prehistorians have theorized, in fact, that the state, one of the salient characteristics of civilization, came into being as a means of controlling and expanding irrigation systems. Others reject this notion flatly—or as flatly as a prehistorian is capable of rejecting anything—saying that the state appeared well before large-scale irrigation. But "the state" is a term that can stand a lot of definition—and so is "large scale." Certainly there seems no doubt that some kind of irrigation came on the scene well before anything that can reasonably be described as a state. We do not, to be sure, know precisely when or where irrigation began, the more so in that it undoubtedly started small, like so many human institutions. At the beginning it likely involved nothing more than carrying leather buckets or pots of water from a stream to a dried-up garden a few yards away; "pot irrigation" of this sort is still carried on in parts of Mexico, and on a fairly large scale at that. But though pots have survived in many places, evidence of their use as watering cans has not.

The first place in which we are reasonably sure that irrigation was used is Jericho—long before Joshua and his trumpets came that way. The site is located well down in the Jordan Valley, below sea level and deep in the rain shadow of the Judean hills. Its climate is dry and tropically hot, but a perennial spring gives a plentiful supply of water. Jericho seems to have gotten its start as a seasonal camping ground, and perhaps a shrine, of nomadic hunters around 8000 B.C. Only a thousand years later it had grown to a sizable community covering some ten acres, fortified with a ditch 9 feet deep backed by a 20-foot stone wall, the whole dominated by a stone watchtower at least 30 feet high.

Now the Jericho of 7000 B.C. was certainly a permanent settlement. No nomadic people is going to bother building fortifications on that scale against their enemies; it's much simpler to pull up stakes and run away. Moreover, it was a settlement whose inhabitants had enough leisure—and food—to engage in fairly elaborate construction projects. Part of the community's wealth may well have come from trade; later on Jericho was unquestionably an important center on the trade route along the Jordan from Palestine and Syria to the Red Sea. But trade alone doesn't seem sufficient, the more so in that the community had little to trade itself. It is also conceivable that the Jerichoites, with what amounted to a fort controlling a major trade route, were engaged in the first known case of the protection racket, levying tribute on traders. But this, too, seems implausible unless we assume that the community possessed food resources to avoid being starved out. The spring by itself could not have done much for the crops—unless the people had developed means of "spreading" the spring water over their fields, which is to say, of irrigating, thereby raising a plentiful harvest of grain and vegetables in the rich soil and tropical climate.

But it is, after all, not in the Jordan Valley but in Mesopotamia to the east that true civilization began. By 5500 B.C. at the latest, irrigation was well into its stride in the Mesopotamian lowlands; farming communities of that date have been found in locations where the rainfall is, and was, too sparse for dry farming. Only a few centuries later, we begin to get profuse evidence of what this new form of climate control was doing for human culture. Irrigation agriculture was opening up hundreds of square miles of otherwise arid land to human habitation. But that is the very least of it. It was also opening up land which, for a variety of reasons, was far more productive than the dry-farming hill country.

First, Mesopotamia proper—the "land between the [Euphrates and Tigris] rivers"—is markedly warmer than the hills around it, warm enough to experience nothing that most of us would recognize as winter. There is a cool season, comparable to early fall in New York (or midsummer in

London), but seldom if ever cold enough to damage growing crops. And if the temperature is suitable for agriculture all year, so is the available moisture. Once man had learned to tap the water of the great rivers, there was nothing in the climate to keep him from raising two or even three crops a year. Nothing, that is, except soil exhaustion—and between them, desert and river managed to take care of that too.

If Equatorial rain-forest soils are among the poorest in the world, because the intense precipitation leaches most of the nutrients out of the earth, then desert soils, other things being equal, are among the richest. Desert soils next to great rivers, as in Mesopotamia, Egypt, and the Indus Valley, are the richest of all because they are renewed every year. Annually, the brown flood-waters lap over their banks and, spreading in quiet sheets across the land, drop their load of fine silt to revivify the soil. Once in a while the floods will fail and the next year's crops will suffer; sometimes, too, the floods will be so severe as to wash away human habitations. But by and large the annual floods were a blessing, not a curse. Thus, while villages in the hill country had to be abandoned, because of soil exhaustion, after two or three centuries at most (and even this tenure must have required pretty sophisticated agricultural techniques), the valley lands could and did yield crops for thousands of years. (Ultimately, much of Mesopotamia did revert to desert, but for reasons having nothing to do with soil exhaustion. The story will be told in Chapter 29.)

With no winter, virtually unlimited water, and constantly renewed soil, the farmer's labor could yield much more than mere subsistence. This higher productivity implies either more leisure, which can be turned to the production of "luxuries," or surplus food, which can be traded to other communities for whatever desirable commodities they might have in stock. In fact, relics from the period between 5000 and 4500 B.C. include stone bracelets, shell necklaces, copper beads, semiprecious stones, carefully polished stone bowls and cups, and, especially, increasingly graceful and ornate pottery. (Pottery itself had been invented a thousand years or so earlier.) The elaborate geometrical ormanentation of some pots so closely resembles textile patterns as to strongly suggest that somebody had also invented weaving. The interior walls of houses, too, began to be plastered and painted.

Naturally all this did not happen at once, nor did it happen only in the irrigated lands. Even in dry-farming areas, agricultural techniques continued to improve, and the domesticated strains of plants and animals grew more productive. At least one dry-farming settlement in southern Turkey reached the remarkable area of thirty-two acres. It may have owed its prosperity partly to a monopoly of the local trade in obsidian (a volcanic glass which makes superb stone tools), and may prove, when excavations have been completed, to have been a manufacturing center as well. But even the most complete monopoly and the most skillful manufactures are worthless unless your neighbors have surplus food or other valuables to trade for them.

If the Middle East as a whole was booming, it was in the valley lands of Mesopotamia that the boom was at its highest, since it was there, thanks to irrigation, that land and human labor were most productive. But even as man's new venture in climatic control was beginning to bring him modest riches, it was also providing a considerably more ambiguous gift: human inequality.

In a nomadic tribe of hunters, a man's possessions are limited to what he can carry on his back; nobody can be much richer than his fellows. Even in the primitive highland agricultural communities, wealth seems to have been fairly evenly distributed, judging from the fact that the houses were pretty much of a size. The village fields, in particular, were most likely owned and cultivated in common, as remained the case in many such communities right through historic times. But the increased productivity made possible by irrigation and other agricultural improvement produced something new: a reliable food surplus. Every bit of human history since has proven that, given a surplus, the man who is cleverer or less scrupulous or stronger or simply more persuasive than his fellows will somehow manage to get hold of a disproportionate part of it.*

Other agricultural developments fed this trend: the introduction of the vine, the olive, and the date palm, all of which take years to yield fruit. When a man has planted a tree and tended it for five or ten years, he is likely to take a distinctly proprietary attitude toward it—and to the land upon which it grows. It is no coincidence that this period of growing prosperity also produced a new kind of artifact: pieces of baked clay with different ridged patterns on one side. Judging from rather similar objects known from later times, there can be little doubt that these were seals, used to impress their pattern on the layer of mud which covered and made airtight the lid of a pot containing grain or oil. They are the visible evidence of individual ownership.

Inequality developed in another way too. The peoples of the irrigated lands were necessarily richer than the remaining tribes of nomadic or seminomadic herdsmen who still inhabited the higher or dryer regions. Nomads, by and large, are a good deal tougher than the average farmer; they have to be. The result, of course, must have been raiding expeditions by the nomads (remember those fortifications at Jericho?) when food was short, or even when it wasn't. Some nomads were very likely able, on occasion, to go beyond the swift swooping raid that carried off grain, cattle, and an attractive girl or two and actually take over the village as permanent overlords. Their toughness and military skills would enable them to enjoy the pick of the crops—and at the same time earn their keep, after a fashion, by fighting off attacks from their still-nomadic cousins. The distinction between robber and cop, between extortion and taxation, has

* Rather later, when irrigation had developed to the point where it embodied the building of canals and similar constructions, land was no longer just "land," with one piece about as good as another; some was "improved land," meaning that there was a distinct advantage in owning this field rather than that.

been blurred at many times in human history, and never more so, we may imagine, than in early Mesopotamia.

The need for a "defense establishment," whether one imposed by the nomads or home grown, would have been aggravated by another factor: growth of population. Mankind has always bred up to, and usually beyond, the available food supplies, and an increase in food has almost invariably been followed by an increase in population. Result: pressure on the available land, which, bear in mind, was still concentrated along the river channels (large-scale irrigation canals, carrying water for several miles inland, were still in the future). Further results: disputes and fights among landowners and between villages.

Exactly when the military specialist made his appearance in Mesopotamia can only be guessed at. We can be sure, however, that the same food surplus that made his existence possible also made possible the development of other specialists: the potter, the weaver, and especially the smith. The intricate technology needed to smelt copper and tin from their ores and to fabricate them into bronze tools and weapons could hardly have been carried on by amateurs; it implies a full-time devotion to the craft. Moreover, since metal ores are not found in the alluvial soil of Mesopotamia, metallurgy also implies some sort of organized trade with mining and smelting communities in ore-rich Iran and Asia Minor. Some nomad groups may well have served as intermediaries in the metal trade, thus evolving into professional or semiprofessional merchants.

All these specialists were in a sense variations on a theme already established by the very first specialists, the priests. Before the smiths, before even any evidence of a military "ruling class," we find the remains of temples, and fairly elaborate ones at that. (As noted earlier, the full-time priest–medicine-man–shamen may well date back to the hunting tribesmen of the last glaciation.) We can believe that as the riches of society grew, so did those of the priests. Man has always been eager to propitiate whatever forces he believes control his world and to support in fine style the men thought to have special knowledge and influence with those forces. He still does so, though the modern priesthood includes far more scientists than religionists. Nor was the growth of the priesthood mere superstition, even six or seven thousand years ago. Whatever the value of their spells and chants in propitiating the gods, there can be little doubt that they performed very necessary social and economic functions. One of the more important of these had to do with climate, and as such is worth a chapter of its own.

28. OF TIME AND THE RIVERS

Seasons, Priests, and Calendars

Among writers—including some scientists—who deal with the history of time reckoning, it is almost a matter of faith that man became aware of the month long before he began reckoning time by years. This theory is based on the presumed early history of the calendar in Mesopotamia and Egypt, as well as on very selective, not to say dubious, evidence from contemporary anthropology. It has been rationalized on the grounds that to a hunting people moonlight is of special importance, whether for hunting itself or as a means of illumination for other activities.

This is the sort of thing that makes you wonder whether writers really read what they write—or if they do, whether they think about it. The picture of primitive man hunting by moonlight staggers the imagination. I walk in the country on moonlit nights—on roads, not in fields or woods—but I stop short of running, as hunters must often do, because it is far too easy to trip or turn an ankle over some dimly seen obstacle. And this, mind you, in country where there are no leopards or tigers, wolves or bears. Had primitive man conducted his foraging expeditions during the night hours, he would have ended up as prey, not predator. But quite apart from this fantastical overvaluation of moonlight as an aid to man's activities, the partisans of a primitive lunar "calendar" neglect one of the most obvious facts about human ecology: the natural cycles important to men were, and are, governed chiefly by climate—which means, of course, that they are annual cycles, not monthly ones.

In the Equatorial rain forest and in the desert (neither attractive to primitive man), climate varies little from one year's end to the next. But everywhere else, the rise and fall of temperature, the arrival and departure of the rains, produce major seasonal variations in plant and animal life—variations which any primitive must take account of if he cares about eating regularly. It is no surprise, therefore, to find that even the most primitive peoples on earth today keep track of the seasons in a rough way. They do so by watching the stars.

Living as most of us do in cities, in which starlight is dimmed by haze and skyglow, we are little aware of the extraordinary spectacle that the stars present on a clear night in the country. But if, as I have done, you happen to find yourself well away from city nights on a clear summer night, spend a half hour or so gazing up into the diamond-dust-sprinkled sky and imagine its effect on primitive man: the wonder, the growing familiarity, the singling out of particular groups and the weaving around them of pictures of beasts and gods. From this regular observation of the stars, primitive man knew quite well what devotees of astrology have recently rediscovered: the aspect of the heavens changes from day to day, from season to season. The sun "moves" through the constellations, and as it moves changes the portion of the heavens that is invisible during the day and the remainder that is visible at night.

Much of the lore of primitive star reckoning has been collated by anthropologist Martin P. Nilsson. He points out, among other things, that primitive man was doubtless particularly aware of what the astronomers now call heliacal risings and settings. Waking at dawn, his eye would naturally be drawn to the eastern sky and would note the last stars to rise there before sunrise; returning to camp at dusk, he would mark the first stars appearing in the west, to set as the last sunlight faded. We find even today that primitive peoples possess a knowledge of seasonal changes in these heliacal phenomena which is remarkable only to those for whom "primitive" means stupid. A tribe of Australian aborigines —than whom there is no more primitive people on earth —determines the season to hunt for termite grubs by noting the position of Arcturus on the western horizon; the star Vega tells them that the mallee hen has laid its toothsome eggs. Primitive agricultural people are, if anything, even more dependent on time reckoning. Indians in northwest Brazil, for example, watch the Pleiades; the heliacal setting of this star-group warns them it is time to plant yams, since the seasonal rains are due.

In some tribes, star-reckoning seems a matter of common knowledge. In others, even very primitive ones, it is on the way to becoming a specialized skill. Among the Australians, Nilsson writes, "all tribes have traditions about the stars, but certain families have the reputation of having the most accurate knowledge." Among the Kenya people of Borneo, matters have moved a step further; "the determination of the time for sowing is so important that in every village the task is entrusted to a man whose sole occupation it is to observe the signs. *He need not cultivate rice himself, for he will receive his supplies from the other inhabitants of the village* [my emphasis]. His separate position is in part due to the fact that the determination of the season is effected by observing the height of the sun, for which special instruments are required. The process is a secret, and his advice is always followed." Thus, says Nilsson, "the development of the calendar puts a still wider gap between . . . the calendar-maker and the common people. Behind the calendar stand in particular the priests. . . ." Among the Maori of New Zealand, he notes, there was a regular annual "seminar" of priests and high-ranking chiefs, which determined the days on which crops should be sown and reaped.

In Mesopotamia, as in New Zealand, we can be sure that it was the business of the priests to tell the villagers the "time to plant and the time to pull up what is planted," as well as when to expect the annual floods which, while they replenished the land's fertility, were also a source of inconvenience and at times even danger. The primacy of the priests in keeping track of the seasons was undoubtedly solidified by the long-time human propensity to worship heavenly bodies as well as reckon time by them. And the man who knew when particular constellations would appear, or when the noonday sun would reach its summer height, obviously knew more than ordinary folk about the gods' secrets and was the logical man to intercede with them and interpret their will.

As the first intelligentsia, the priests were also the logical ones to plan and supervise the increasingly more complex irrigation works. They were the logical ones to keep tally of the offerings made to the temples and of their disposition, scratching crude picture records on lumps of damp mud. In a few centuries, these pictures would be conventionalized and expanded into a true written language, some of whose early documents give us a vivid sense of the importance of climate control to the Mesopotamians—clearly identifiable by this time as the people we now know as the Sumerians.

God is said to have created man in his own image, but I suspect that more often the reverse has been true. The gods of most peoples are men writ large, and whatever was believed to concern the gods deeply must have been of overwhelming concern to the people. Thus we are not surprised to find in the Sumerian creation myth an account of how the goddess Ninhursag, the "mother of the land," was impregnated by Enki, the water god. The passage mingles the symbols of irrigation with those of human fertility:

> He filled the dykes with water,
> He filled the ditches with water,
> He filled the uncultivated places with water,
> The gardener in the dust . . . embraces him . . .

and then "poured the semen in the womb of Ninhursag."

Nor was Enki the only God concerned with water; there was also Ennuge, the irrigator (his title translates literally as "inspector of canals"). In another fragmentary mud-brick document, we learn how an unnamed god (quite possibly Ennuge) "established the cleaning of the small rivers," meaning, no doubt, the regular dredging out of the natural and artificial water channels which would otherwise have become blocked with silt.

A society in which the gods, through their priestly deputies, act as inspectors of canals and superintend canal maintenance can no longer be classified as barbarous. The rich harvests of the irrigated lands had given birth to a far more complex social structure, with rich and poor, laborers and intelligentsia, professional craftsmen, professional bureaucrats —and professional rulers. For without the priestly-soldierly ruling class, the rapacious rich and the hungry poor would have been locked in endemic civil war, during which the irrigation system, on which the whole society depended, might well have fallen into fatal disrepair.

Man's discovery of how to make the desert bloom was, like so many technological achievements before and since, a mixed blessing. It brought wealth, literacy, the refined skills of the specialized craftsman, the developing science and mathematics of the priestly reckoner of the seasons, and the enormous stimulus to imagination that comes from the jostling together in larger communities of men with different skills and experiences. But all this was purchased at a price: the poverty which made riches possible; the institutionalized superstition of a state religion; slavery, coercion, and the domination of man by man. In eating the fruits ripening along the irrigation channels, man had truly lost his innocence and acquired in exchange the knowledge of social good and evil.

29. VARIATIONS ON A THEME

The Characters of Civilizations

Almost anyone who has had any extensive contact with the culture of another country must at some point have wondered why "foreigners" are different from "us." One of the oldest explanations relates these differences to climate. Italians and Spaniards, living in the hot-summer Mediterranean Zone, are "hot-blooded"; British, Scandinavians, and Germans, their dispositions tempered by Maritime winds and more northerly latitudes, are coolly reserved. The French, lying betwixt and between, partake of both worlds; one can distinguish between the "stolid" Normans and Bretons of the north and the more ebullient Provençals. (Similar climatic-temperamental differences also show up in northern and southern Italy.) One can even carry the argument into the United States, contrasting the rather dour New Englander with the more outgoing Georgian and the flamboyant Texan. And so on.

The trouble with this theory is that it doesn't work very well outside Europe and North America (if it does even there). If the Italians are "hot-blooded," the Thais and Burmese should be positively frenzied, but as everyone knows, they are actually "fatalistic." The Chinese, with a temperate climate, are equally "fatalistic," or were until recently. I forget what they are supposed to be now—apart, of course, from "inscrutable." But the superficiality of these climatic-characterological clichés should not blind us to a much subtler truth: The character of a nation, or of a civilization, is largely the product of its history, and that history, in turn, is necessarily in part the product of the climate in and around that civilization. Some useful examples can be found in the development and fate of the four pioneering Old World civilizations.

Most instructive, perhaps, is the contrast between Egypt and Mesopotamia. Egyptian civilization, as we know, developed in the delta and valley of the Nile, which lie in the heart of the Desert Zone. The flat delta, threaded with the river's wandering mouths, is not much more than a hundred miles wide and the valley much less—some ten to twenty miles. Thus Egyptian civilization was confined to a narrow strip of cultivation—the "black land" of the hieroglyphs—lying between rocky cliffs, beyond which the arid "red land" stretched far past the horizon. To east and west, then, Egypt was guarded by the red desert. During the Climatic Optimum, indeed, the desert had probably been a grassy steppe supporting a fair number of grazing animals and their human hunters, but by the time Egyptian civilization opens, shortly before 3000 B.C., the desert had taken over. Tombs from that period, dug into the desert soil, have been found to contain well-preserved wooden objects—meaning that the red land could hardly have been much moister than it is now. To the north lay the Mediterranean, then as now a commercial highway (Egypt imported timber from Lebanon long before Solomon instructed Hiram to "hew me cedars"). But the technology of the time, though adequate for commercial shipping, was not up to amphibious warfare, meaning that Egypt was very nearly as secure against invasion from the north as from east and west. In the south was a similar story. In Egypt proper, water transport (then the only feasible method of moving goods in bulk) was simple. The prevailing northerly winds, sucked in by the desert heat from the cooler Mediterranean, blew sailing ships upstream; for travel in the reverse direction they simply furled their sails and drifted with the current.* But beginning at the site of the present Aswan High Dam, the smooth flow of the Nile is broken by a series of cataracts (more accurately, perhaps, rapids) which hindered, though they did not prevent, the river traffic. Moreover, south of the First Cataract the desert cliffs press in close to the river, limiting irrigation and cultivation to narrow strips, so that no very powerful culture could exist there.

Thus Egypt was as secure against invasion as climate and geography could make it. The first burgeoning of Egyptian civilization does, indeed, seem to have been set off by merchants or travelers from Mesopotamia; there is archeological evidence of this influence. Some have speculated that Egypt was actually invaded, but this seems doubtful at best. What is certain is that during the first fifteen centuries of Egyptian history (equivalent to the time between the fall of the Roman Empire and the present) the land was never successfully, or unsuccessfully, attacked. Egyptian towns were almost never walled.

But isolation is a curse as well as a blessing. Invasion brings death and destruction, but it also brings new ideas—as do the traders and mere wanderers who flock to less isolated lands. It can be argued, indeed, that isolation makes for conformity even apart from the lack of foreign influences; in more accessible lands, dissidents can fly across the border or take to the hills. In Egypt the only refuge for the dissenter was the desert, which was no refuge at all.

The character of Egypt was shaped by another climatic factor, this one lying far to the south in the Ethiopian and central African headwaters of the Nile: the timing of the summer rains in these areas, which determines the timing of the Nile flood. Now it happens that those rains are

* Their language included two quite different verbs: *Khd,* to fare downstream, and *Khnt,* to sail upstream.

"monsoonal"—caused, that is, by the summer heating of the Sahara, which draws moist air in across Africa from the Indian Ocean, precisely as a sea breeze is drawn inland along our own coasts on a hot July afternoon.* Since the heating of the Sahara, and therefore the monsoons, and therefore the Nile floods, are controlled by the regular course of the earth in its orbit around the sun, it follows that all three phenomena are, by Temperate Zone standards at least, remarkably regular. The August arrival of the Nile flood seldom varies more than a week or two either way. It occurs, moreover, at a time when the harvest is already in, so that the land lies ready for the river's annual fertilizing embrace.

We must not exaggerate the regularity. Though the timing of the Nile flood was predictable, its height was not, and years of a "poor Nile," in which the flood reached only a few miles beyond the river banks, were years of hunger. (The "seven lean years" that Joseph coped with so shrewdly have their prototypes in much older Egyptian documents.) And if the land was seldom disrupted by invasion, there were occasional instances of civil war.

For all that, however, life in Egypt was more predictable, more bound by certainties, less subject to disturbing foreign notions, than almost any other civilization then or since. The result was about what you would expect: after the brief but intense burst of innovation and creativity which launched Egyptian civilization, the culture rather rapidly became almost static, not so much resistant to new ideas as indifferent to them. The government, apart from the rare periods in which internal unity broke down, was almost unimaginably absolutist: Pharaoh was not the deputy of the gods, or the anointed of God, as were kings in other places and times, but a living god himself. His will was law; the Egyptians, unlike other ancient civilizations, never evolved any written code of laws. It is doubtful whether they missed it. They were (or soon became), a parochial, uncurious, and prag-

* This condensed explanation is naturally somewhat oversimplified.

matic people; they had a good thing going—why tinker with it? The Nile rose every year, as surely as the sun-god Ra rose in the sky; Pharaoh was in his heaven-on-earth, in Thebes or Memphis—so what could be wrong with the world? While it lasted, nothing. The common people, of course, had a rough time of it—as always. But for the small upper class, it was "eat, drink, and be merry—for tomorrow there'll be another party." Of all the ancient civilizations, excepting perhaps the Chinese, it was Egypt which produced the most beautiful (by our standards) sculptures and paintings—and the first Beautiful People.

It is a common fallacy that the Egyptians were morbidly preoccupied with death. One source of this error is the fact that they commonly buried their dead—or at least the important dead—in the desert, so that it is the trappings of mortality rather than of everyday life that the desert climate has preserved for us. But even the tombs themselves make clear that the Egyptians saw death as, hopefully, a continuation of life. The tomb paintings, the carefully shaped models of little men and animals, showed what the deceased had enjoyed in life and what, if the magic worked properly, he would continue to enjoy eternally: rich food to repletion, wine and beer to intoxication, servants to take care of his physical needs, sports, games, and dancing girls to satisfy less tangible requirements. Egyptian upper-class life, before and after death, can be summed up in a common phrase from the tomb inscriptions: *het nebet nefret,* which can be translated as "the best of everything."

But for a civilization, as for an individual, the best can sometimes be a bit too good. For fifteen hundred years, Egypt drowsed between its desert ramparts, secure in its own traditions and power, ruled by the Pharaohs who symbolized the power, and by the priest-bureaucrats who guarded and enforced the tradition. But those who dance to the tune of "what was good enough for great-great-grandfather is good enough for me" must eventually pay the piper, when something happens which great-great-grandpa didn't anticipate.

What happened was, in the first place, an invasion from Palestine, along the coast of Sinai and across the Isthmus of Suez, by a hard-to-identify people whom the Egyptians called the Hyksos (the name, uninformatively, means merely "rulers of foreign countries"). How it happened is a good deal less clear. It appears that the conquest was aided by one of Egypt's periodic descents into civil war, but this doesn't really explain much. Up to the Hyksos invasion, about 1700 B.C., the desert barrier had isolated Egypt whether she was strong or weak. After it, there was almost continuous land traffic between Egypt and Palestine. When Egypt was weak, she was invaded from the east; when, less frequently, she was strong, she invaded in turn, at one point pushing her frontier almost to the upper Euphrates.

It may be that the Sinai-Suez desert barrier had been, as it were, eroded by the gradual establishment of roads and sinking of wells for drinking water. By themselves, these would have perhaps been adequate for a fast-moving military expedition or merchant caravan. Less easy to explain is the fact that nomadic herdsmen from Palestine and Sinai were able to bring their flocks overland into the delta in times when pasturage elsewhere was scanty (this is attested to not only by the Biblical story of Joseph and his brethren but by Egyptian documents of an earlier period), which would argue not merely water but also some degree of pasturage en route. It is just possible that the climate had changed slightly—specifically, that the Stormy Zone had dipped far enough south to bring a little rain to the coastal strip. By Roman times, certainly, something of the sort seems to have happened; writings of the period indicate that the Egyptian coast then enjoyed some rain not merely in the winter but also in the summer. It is uncertain, however, how early this trend became sufficiently marked to play a part in facilitating traffic between Gaza and the delta.

Whether the cause was cultural, climatic, or both, Egyptian isolation was no more. But the new contacts with foreign lands, though they superficially affected Egyptian culture, do not seem to have changed it in any fundamental way; it remained as rich as the Nile Valley—and as narrow. If anything, its narrowness and rigidity increased; the priestly bureaucrats clung to and enlarged their powers (a religious and apparently governmental revolution led by the pharaoh Akhnaton lasted no more than a generation). Imperialism abroad led merely to a swollen "defense establishment"; conquest from abroad engendered no renascence but merely an intensified preoccupation with the past—manifesting itself among other things in detailed copying of the painting and sculpture of the earlier period when the unbreached desert wall held the Nile's bounty secure. At last the Gift of the Nile became the prize of any conqueror who happened to be in the neighborhood: Libyans, Ethiopians, Assyrians, Persians, Macedonians, and Romans (eventually, Arabs, Turks, and English). A great, if isolated civilization for fifteen centuries, and a great power for much of the succeeding five, she became, and for some three thousand years has remained, a social mummy, watched over by the colossal monuments of her former greatness, a case study in the principle that those who cling too firmly to the past, however rich, glorious, and protected, will ultimately be crushed by it. Egypt was perhaps the first civilization to mistake the special advantages of climate and geography for the special favor of the gods; she was not the last.*

The Mesopotamians were not at all sure that they were favorites of the gods. For one thing, their situation was distinctly insecure. To the southwest, indeed, the desert protected them almost as absolutely as it did Egypt, but to north and west lay Syria and Asia Minor, both supporting fairly dense populations, while the mountains all along the northeast were infested with tribes of tough hillmen and, on

* Recently *The New York Times* quoted an Egyptian journalist as lamenting that his countrymen "are too easygoing. For thousands of years we have lived on the gifts of the Nile, knowing that the flood would bring us new riches each year and that we were protected by deserts on both sides." Even after five thousand years, climate continues to do its work!

occasion, with even more formidable enemies lapping over from the Iranian plateau. The result was that during the centuries when Egypt was evolving its placid and hedonistic culture between the desert ramparts, Mesopotamia was invaded by Elamites, Guti, Mitanni, and Kassites—among others. When the Mesopotamian cities were not fighting against foreigners, they were likely to be fighting among themselves. While in Egypt the very narrowness of the Nile Valley made it easy to control and therefore unify, the Euphrates-Tigris Valley rambles over a far wider expanse, the rivers and their subsidiary channels rambling with it. The complexities involved in unifying and governing this territory probably help to explain the early appearance there of written law codes.

Given the prevalent state of insecurity, we are not surprised to find the remains of massive walls around the sites of ancient Mesopotamian cities. The great city of Ur, in the days of its glory around 2200 B.C., was girded with a mudbrick rampart two and one-half stories high, topped with a wall of burnt brick "like a mountain."

Mesopotamian life was unsettled in another way, stemming from the nature of the rivers that were its life. The Euphrates in particular carries an enormous load of silt, far greater even than the Nile; the result is that it has over the millennia built up its bed and banks (the "natural levees") well above the surrounding plain—at the site of Ur, the *bed* of the river is six feet higher than the railroad line alongside.

Water for irrigation could be obtained simply by cutting through the natural levee. But if the levee crumbled away in flood time—watch out! The flood water could and did spread for miles across the almost totally flat plains (the Nile Valley, by contrast, is slightly dish-shaped in cross-section). A severe flood, of which there are numerous traces in Mesopotamian diggings, might well seem to have inundated the entire world; it is not surprising that the Deluge of the Bible harks back to the Babylonian, and ultimately Sumerian, legend of Gilgamesh, in which "the windstorms, exceedingly powerful, attacked as one; at the same time, the flood sweeps over the cities." Nor was the timing of the annual flood any more reassuring. The Euphrates and Tigris flow from the north, opposite to the Nile (the Egyptians, indeed, sometimes referred to Mesopotamia as "the-land-where-rivers-run-backward"). Thus the flood season depends not on the climatic cycle of tropic Africa but on that of temperate Asia Minor: specifically, on the melting of the winter snows there. This, in turn, is likely to depend less on the regular timed advance of the sun than on the happenstance of the last rainstorms of spring. Thus the annual floods in Mesopotamia can arrive any time between the beginning of April and early June, and with little warning at that.

The floods' onslaught, finally, came at a uniquely inconvenient time—not, as in Egypt, when the harvest was already in, but three to four months earlier (the difference between the spring rain of Asia Minor and the summer rain of tropical Africa), at a time when the winter crop was usually not yet ripe and the summer crop was still in the ground. To prevent the crops from being destroyed by the flooding waters, the farmers and gangs of temple laborers had to quickly turn from husbandry to flood-fighting, strengthening the river banks with reed matting and doubtless building up weak places with dirt.

In Egypt, the god of the inundation, Hapi, was a wholly benevolent deity, and the priests could sing, "Praise to thee, oh Nile, that . . . comest to nourish Egypt, that waterest the meadows . . . that givest drink to the desert places." The Mesopotamian attitude toward the rivers, and to Nin-Girsu and Tiamat who ruled the waters, was considerably more ambivalent. They were well aware of their dependence on the rivers; their mythology, cited in Chapter 27, makes that clear. But they were equally well aware that the rivers and their gods were capricious and even dangerous.

It is from Mesopotamia, not Egypt, that we get the literary prototype of Job, the Righteous Sufferer, the man who, though virtuous, is yet smitten by the gods with misfortune.

> I am a man, a discerning one, yet he who represents me prospers not;
> My righteous word has been turned into a lie . . .
> The man of deceit has covered me with the South wind * . . .
> You [the god] have doled out to me suffering ever anew . . .
> My herdsman has sought out evil forces against me . . .
> My companion says not a true word to me . . .
> The man of deceit has conspired against me
> And you, my god, do not thwart him. . . .

It is true that early Egypt also had its bad times (notably a period of civil disorder around 2200–2050 B.C.**) and that these produced some creditable literary lamentations. But what one hears in them is less the anger of a man smitten with evil than the voice of a child that has been hurt and bewilderedly asks why. In a sense the early Egyptians *were* children—precocious, talented and charming, but also overprotected and rather spoiled by their climatic situation, so that they could not cope with either success or failure. Mesopotamian civilization, inured by climate from birth to the reality that life is a difficult and chancy business, was less charming but also more durable. For Mesopotamia, despite some special difficulties resulting from climate control, despite periods of decadence and eclipse, did endure as a major center of civilization for some four thousand years. It was the heartland not merely of the Sumerians and Babylonians but of the Assyrians, their successors the Neo-Babylonians, the Persians, the Seleucid Greeks, the Parthians; as late as the early medieval period it nurtured the sunset brilliance of the Caliphs of Bagdad.

* I.e., the hot humid wind from the Persian Gulf.

** Which may have been set off by a long drought in Central Africa, bringing a run of low Niles.

Not only did Mesopotamia endure, it also influenced other civilizations including, ultimately, our own far more than did isolated Egypt. For centuries, Babylonian cuneiform was the recognized medium of diplomatic correspondence throughout the Near East (as French was in nineteenth-century Europe); documents (i.e., seal rings, tiles, mud-baked bricks, clay tablets, etc.) written in it have been found in Palestine, Asia Minor, and Egypt itself. Babylonian astronomy and mathematics were the foundation of Greek science; some authorities believe that the Babylonians discovered the Pythagorean theorem ("the square on the hypotenuse of a right triangle . . .") well before Pythagoras. Our own system of measuring angles and time in minutes and seconds goes back to Babylonian arithmetic, which reckoned by sixties as well as by tens. It was the Mesopotamians, not the Egyptians, who invented written law codes, large-scale "private enterprise" commerce, and the whole apparatus of banking, including the mortgage and letter of credit. It was they who gave the Hebrews the story of the Deluge. And in the lament of the Righteous Sufferer, we see foreshadowed not merely Job but also that terrible cry from the Cross, "My God, my God, why hast thou forsaken me?" When the Egyptians were spinning their charming theological nursery tales, the Mesopotamians were grappling with the problem of evil and human suffering. They never solved it—but then, neither have we.

More durable than Egypt, Mesopotamian civilization was still not immortal. Its decline and fall, if slower, was as sure. Iraq, which includes most of ancient Mesopotamia, is one of the poorest countries in the Middle East—and but for its oil would be poorer still. The reason for this decline seems less related to the quality of Mesopotamian civilization than to the decline of the land in which it once flourished. Mesopotamia, indeed, is an early and sobering example of a principle whose full force we are only now beginning to comprehend: when man sets out to tinker with the climate (or with any other aspect of his environment) he is likely to run into consequences he never expected and would much rather have done without.

The problem with Mesopotamia was its almost pool-table flatness, which we have already contrasted with the slightly "dished" cross-section of the Nile Valley. Moreover, the "fall" of its two rivers is minute: a little more than one inch per mile, half that of the Nile. The result was that while the Egyptians employed a system of "basin irrigation," in which water was let into fields from the Nile at one end and then drained off into the river or another field downstream, the Mesopotamians, perforce, let the water remain on the land until it sank in or evaporated. Now all rivers contain some proportion of dissolved salts (chiefly "common salt"), the Tigris and Euphrates rather more than most. When the river water evaporates, the salt is left behind, to build up over the centuries. Moreover, irrigation itself raises the underground "water table," so that even salt that has been leached below the topsoil is brought up again to poison it. We now know that in such situations periodic drainage, to flush away the salt, is as important as irrigation. The ancient Mesopotamians did not know. As early as 2400 B.C., we find Sumerian records referring to the salinization of soils in southern Mesopotamia. The first result was the gradual replacement of wheat by barley, a more salt-tolerant crop, but later even the barley crop began to fail. The process is thought to explain at least in part the collapse of the pioneering Sumerian culture. With the decline of agriculture in the south, the center of Mesopotamian civilization moved from such Sumerian cities as Ur and Lagash to Kish and Babylon, in the more northerly land of the Semitic Akkadians. The northward movement continued to Assur, a city of the Assyrians, and by 1000 B.C., with Ur already a "lost city," Mesopotamian culture is dominated by the northern cities of Nineveh and Nimrud.

The final collapse of Mesopotamian civilization some 2,000 years later is often ascribed to invasion by the Mongols, who are said to have destroyed the irrigation works out of sheer malice. This seems a rather dubious theory; to destroy irrigation works by hand (i.e., without modern explosives and bulldozers) is a task which even the most brutal invader would hardly undertake for mere sport. The true picture seems to be less sensational: Over the centuries, the annual silting over of the fields had gradually raised their level above that of the canals, so that the water could no longer reach the land.

This was not the simple problem of the canals themselves silting up—the Mesopotamians had coped with this successfully ever since the gods first superintended "the cleansing of the small rivers"—but a matter of building a whole new irrigation system. And the ramshackle government of Mesopotamia, thanks to both earlier Turkish and later Mongol invasions, was simply inadequate to a public works project of such magnitude. The result was one of those vicious social circles: The decay of the irrigation system led to a decline in agriculture and therefore population, which led to an even more dilapidated irrigation system, and so on. Thanks to salinization and silting combined, the climate control which had launched human civilization had produced its opposite—an impoverished, almost barren land, which only within the past generation or so has begun tackling the formidable job of drainage and canal-building necessary to reclaim the soil for agriculture.

30. STILL MORE VARIATIONS

China and India

Civilization developed in the Far East considerably later than in the Near East—by present reckoning, around 1500 B.C. As already noted, it arose first in north China, where the inconvenient and seasonal rains provide a powerful incentive for irrigation, while the unruly Hwang Ho (Yellow River), its waters laden with the silt that gives it its name, furnishes an equally pressing motive for flood control.*

The course of Chinese civilization, no less than that of Mesopotamia or Egypt, has been heavily influenced by its climatic surroundings. It was, and up to a century or two ago remained, in the peculiar situation of being both isolated and vulnerable.

Starting in the northeast and moving counterclockwise around the boundaries of the Chinese "heartland," we find first the frigid Siberian taiga, then the steppe/desert of Mongolia and Sinkiang, less cold than Siberia but ranging from dry to arid, then the Tibetan plateau, which is also cold and dry, thanks both to the rain shadow of the Himalayas and its altitude (the "lowlands" of Tibet are nearly as high as our Rockies), then the corrugated ranges between China and modern Burma, which World War II flyers knew as the dreaded "Hump," and finally the relatively narrow coastal strip which we now know as Vietnam. None of these areas, apart from portions of Mongolia and Sinkiang, were as hostile to human habitation as the desert surrounding Egypt, but on the other hand, none of them, except for Viet-

* Sometimes it is called "China's Sorrow."

nam, could support a dense population or the civilized society that only a dense population makes possible. Thus China, once her civilization had expanded to fill her natural climatic boundaries, was and remained unchallenged, and almost uninfluenced, by any other civilization. She could be, and was, invaded at various times from both Mongolia and central Asia, but never conquered, in the sense of being made subordinate to any other country. The invaders, however militarily irresistible, were, because of the inhospitable lands they had come from, few in number. The Mongols, perhaps the most formidable, never numbered more than about 3 percent of the total Chinese population. For this reason, and because they were usually of a lower cultural level than the Chinese, they were assimilated into Chinese civilization with few traces.

(Vietnam, whose civilization came later than that of China and was in part an offshoot of it, was never large or powerful enough to even consider invading her immense neighbor. The shoe was rather on the other foot: China for many centuries controlled Vietnam—a fact of which the Vietnamese, both Communist and anti-Communist, are well aware. Much the same considerations applied to Japan, though her oceanic situation saved her from invasion.*)

Thus Chinese culture maintained a continuity unmatched anywhere else in the world. The cuneiform characters invented by the Sumerians, the Egyptian hieroglyphs, died out nearly 2,000 years ago. But the Chinese writing system is essentially the same today as it was in 1500 B.C.—a fact of no small importance to Chinese schools, since it is estimated that the complex and beautiful characters take at least twice as long to master as our own alphabetic script.

China, whose culture was more durable than that of Egypt or even Mesopotamia, and who in numbers, technological development, and prosperity eventually outshone them

* On one occasion, a Chinese-Mongol invasion fleet was destroyed by a typhoon, the "divine wind" or *kamikaze*.

both, remained singularly parochial. Most civilizations have tended to regard "foreigners" as barbarians; in China that tendency was reinforced by the fact that nearly all her neighbors *were* barbarians. For most of her history, her civilization has been more advanced than any other with which she was in even moderately close contact; at times, especially after the fall of Rome, more advanced than any in the world.

Only once did China attempt to break away from the isolated existence to which her climatic situation had conditioned her. Beginning in the twelfth century A.D., the merchants of south China embarked on a remarkable period of voyages which took them far into the Indian Ocean, even to the East African coast. But in 1424 these operations were suppressed by imperial edict. The reasons are complex and not wholly understood; the threat of a renewed Mongol invasion from the northwest seems to have been one of them. But one cannot help seeing as perhaps the most fundamental cause the habit of mind which by that time had become ingrained: that China was, or should be, sufficient unto herself; that other lands had nothing to give or teach her. In any case, China has since shown little interest in maritime or naval affairs—a fact worth pondering by those earnest souls who envision Chinese armadas invading Hawaii and California.

Psychological and ideological habits established over thirty-five centuries do not die quickly, not even amid the military and social convulsions which have shaken China for the past sixty years. Scanning the dreary and mendacious polemics that have been rocketing back and forth between Peking and Moscow, I cannot help feeling that the Chinese are in part merely reorchestrating a very old tune. When they intimate that they are the most Communist country (and therefore, of course, the most advanced country of any sort) in the world, they are saying nothing that their ancestors did not say many times before them. And Mao's recipe for instant communism, the spectacularly unsuccessful "Great Leap Forward," amounted among other things to the declaration, "It is we Chinese, not any foreign devils, who know how things should be done." Chinese ways and Chinese thinking have changed enormously over the past generation, but in some of the basics, shaped by geography and climate, they have remained pretty much the same.

When we turn from Egypt, Mesopotamia, and China to the Indus Valley, we move into murky territory. We know civilization arose there some time before 2500 B.C., that it waxed rich and powerful, extending over (or at least dominating) a far larger territory than either Egypt or Mesopotamia, and that it disappeared utterly between 1700 and 1500 B.C. Few of its writings have survived, and none that we can read; the first legible Indian inscriptions date from nearly a thousand years later and are none too easy to interpret. It is as if our entire knowledge of the Roman Empire was based on purely archeological findings, plus a few legends first written down by the Italians of Leonardo's day. But there is one climatic question connected with the Indus Valley civilization that is worth exploring, partly because the possibilities it raises are, in light of our present knowledge and activities, peculiarly ominous. The Indus Valley civilization may have committed ecological suicide.

To understand how this may have come about, we must first dispense with the term "Indus Valley." Many settlements of the civilization, including its two chief cities (Harappa and Mohenjo-Daro) have indeed been found along the Indus and its tributaries. But it stretched east as far as the upper reaches of the Ganges river system, west along the coast almost to the present Pakistan-Iran border, and on the southeast, covered the Gujerat Peninsula (between the Gulf of Cambay and the Rann of Cutch) with a thick network of villages and at least one sizable town. A better term, therefore, is "Harappan."

Of the many mysteries about the Harappan culture, easily the biggest, is why it vanished so completely. Traditionally, its destruction has been laid to the Aryans—a group of semi-nomadic farmer-herdsmen speaking a language of the

Indo-European group, which includes nearly all the modern European tongues and those spoken in present-day Pakistan and northern India. (The Harappans very probably spoke a language of the Dravidian group, whose modern representatives dominate the speech of southern India.) That the Aryans invaded India is incontestible. For one thing, they are there now; for another, their legends clearly imply a bellicose past (their god Indra is called the "fort destroyer," who "rends strongholds as age consumes a garment"); and finally, the upper layers of Mohenjo-Daro contain several contorted groups of skeletons which give clear evidence of sudden and violent death. But proof of invasion and massacre is not proof of who did it, nor of what the overall effects were. Egypt, Mesopotamia, and China were repeatedly invaded over the millennia; indeed, at just about the time when the Aryans were supposedly tangling with the Harappans their Indo-European cousins, the Kassites, were thrusting into Mesopotamia. But through all those invasions, civilization continued, for very good reasons. First, the most barbarous of the invaders were generally the fewest in number because they came from relatively inhospitable regions (in the case of the Aryans and Kassites, from the dry plateau country of Iran and Afghanistan). Second, the whole point of the invasion was to enjoy the riches of civilization, and you do not enjoy by destroying the source of that wealth, but by "taking over the business" and keeping it going.

The distinguished British archeologist Sir Mortimer Wheeler, one of the leading experts on the Harappans, is convinced that the end of their civilization is "rooted in deeper causes of decline." Among other evidence is the drop in the quality of artifacts in the latter stages of the civilization. Intricately painted multicolored pottery is replaced by plain, unpainted ware; the delicately carved animal figures on soapstone seals are discarded in favor of crude geometric designs. What could the "deeper causes of decline" have been? The archeologist George F. Dales, among others, has suggested flooding. Certainly there is plenty of evidence that

Mohenjo-Daro, and the whole lower Indus Valley, suffered from periodic floods during the Harappan period. But floods, and pretty severe ones, were no strangers to either Mesopotamia or China. Dales believes, however, that the Indus floods were super-floods, resulting not from excessive rainfall upstream (aggravated, perhaps, by the destruction of forests for fuel; the Harappans used enormous quantities of baked brick), but from earth movements which periodically threw up dams of rock or, more likely, mud across the lower Indus, thus drowning the valley lands beneath a great lake. Only in the course of perhaps a century would the dam be eroded away and the waters depart.

There are too many things wrong with this theory. For one thing, Dales sees this flooding cycle as having been repeated at least five times; if the civilization could flourish despite four such floods, why did the fifth destroy it? For another, on his own showing the floods could not possibly have affected either the upper portions of the valley, which include Harappa itself, a city quite as large and prosperous as Mohenjo-Daro, or the Gujerat, whose present total of known Harappan settlements exceeds that of the lower valley. Dales suggests that the evident decline of the Gujerat settlements was produced by an influx of refugees from the flooded lands to the west, but this strikes me as a bit of circular reasoning: Having assumed that there were refugees, who must have gone somewhere, Dales then assumes that they went to the Gujerat; but in fact, since his hypothetical lake would have gradually spread upstream from his hypothetical dam, the obvious line of retreat for the refugees would also have been upstream, toward Harappa, not the Gujerat. Floods of some sort there certainly were. But the flood theory, explaining the destruction of the far-flung Harappan civilization by a catastrophic but relatively local event, amounts to saying that things fell apart because the center could not hold. Poetically appropriate, perhaps, but not convincing.

A peculiar fact about the Harappan civilization is that many of its settlements lie in places that are today distinctly inhospitable to man. The lower Indus Valley itself is near-desert, except for its irrigated portions, and not a few of the Harappan remains there lie at a distance from the river, or at heights above it, that would make irrigation difficult or impossible. The Gujerat, though somewhat moister, is no Garden of Eden—and moreover, has no major rivers to supply irrigation water. Even more remarkable is the existence of more than forty settlements in what is today unrelieved desert—in the empty lower valleys of rivers that once flowed into the Indus but which today sink into the desert sands long before reaching it. This and a good deal of other evidence makes it clear that the climate of the Harappan homeland today is markedly dryer than it once was. And there are strong reasons for believing that the change in the climate, which by itself could have destroyed the Harappan civilization, was the direct result of that civilization.

This theory is particularly associated with Reid Bryson of the University of Wisconsin. Parts of it are a bit too technical for me to follow, but in essence it runs something like this. The present climate of the Indus Valley and adjacent regions is a "problem climate," meaning that there is no very good reason, in terms of normal atmospheric processes, why it should be as dry as it is. In winter, to be sure, there is no problem; like most of India it is covered by the Desert Zone's belt of descending dry air, and is dry. In the summer, however, the whole Indian subcontinent is dominated by the southwest monsoon, a flow of warm, moist air from the Indian Ocean. To most of that region the monsoon brings rain, usually copious and sometimes torrential, but to most of the northwest portion, only occasional sparse showers at best. This lack of rainfall implies that the upper layers of the atmosphere must be moving downward. This might be caused by some "outside" force (i.e., the General Circulation), but under these conditions air temperatures would increase with decreasing altitude. Since this does not seem to be the case, it appears that the downward motion is being maintained by some compensating mechanism for cooling the atmosphere.

Some years ago, the Indian meteorologist P. K. Das estimated the amount of downward motion of the desert and calculated the amount of atmospheric cooling that would be needed to sustain it. He then calculated how much cooling *ought* to be taking place considering merely the composition of the atmosphere (i.e., its percentage of carbon dioxide and water vapor, the substances responsible for the greenhouse effect). This second figure turned out to be only about 75 percent of the "needed" figure. A year or two later, Bryson and some associates set out to discover the source of the "missing" 25 percent. They began by measuring, with instrument-carrying balloons, the amount of cooling actually taking place over the desert and found it very close to the "needed" cooling that Das had calculated. The gap between this figure and the calculated "greenhouse" cooling was accounted for by a dense layer of atmospheric dust. Air sampling from planes showed that the dust extends as high as 30,000 feet, and photos taken in orbit by U.S. astronaut Gordon Cooper reveal that at its worst it completely obscures the ground. The effect of the layer is to step up the loss of atmospheric heat (by radiation) some thousands of feet up, and thus provide the additional cooling to keep the air moving downward.

Thus it appears that the dust sustains the desert—and the desert, of course, produces the dust. But if, as seems clear, the desert was not always there (or at least was not always so extensive), where did it, and the dust, come from?

An obvious answer is some general, world-wide climatic change. The trouble with this is that there have been considerable changes in world climate over the past thousand years, but no evidence of much change in the Indian Desert. Prior to that time, the area seems to have gone through several cycles, in which evidence of a fairly dense population (presumably corresponding to a moister climate) alternates with small and scattered settlements (corresponding to desert or near-desert). Significantly, the first known shift to desert conditions seems to have occurred around 1500 B.C., just about the time that the Harappan civilization was falling apart. We may have done the Aryans an injustice! Warlike though they unquestionably were, the civilization which they allegedly tore down may have already collapsed before they arrived.

There is still no conclusive evidence on why the desert advanced at just that time. Bryson's proposed explanation is based on parallels from the recent history of India. As population has increased, he points out, "the forests have been destroyed for timber and fuel, so that large areas . . . are almost completely cleared of trees." This by itself alters the "effective" climate, especially in hillier areas, by increasing runoff and soil erosion and thus lowering soil moisture. But also, says Bryson, with the trees, gone, "cow dung, which should be used as a manure for fields, is burned as fuel. The consequent loss of fertility can in turn promote overgrazing. . . ." (This, though he does not say so, is aggravated by the Hindu taboo on killing cows. We do not know whether the Harappans also held the cow sacred, but a seal engraving of a bull, draped with what seems to be a garland, suggests that they may have.) Overgrazing leads to further soil destruction, to dust, and to deterioration not just in the effective climate, but in the *real* climate—not merely more runoff but less rain. Once the process gets well underway, a few generations could produce a self-perpetuating desert.*

If one accepts this very plausible theory, Harappan civilization would have differed from that of Egypt or Mesopotamia in being based only partly on irrigation agriculture. Assuming greater rainfall in the region than at present—which would help explain why the civilization used such a high proportion of burned brick: unfired mud brick would

* Bryson has also speculated that the dust may have further decreased rainfall by "overseeding" potential rain clouds—a mechanism described more fully in Chapter 40. He now suspects, however, that while the initial shift toward desert conditions was the product of human improvidence, the *coup de grâce* may have been delivered by a general climatic change around 1500 B.C. "When man is pushing the environment to the limits," he points out, "he is especially vulnerable to natural changes."

have been washed away—the people could have carried on extensive stock raising beyond the irrigated lowlands and, in the moister places, some dry-farming as well. But the consequent "population explosion," by promoting overgrazing—and also, no doubt, by pushing dry-farming and stock-raising into marginal areas—would ultimately have destroyed the favorable climate which made the whole civilization possible. Agriculture would have been pushed back into the river valleys, and some of these would have withered as stream flow diminished. At last, the civilization would have perished from its own success.* Recovery, or partial recovery, of the land (as a result of depopulation) may have taken a thousand years or more; there followed another period of expanding population, of misuse of the land, and then the renewed advance of the desert we see today.

But, says Bryson, if history suggests that man played an important role in making the Indian Desert, science suggests that he may be able to unmake it. He notes, for example, that experiments in quite arid parts of the region have shown that "it is only necessary to exclude animals for a year or two in order for a fine stand of native grass to spring up." If this was done on a large enough scale, "there would be little blowing dust"; high-level atmospheric cooling, and therefore subsidence of air, would be reduced, which would mean more summer showers, which would mean more grass, and so on. All this would require major changes in living habits among the peoples of the region, which is, says Bryson, "not a job for meteorologists but for social scientists." He is right, of course, but the social scientists won't appreciate the assignment. Changing human patterns is not much easier than changing the climate directly—and a good deal more thankless.

* Recent pollen studies have given further confirmation to this theory. Vegetation during the Harappan heyday points to a considerably moister climate at that time; subsequently, a shift toward desert conditions was followed by a drastic decline in grain pollen, implying a sharp restriction in cultivation.

The story of the Harappan civilization, fragmentary and uncertain though it is, is worth bearing in mind today. All over the world, expanding populations, along with human thoughtlessness and greed, are engaged in destroying our environment, even as the Harappans seem to have destroyed theirs. And once an environment is badly damaged—by man-made desert, by erosion, by pollution—it becomes much too much like Humpty Dumpty and the wall. We *may* be able to put it together again—but we will need all the king's horses and men to do so.

31. HIGHWAYS AND BARRIERS

Civilization in New Climes

If the development of the various ancient civilizations was influenced by climate, the spread of civilization from these pioneer centers was no less shaped by the climatic regimes, favorable or otherwise, that it found in its path. Consider first the region to the east of Mesopotamia.

The Iranian plateau is high and dry. Encircled by rainshadowing ranges, it is suitable under natural conditions for nomadic herdsmen and, in the higher areas, a relatively sparse agricultural population, dependent for their crops on hill showers and perhaps on a bit of crude irrigation from mountain torrents. Agriculture must have reached Iran soon after its development to the west, but for several thousand years the climate inhibited its development. The tough hillmen of the region could and did raid and even conquer adjacent, richer lands, but Iran itself remained a backwater.

Some time around 800 B.C., some clever Iranian farmers developed a new method of climate control. Our earliest account of it comes from a report of the Assyrian king Sargon

II. During a military campaign, presumably against Iranian raiders, he found near Lake Urmia a system of tunnels for tapping underground water, the first known example of what later came to be called the *qanat*. A qanat is, in effect, a subterranean irrigation aqueduct. It draws its water from aquifers—water-bearing strata—in hilly country and transports it to the lowlands via a tunnel. Unlike river irrigation, it does not depend on surface water (scarce in Iran); moreover, by transporting its precious liquid underground, it prevents losses from evaporation in the dry plateau air. The remarkable utility of the qanat is shown by the fact that there are today some 22,000 of them in Iran, with 170,000 miles of tunnels; they supply three-quarters of all the water consumed there.

Less than a century after the first report of a qanat, Iran emerges as a powerful, civilized state. The Medes, the first to unify Iran, marched west to sack Assyria with the help of the Babylonians, and then took over a good deal of Asia Minor. In another century or so, their successors, the Persians, controlled an empire stretching from the Indus on the east to the Nile Valley on the west, including all of Asia Minor. Their grab at Greece was stopped only by the heroic resistance of the Greek warriors at Thermopylae, Salamis, and Platea. The Iranian outburst, coming hard on the heels of a new technique of climate control, may be coincidental, but I don't think so.

The repercussions of the Persian conquest spread even farther than Persian armies. In particular, they seem to have stimulated the second emergence of civilization in India—this time in the rich Doab, the "land of two rivers" centered on the Ganges and its tributary the Jumna. The rains in this region, though still highly seasonal, are far more copious than in the Indus Valley—which, significantly, never again became an important center of civilization.

By 1000 B.C. or soon after, the Aryan invaders had settled down in the Doab and were beginning to build sizable towns; in another few centuries, Persian and other Middle

Eastern influences had contributed both writing and a crude form of coined money and had propelled them into full urban life. The further expansion of civilization into southern India seems to have been largely a matter of alternating diffusion and conquest from the center along the Ganges. Much the same is true of eastern India, though here the process was slower, due to the even wetter climate and consequently thicker forest cover. The slow development of agricultural techniques adapted to this climate, plus the sheer physical problem of hacking out settlements in the jungle, seems to have kept the area rather thinly settled almost until the Middle Ages (today it is one of the most densely populated regions on earth).

We turn now to the west and north.

Agriculture came to Palestine-Syria at the same time as it came to the rest of the Middle East. Full civilization, however, was delayed. With a Mediterranean climate, the residents could harvest a crop of grain soon after the winter rains, and pasture their flocks on the stubble before herding them up into the hills to graze during the long hot summers. But irrigation, outside the relatively narrow valleys of such streams as the Jordan and Orontes, was impractical, so that subsistence agriculture alone could not support the dense population that civilization requires. One answer was agricultural specialization. The vine and the olive tree are both native to the Mediterranean and thrive on the summer sun (in France today, a damp rainy summer means a bad year for vintners.) Either can produce a crop on land that is too hilly, or too stony, for anything else but rough grazing.

However, nobody can live on olive oil—or wine (though some have tried)—so that agricultural specialization meant trade. Wine, indeed, was produced in both Mesopotamia and Egypt, but we can guess that aristocratic topers in both lands often preferred "imported" to "domestic," if only because it was more expensive. The main export crop, however, was oil, which was used not only for cooking, but also in lieu of soap (not yet invented), as the base for what must have been distinctly powerful perfumes, and as fuel for domestic lamps.

Once propelled into trade by the limitations of their climate and geography, the Levantines found other goods to load their ships and caravans. The Lebanese hills yielded tall cedars, while the sea brought forth fish which, when dried, could be shipped inland to vary the diet of oil-raising farmers. (The difficulty of preserving shellfish in any safe, or even palatable, form may well explain the Mosaic ban on the consumption of these animals. The Jews lived far enough inland to ensure the spoilage of any shellfish that reached them from the coast.)

Of course trade must go both ways. The nearby island of Cyprus yielded copper (and also gave the metal its name); Egypt could supply linen, papyrus, and the always welcome gold; Mesopotamia, dried dates and asphalt for calking ships; Asia Minor, minerals such as obsidian and emery and, as middleman for miners in the Caucasus, silver and tin. Even coasts with nothing to sell could be raided for slaves. Trade, of course, encouraged manufacture: pottery (to ship the oil and wine in, if for no other purpose), textiles, dyes from the *Murex* snail whose precious purple ultimately became the symbol of royalty, jewelry in gold and ivory, as well as the more utilitarian tools and weapons. By approximately 1800 B.C., this process had produced along the Mediterranean coast the world's first truly commercial civilization, populated by merchants, craftsmen, and sailors. It was the latter who carried goods, and ideas, between the various civilized or quasi-civilized lands of the Near East and who, faring westward, bore the enticements and vices of civilization to new lands.

One does not usually think of climate as affecting the sea as well as the land, but of course it does. And from this standpoint, the Mediterranean could hardly be better suited to the interlinked maritime professions of commerce, exploration, and piracy. During the summer there are few storms; moreover, the clear, starry skies were a boon to navigators who knew neither chart nor compass. Fogs, a serious hazard even with today's radar, are rare. The winds are reasonably variable, no small matter to sailors who had not yet learned the technique of beating to windward; though predominantly from the north, the airs shift sufficiently to provide motive power in most directions sooner or later. Moreover, even when overall winds are unfavorable or feeble, the sailor can take advantage of the breezes set up by contrast between land and ocean. Heating up rapidly during the day, air over the coastal lands rises aloft, sucking in a sea breeze from the cooler waters; at night the reverse process creates a land breeze blowing offshore. Either one could move a ship lying as much as twenty miles out to sea.* The Mediterranean, moreover, is wholly free of the really violent marine storms, the hurricane and the typhoon, that ravage the Caribbean and the western Pacific. To Syrian or Phoenician sailors, to be sure, the distinction would have been trivial; an ordinary gale could, and often did, sink them or drive their vessels ashore—as St. Paul discovered and modern skin-diving archeologists have confirmed. But where a gale can sink a ship, a hurricane can lay waste an entire port city.

By 2000 B.C. or even earlier, Levantine sailors had carried the germs of their commercial civilization to Crete—or so it is thought. Certainly the Cretans were great traders, and the tentative and fragmentary readings of the earliest Cretan script (the "Linear A" dating from around 1800 B.C.) suggest that their language was akin to Phoenician. In any case, the Cretans in turn supplied both the impetus and the script for the first truly European civilization, which arose, albeit temporarily, in the person of the Myceneans. Some hundreds of

* Rather later—i.e., shortly before the Christian era—crafty mariners in the Indian Ocean learned to take similar advantage of the Indian monsoon winds, which are in some respects sea and land breezes expanded from coastal to continental scale and alternating from summer to winter rather than from day to night. During the summer, the southwest monsoon (the "sea breeze" sucked in by the heating of the Indian peninsula) carried them straight from the mouth of the Red Sea to the west coast of India, while the northeast monsoon (the "land breeze") powered the return trip in winter.

years later still, another outburst of Phoenician free enterprise founded Carthage, whose North African civilization dominated the western Mediterranean until it was smashed by Rome.

The peculiar civilization of the Levant, stemming in turn from its climate and geography, produced two major, and permanent, contributions to human culture—one "practical," one moral. With no great river valleys requiring coordinated irrigation and administration, the Levant had no need for a large hierarchy of priestly bureaucrats; considering the perennial attitude of commercial interests toward "government interference," it undoubtedly had no desire for them either. Thus literacy never became the near-monopoly of the scribes and priests that it was in Egypt and Mesopotamia. Moreover, a busy merchant simply didn't have time to master the complicated symbolism—syllabics, consonantals, bi-consonantals, ideographs, determinatives, and so on—of Egypt and Mesopotamia. Thus the first Levantine script of which we have knowledge (that of the port city of Ugarit) was written in the cuneiform characters of Mesopotamia, but it was an alphabetic cuneiform. Not a perfect alphabet—the vowels (which are quasi-expendable in Ugaritic and the other Semitic languages) were omitted—but for all that establishing for the first time the principle of a single sign for a single sound. The ancestor of our own alphabet was devised a little later by the Phoenicians, though it was based ultimately on Egyptian rather than Mesopotamian originals.

For the same reasons, the Levant had next to no political unity, a fact that will come as no surprise to Bible readers who recall the innumerable tribes and cities with whom the Jews alternately fought and allied themselves. The Hyksos invasion of Egypt was, indeed, carried out by what seems to have been a coalition of Palestinian cities and perhaps Arabian tribesmen, but with their expulsion from Egypt the coalition fell apart, as such predatory alliances usually do.

Thus the Levant, small, rich, and disunited, lay between Egypt and Mesopotamia, both of them large and rich. The consequences require no special prescience: the Levant was periodically conquered by one or the other, and occasionally served as the battleground for both. For the coastal trading cities, these perennial invasions generally meant little more than a shift from one tax collector to another; beyond this, it was business as usual—the more so in that few conquerors cared to meddle too strenuously with the commercial goose that yielded such handsome eggs. But for the backcountrymen, with little ready money to hand over for protection, invasion was likely to mean the ravaging of his fields, the "liberation" of his cattle, and on occasion the carrying off of his family and himself into slavery—with a rape or two to seal the bargain.

One of these peoples brooded about the situation. The Israelites had been exposed to literature on the problem of evil in Mesopotamia (Abraham hailed from the city of Ur). Settled in Palestine and periodically clobbered by armies from Mesopotamia or Egypt, they had a great deal more evil to speculate about. One possible response would have been a mere intensification of ceremonial observances to woo the gods who unpredictably visited their people with evil; neighbors of the Jews carried this propitiatory technique to the point of regular human sacrifice. The Jews, for reasons we can only guess at, took a different path. Asking the perennial question of the man, or community, in trouble, "What did I do wrong?" they began seeking the answer, not in terms of ceremonial observances, which were already pretty elaborate, but of man's conduct toward his fellows. "What," they asked, "does the Lord require?" The answer of one of their prophets—"to do justly, and to love mercy, and to walk humbly"—has, in my view, never been improved upon.

By inventing alphabetic writing, the coastal Levantines shifted the focus of learning and literacy from the temple to the individual citizen; the dour Jewish backwoodsmen did the same for morality.

Asia Minor, like the Levant, had shared in the initial development of agriculture; recent excavations, indeed, suggest that cattle may have first been domesticated there, but, as in the Levant, further development was delayed. Along the coastal strip, farming must have been much like that of Palestine and Syria to the south; in the interior—higher, dryer, and more rugged to boot—conditions were even less conducive to a dense population. Again, as in the Levant, it was trade that turned the scale—here predominantly in minerals, those extracted in the area, such as copper, and others obtained from still-savage tribesmen farther north.

Soon after 2000 B.C. an invading people, the Hittites, conquered eastern Asia Minor and organized it into a powerful civilized state that kept records in cuneiform script and in locally-invented hieroglyphs. Not much later, the Hittites, or one of the mountain tribes they controlled, discovered how to smelt iron, thus providing a new and important source of wealth which could be traded for agricultural and other imports. The Hittites kept the process to themselves as long as they could; archeologists have discovered a long correspondence between one of the pharaohs and a Hittite King, in which the Egyptian repeatedly urges his "royal brother" to send him more iron, while the Hittite, blandly ignoring the plea, forwards only a few ceremonial weapons of the precious metal. But like all secret weapons, the techniques of iron-working eventually leaked out, and Hittite power leaked away with it.

We turn now to Europe.

It is often said nowadays that Western people devote too much attention to their own culture and too little to that of others. I will not quarrel with this; the arguments for improving our understanding of how other people live and, so far as we can grasp it, how they think, are unanswerable. Having made this concession, however, I am still going to devote a major portion of the remaining chapters to climate in its interrelations with Europe and Europeans, including, of course, the transplanted Europeans who populate so many other regions. I make no apologies; my own heritage is European, and there you have it. Beyond this, however, there is the incontestable fact that European civilization, for good or evil, has been far and away the most successful in the world. All other cultures have been, at best, regional; European culture (including its American and other offshoots) is a truly world culture and has, incidentally, adapted itself to a greater variety of climates than any other. Bigger, to be sure, doesn't mean better (personally, I am not at all sure what makes one culture "better" or "worse" than another), but it certainly means a greater impact on the human race. One may like European culture or dislike it but one certainly can't ignore it.

The agricultural beginnings of civilization reached Europe in the portion nearest the Middle East, on the Greek coast that faces Asia Minor across the Aegean. They got there, moreover, quite quickly—by around 6500 B.C. or even earlier. But though geographically Greece is European, climatically it is part of the Mediterranean Zone. The further infiltration of agriculture to the north and northwest faced some climatic problems.

The Mediterranean Zone in Europe is divided quite sharply from the climatic regions to the north. Along its northern boundary lie the Pyrenees, the Alps, and the Balkans, which together form a considerable barrier to the penetration of cold air from the north—or the passage of warm air in the opposite direction. There are gaps, of course, as in Yugoslavia, where cold air can funnel down into the Adriatic as the tempestuous Bora, and in the Rhone Valley, which opens Provence to the even more violent mistral (described by Cyrano de Bergerac as the only wind that could implant a cold in his magisterial nose). But elsewhere the separation is quite abrupt, so that in traveling north from coastal Greece into Bulgaria, Rumania, or Hungary, one passes within about a hundred miles from a climate like that of southern California to one resembling that of Iowa or eastern Nebraska. During the Climatic Optimum (the period under discussion), the entire area was somewhat warmer than at present, but the contrast was quite as marked as it now is. This seems to me as good an explanation as any for why agriculture took so long—some 1,500 years—to move from Greece into the Danube Valley. Considering that Iowa and Nebraska are today in the heart of the North American wheat belt, this may seem a strange statement. But at that time, wheat and barley were 7,000 years closer to their ancestral habitat, which was subtropical. It is almost certain that it took time to evolve strains which could thrive in the more severe climate.

There was also the matter of agricultural methods. Mediterranean dry-farmers typically sowed their grain in winter, to take advantage of the seasonal rains. Farming to the north, however, required the learning of a different pattern —either sowing in late fall, before the ground froze, or, probably, in spring, as soon as the danger of frost, which could kill the young shoots, was past. Then there were the problems of soil and natural vegetation. The Danube Valley receives appreciably more rainfall than the Mediterranean and, being considerably cooler, also loses less by evaporation. The added moisture produces a thick forest cover over most of the region, in contrast with the Mediterranean, where the lowlands (i.e., the agricultural areas) tend to be grassland or sub-tropical savanna. Obviously, before you can plant a crop in forest country you must hack out a clearing. This by itself was no easy matter for the first Danubian farmers; most of the trees were oaks and other hardwoods. (If you have ever tried to fell an oak with a modern steel ax, you can imagine what it was like with a stone one. And these oaks, remember, were virgin timber, four feet or more in diameter!) Even when a clearing was made, the soil was distinctly thinner and less fertile than that of most grasslands.

The Danubians solved these interlinked climatic-vegetational problems by several techniques. First, they concentrated their crops on plateaus and ridges of loess, the rich soil deposits laid down by the dry winds that had blown south of the recently departed glaciers. These, being well-drained (loess literally means "loose"), were less densely forested and therefore easier to clear. (Elsewhere—notably in southern England—later farmers traveled, and settled, along the well-drained chalk ridges sometimes called downs.*) Second, it is probable that they invented a procedure which was to be reinvented in almost every part of the world where man sought to carry agriculture into forest country: slash and burn. This involved slashing out a clearing, letting the accumulated timbers and brush dry out for a season, and then setting fire to the whole business, thus simultaneously removing obstructions and fertilizing the soil with the ashes. We have positive proof that this technique was used elsewhere in Europe a thousand or two years later; pollen records from Denmark frequently show sequences of tree pollen, followed by grain pollen, followed by weed pollen—and they are weeds of species known to flourish after fires. There is every reason to suppose that the technique was old at that time.

Whatever the precise process, the Danubians over the centuries succeeded in developing agricultural techniques well suited to the climate of their homeland. With much better food resources, their population increased, and this meant expansion—the more so in that slash-and-burn agriculture itself supplied a powerful motive. Shallow forest soils, despite a top dressing of ashes, lose fertility rather rapidly. They could, of course, have been maintained by a system of manuring and fallowing, as was the case along the Mediterranean, but in Europe there was virgin land for the taking.

Ice Age Europe maintained a considerable population of hunting peoples. But as the glaciers retreated, pines and birches invaded the tundra and grasslands which had nourished the great herds of reindeer and mammoth. With the coming of the Climatic Optimum, things got worse; decid-

* There are four such structures in southern England: the North Downs, South Downs, Somerset Downs, and Chiltern Hills, all of them radiating out from the chalk plateau of Salisbury Plain. It is surely no coincidence that the prehistoric English farmers chose this central location for their great temple, Stonehenge.

uous trees such as oak, elm, and beech took over, further shading the ground and drastically diminishing the grass and undergrowth available for deer and wild cattle. The response of the hunting peoples—those who survived this almost catastrophic climatic shift—was to concentrate along the seashore (or in some cases, along lakes and a few rivers), where fish, shellfish, and waterfowl could provide substitutes for the increasingly hard-to-get mammalian quarry.

When the Danubians began moving, around 4500 B.C., nourished by their loess-grown crops and propelled by their slash-and-burn technique, they found an almost empty continent before them. The result was an almost explosive expansion—east into Russia, north into Poland, and northwest into Germany. By 3000 B.C., they—or somebody—had carried agriculture as far west as the British Isles and as far north as the Scandinavian peninsula. How far they got depends on who you think they were—and who you think they were involves putting together some rather complicated evidence, archeological as well as linguistic.

Some people (I for one) identify the Danubians with the original Indo-Europeans, that remarkable and turbulent folk whose cultural and linguistic descendants dominate Europe, the Indian subcontinent and both Americas, not to mention northern Asia, Australia, and New Zealand. Many archeologists disagree. The first Indo-Europeans, they believe, were a much later people, the so-called Battle-Ax Culture, whose characteristic artifacts (notably, of course, certain types of battle axes) have been found around 2500 B.C., or some twenty centuries after the Danubian explosion. Now there can be little doubt that the Battle-Ax folk were Indo-Europeans; at the times and places they lived, they could not well have been anything else. But were they *the* Indo-Europeans? I think not. The reasons lie chiefly in what we can infer about the "original" Indo-European culture—including, of course, the natural environment in which that culture evolved.

Words are the key. Our knowledge that the various Indo-

European tongues are related is based on similarities in words as well as grammar. From these word resemblances, philologists have been able to reconstruct a sizable portion of the original Indo-European vocabulary. Knowing the things these people talked about, we can make some good guesses as to how and where they originally lived.

Thus when we find the words *arðr* in Icelandic, *aratron* in Greek, *aratrum* in Latin (the source of our "arable"), *arathar* in Irish and *arklas* in Lithuanian, every one of them meaning "plow," we can be pretty certain that the Indo-European ancestors of all these peoples used plows. (English has the archaic "eär," meaning "to plow.") The presumption that they were farmers becomes a near certainty when we find comparable similarities with "barley," and in words having to do with sowing and mowing. Likewise, the philological relationships of "door" and "thatch" indicate that we are dealing with a people who built substantial houses, and hence were relatively sedentary. All this points to the Danubian farmers rather than the Battle-Ax folk. The latter, dwellers on the grassy steppes, are described by one authority as "nomads, cattle rangers and perhaps horse rangers, stock keepers with little or no knowledge of agriculture."

The names of plants and animals provide further evidence. "Beaver" and "sow" come from Indo-European, and so do "birch," "beech" and, probably, "oak." This sounds like forest country, not grassland. The beaver, of course, requires terrain moist enough to provide numerous streams and forested enough to furnish saplings to dam the streams. The pig, too, is a forest, not a steppe animal—and is, incidently, utterly unsuited to the requirements of nomadic herdsmen, as we know from folk tales discussing the difficulties of driving the animals even a few miles to market.

A second linguistic point has to do with population. Mixed farming—agriculture plus herding—is intrinsically a good deal more productive per acre than pure herding. Thus a nomadic people will almost certainly not be as "thick on the ground" as a population of farmers. The hardihood and martial skills fostered by the nomads' wandering life may well enable them to conquer the farmers—but in the process they will likely be absorbed linguistically by the more numerous conquered population. Thus nomadic Indo-European peoples have repeatedly made themselves masters of agricultural districts in the Near East, but the only place their tongues survived was in such regions as Iran, where, because of unfavorable climate, the farming population was thin to begin with.* A more recent example is the Norsemen, who were in a sense sea-going nomads. They conquered Normandy soon after 900 A.D.—but as early as 1066, when they grabbed England, they were speaking French and not their original Norse dialect. In another couple of centuries, the Norman conquerors had shifted again, to English.

We cannot, of course, prove that "Battle-Ax Indo-European" was submerged in "Danubian Indo-European," since neither people left written records of its language, or languages—but it seems likely on population grounds. Moreover, the contrary assumption—that the Battle-Ax folk conquered the descendants of the Danubians linguistically as well as physically—leads to still other problems. One would expect, in that case, to find some traces among modern European languages of the extinct (non-Indo-European) Danubian tongues. But there aren't any, not even in the very tenacious names of geographic features. Conquering populations normally take over many of the names that the conquered have already given to hills, lakes, and rivers. By Kennebec and Monadnock, by Allegheny and Mississippi, we would know with no other evidence that America was seized from people who spoke tongues very different from English. (Prairie du Chien and Fond du Lac, Santa Fe and Sangre de Cristo, testify to other linguistic predecessors.) The map of eastern and southern England is sown with names originally

* The qanats, of course, were developed when the Indo-Europeans were already in residence.

Latin or Celtic, though neither Roman nor Celt has lived there for some fifteen centuries. The Danubians, however, took over lands from which most of the "native" hunters had already been driven by the advancing forests of the post-glacial era. Naturally, then, they would have made up their *own* names for the geographical features they encountered, and since the surviving names are Indo-European, it would seem that the Danubians must have been also.*

For all these reasons, then, I see the Danubians' adaptation of agriculture to the climate of non-Mediterranean Europe, and the resulting population explosion, as one of the most momentous events in human history. It was the beginning of a process which, over the next six thousand years, would carry Indo-European tongues to the Indus and Ganges, to the Mississippi and Columbia, to the Amazon, the Plata, the Darling, and the Murray.

Can a culture acquire a bias that lasts six thousand years? Who knows! Can one even speak, today, of a modern Indo-European culture? Probably not. Yet the fact remains that of the three-billion-odd people on earth today, about half speak tongues descended from the language of those long-dead Danubian pioneers. Having begun as expansionists, the Indo-Europeans never stopped, except where climatic barriers or the resistance of other peoples blocked them. They were, and are, the outstanding historical refutation of the theory that the meek shall inherit the earth.

* Since writing this, I have been pleased to find that Calvert Watkins of Harvard, a leading authority on Indo-European, agrees with me in dismissing the Battle-Axe people as the Indo-European founding fathers, on quite different evidence. He cites the two earliest Indo-European languages known, Hittite and Mycenean Greek, for both of which we have documents from around 1500 B.C. And the differences between them, he says, are much too great to have developed in the mere thousand years that had elapsed since the Battle-Axe Folk began their dispersion.

32. BARRIERS AND HIGHWAYS

Some Bad Climatic Scenes

Expansionist though they were, the Indo-Europeans did not have matters all their own way. Pushing into western Europe, they met other farmers who had learned about agriculture by a different route, along the Mediterranean coast and up through southern France. Their advance, however, was merely slowed, not stopped; peacefully or otherwise they absorbed the indigenous agriculturists and by around 1500 B.C. dominated all of western Europe except for southwest France and the Iberian Peninsula. These areas, presumably, were populated by the ancestors of the present Basques, whose curious language has no resemblance to Indo-European—or, indeed, any other known tongue.*

To the north the obstacles were more climatic than human—and more serious. In western Europe, with its mild Maritime climate, wheat and barley, introduced from the east or the south, did well enough. But on the plains facing the North Sea and the Baltic, and in Scandinavia, conditions were more difficult. Summer temperatures, indeed, were no lower than to the south, but winters were distinctly more severe (average January temperatures today: 37° in Paris as against 30° in Berlin and 25° in Warsaw and Oslo). This meant a shorter growing season—in a cool year, probably too short to raise much of a wheat or barley crop.

Some time before 500 B.C. this problem was partially solved by the gradual adoption as food crops of two former weeds in the grain fields, oats and rye. Their shorter growing

* One archeologist has classified it as Modern Cro-Magnon—not altogether facetiously. The suggestion, while unprovable, is at least plausible.

season made them a distinctly more reliable crop, and rye, rather than wheat, bread remains the staple today on both sides of the Baltic (the Scots, also situated to the north, are no less addicted to oatmeal and oatcake).

Hardly had the new crops been adopted, however, than a new problem arose. The climate, which for some 2,000 years had been much like the present, took a turn for the worse, with even colder winters and, probably, cooler summers as well. We are not surprised, then, to find that the north Europeans—the ancestors, presumably, of the Teutonic, Slavic, and Baltic peoples—as late as the beginning of the Christian era were forced, at least occasionally, to supplement their crops with the seeds of hardy wild plants (wild now, anyway; some of them may perhaps have been cultivated then). Memories of this cold and difficult era are thought by some to be the factual basis of certain Teutonic myths: the Twilight of the Gods or Fimbulwinter, in which men and gods alike perish amid eternal frost. Whatever the truth of this, the contrast between the hells imagined by Europeans—freezing cold in Scandinavia, burning hot along the Mediterranean—provide an interesting sidelight on climatic contrasts. "Some say the world will end in fire, some say in ice. . . ."

Thus while western Europeans were developing the rich if barbaric Celtic cultures, the north Europeans remained as impoverished materially as they were climatically. And even western Europe probably lacked the resources, as it certainly lacked the desire, to advance into full civilization. It is an ironic fact that the region which was eventually to thrust its own civilization down the throats of so many "inferior" races was itself initially civilized at sword's point, by the Romans. North Europe, of course, remained barbarian until the beginning of the Middle Ages.

If civilization had hard and slow going in northern Europe, it had even harder and slower going in Africa below the Sahara, where the climate posed problems which made those of north Europe look simple. To begin with, there was the factor of isolation. The drying up of the Sahara, which during the Climatic Optimum was largely steppe grassland, threw an enormous barrier across the continent, broken only by the very narrow corridor of the Nile Valley. Thus the flow of new ideas, sluggish enough in semi-isolated Egypt, dwindled to a trickle as it oozed southward. Moreover, of the ideas and techniques of civilization that managed to make their way into tropical Africa, few were initially of much use to the peoples living there.

Wheat and barley, the staples of human progress in the Near East and Europe, do poorly even in the dryer savannas below the Sahara; in the moister regions (moist Savanna, Equatorial) they will not do at all, since they are wiped out by fungus diseases which thrive in a perpetually warm and moist climate. The animal "staples" were hardly more successful. Neither sheep nor European cattle thrive in the tropics; the high temperatures, often aggravated by high humidity, cause them to feed more sluggishly and also disorder their metabolisms so that a given amount of feed produces less meat or milk.

There is also the fact that both tropical man and his animals must have suffered, as they still suffer, from parasites —notably, insects and other arthropods, and the various disease organisms they transmit. Insects can be a nuisance even in Polar climates, as explorers in the Canadian muskeg know only too well. In the Arctic, though, their assault on man and beast is unchecked only in summer; cold or even chilly weather reduces their numbers to the bare minimum needed to maintain the species. In the tropics, however, there is little or no letup.

(It may be asked why animal or plant species native to the tropics are not decimated by these plagues. The answer is that having evolved in association with the local diseases and parasites, they have over the millennia also evolved resistance to them. Thus African antelopes can act as healthy hosts to sleeping-sickness microbes which, transmitted to domestic, i.e., introduced, cattle by the tse-tse fly, prove quickly fatal; species of wild potato in the Mexican highlands survive blights that can blast imported species in days.)

Finally, many tropical climates have distinctly damaging —and under certain circumstances, catastrophic—effects on soil fertility. The wetter regions in particular (the Equatorial Zone is the extreme case) have only a thin top soil because of the rapid decay of organic matter; its fertility is further reduced by the leaching away of nutrients in tropical downpours. Both processes are greatly accelerated when the forest cover is cleared off for planting. Thus a rain-forest clearing will yield only one or two crops, after which it must be left fallow for fifteen to twenty years before it will regain anything approaching its original productivity. (A temperate-forest clearing, by contrast will yield at least twice as long and recover in about one-third the time.) *

When large tropical areas are cleared, the land may not recover at all. Destruction of the natural vegetation cover aggravates leaching and opens the ground to the scorching sun, which accelerates the oxidation of soil components by the atmosphere. This process, called laterization, loads the soil with oxides of aluminum (bauxite) or iron (hematite and limonite), which may be valuable as ores (to our technology) but are useless for agriculture. Worse, the oxides often bake into a surface layer of bricklike laterite rock, which has been used since ancient times for road-building and other kinds of construction. The "lost cities" of tropical Cambodia are partly built of laterite, and it is thought that the same process that produced this construction material may also, by ruining agriculture, have destroyed the civilization that built with it.

With all these climatic handicaps, the wonder is not that civilization in Africa was delayed, which it was, but that it developed at all. Precisely how the Africans managed to overcome their peculiar difficulties is still poorly understood.

* The problem of soil fertility in Africa was further aggravated, of course, by the lack or dearth of manure from domestic animals.

The study of African civilization has been handicapped both by the climate itself—which in many places is even more insalubrious for white scholars than it is for black farmers—and by the conviction, held until recently by most white savants, that the Africans could not have possibly produced any useful inventions. Thus it became almost an article of faith that the natives' advance beyond the hunting and gathering stage (from savagery into barbarism, scientifically speaking) could only have taken place some time after the beginning of the Christian era when voyagers from southeast Asia introduced crops truly suited to a moist tropical climate. These plants—rice, yam, banana, and the starchy taro root—were thought to have been brought across the Indian Ocean to East Africa, whence they spread westward. More careful and less biased scholars have since noted that in West Africa, for example, the most widespread species of yams are native. Their domestication occurred on the spot, and long before any Asian influences could have manifested themselves (the date may have been as early as 3000 B.C.). The sorghums and millets—until the introduction of maize the dominant grain crops in the dryer areas of Africa—were also domesticated by Africans, and at some unknown date West Africans domesticated an indigenous type of rice. But in Africa as in the Middle East, mere domestication was not enough; a really secure food supply required in addition the upgrading (sometimes called "ennoblement") of the wild species—which surely must have required centuries.

The African food problem was further aggravated by the fact that yams and bananas are native to hot moist regions and as such reflect in their internal chemistry the deficiencies of their native soils. Both are almost totally composed of carbohydrates, low in vital minerals and almost chemically free of protein. To supply the missing dietary ingredients, Africans were long compelled to continue as at least part-time hunters. Only the gradual acclimatization of goats and the introduction of "pre-acclimatized" zebu cattle (natives of tropical Asia first domesticated by the Harappans) solved the problem.* Some white scholars, indeed, have claimed that the Africans "should" have domesticated indigenous species (e.g., some of the many types of antelope), and the fact that they did not indicates their "lack of initiative," or "laziness," or some such. No doubt. These strictures on Black African lack of industry would be more convincing if we did not know that wherever the industrious whites run things in Africa, they arrange matters so that all the really hard work is done by Blacks.

In seeking to introduce exotic species into their own very different climatic conditions, the Africans were doing precisely what the Europeans had done before them in Europe—and would do after them when the colonization of South Africa began in the seventeenth century. The ancestors of today's Boers brought European cattle with them; only after many generations did they learn the value of adding the zebu strain to increase their herds' tolerance to heat.

This is not the place to recount in detail the rise and fall of Black civilization in Africa. Suffice it to say that despite climatic and other ecological difficulties, true civilizations arose in West Africa in the centuries after 700 A.D. and in East Africa not much later—at precisely the same time northern Europe was also becoming civilized. In both north and south, interestingly enough, the civilization represented a fusion of indigenous cultures with elements contributed by Mediterranean peoples—for the Europeans, the Greeks, and Romans; for the Africans, the Arabs. What the African civilizations might have become if they had been left alone is an interesting but pointless speculation. They weren't. In West Africa, civilization was destroyed by Arab and European slavers, assisted by unscrupulous African rulers eager for a share in the profits. In East Africa, it was "pacified" into fragments by the Portuguese, in a process not without modern parallels. By ingenuity and effort, the Africans had

* Except for the areas where tse-tse flies lived; there the problem has not been solved yet.

managed to overcome the hostility of their natural environment, but hostile nature and hostile man together were too much.

In the Americas, the spread of civilization was also severely hampered by climatic barriers. In South America, the Andean protocivilization was hemmed in by the Amazon rain forest on one side and the arid coastal desert on the other. In Middle America, civilization was less consistently limited to the highlands; Olmec civilization was based in the jungles (via slash-and-burn agriculture) and their great successors, the Mayans, certainly flourished there for a while. But these truly tropical civilizations ultimately disintegrated —quite likely because of the same unfavorable climatic factors already cited for Africa—and the center of civilization shifted to the Mexican highlands, where it was hardly less hemmed in than that of the contemporary Andeans.

Mexican agriculture, based on the cultivation of corn, beans and squash, did indeed gradually diffuse northeastward into the Mississippi Valley and thence into most temperate parts of North America east of the Rockies. But the distances involved, plus the semidesert barrier of northeast Mexico, prevented the spread of true civilization by land.

Nor was the sea any substitute. Superficially, in size, shape, and prevalence of islands, the Caribbean is not unlike the Mediterranean. But it lies much farther south than that great human highway, well within the trade-wind belt, so that for something like 90 percent of the time the winds blow in the face of any mariner seeking to fare eastward from Mexico. Just how troublesome this can be emerges from an incident of Columbus' last voyage, during which he cruised eastward along the coast of what is now Honduras and then southward along the "Mosquito Coast" of present-day Nicaragua. The angle at which these two shores meet he christened Cabo Gracias a Dios ("Cape Thank-God"), a name long considered as merely another example of Spanish piety. Samuel Eliot Morison learned its real significance when he cruised those same coasts under sail while researching his biography of Columbus. Like the great admiral before him, he spent days beating back and forth into the trade winds along the Honduran coast; when at last he worked his way around the cape, and could run free to the southward, it was "Thank God!" indeed.

The troublesome trades (not to mention Caribbean hurricanes) are certainly part of the reason why none of the Mexicans developed ocean-going ships. They also explain why Cuba, less than 200 miles east of the Yucatan peninsula (a great Mayan center), was populated by Arawaks and Caribs from South America, who had island-hopped from the southeast along the 2,000-mile chain of the Antilles.

The complexities that have assisted or hindered the spread of civilization from one climate to another imply a generalization worth keeping in mind: The attempt to transfer techniques of civilization mechanically from one natural environment to another is likely to run into difficulties at best and, at worst, to severely damage both the environment and the societies that depend on it. Programs for helping "underdeveloped" areas which ignore this principle, as some have, may well end up paving the road to an environmental hell with good intentions—or with blocks of laterite.

PART FOUR

Climate and History

33. GREECE AND ROME

Two Climatic Puzzles

A little knowledge is a notoriously dangerous thing, and to no one, I suspect, is it more dangerous than to the academic. Knowing as he does that in his own field—usually a fairly narrow one—his knowledge is far above average, he can too easily transfer his professional self-assurance to areas in which his competence is average or below. The anthropological theories of Robert Ardrey received a cool-to-frigid reception among most anthropologists—but not a few professors of English and political science found them remarkably plausible.

In climatology especially, I think, a little knowledge is peculiarly perilous. The historian who skims through climatological literature in search of data that will fortify some pet theory may well fail to grasp the complexity of the problems he is dealing with, not to speak of the tentative, fragmentary, and often contradictory nature of the data itself.

When Rhys Carpenter, Professor Emeritus of Classical Archeology at Bryn Mawr, began delving into the reasons for the decline of Mycenean civilization, he was tackling one of the major puzzles in that field. We know that the Myceneans, whose ancestors seem to have reached Greece from the north before 2000 B.C., derived most of their civilization, including their system of writing, from the earlier maritime civilization of Crete. By around 1500 B.C., they equaled the Cretans in wealth and power, and a century later conquered the island—quite possibly helped by the catastrophic effects (tidal wave, ash fall, and perhaps earthquake) of an immense volcanic explosion on the Aegean island of Santorin.

Some time around 1200 B.C., Mycenean civilization quite suddenly fell apart. Its artifacts show an abrupt decline in both quantity and quality, several of its major cities (e.g., Pylos, Tiryns, and Myceae itself) were sacked, burned, or simply deserted and, most significant of all, the art of writing was lost, being reintroduced into Greece (in a different form) only after a lapse of some five centuries.

What happened?

Later Greek tradition ascribed this abrupt relapse into barbarism to invasion by the Dorians, who in classic times did indeed occupy most of the Mycenean homeland, as well as some of its overseas conquests, such as Crete. Like the Myceneans, the Dorians were Greeks, but spoke a different dialect from the one that John Chadwick and the late Michael Ventris laboriously reconstructed in 1952 from the Mycenean Linear B tablets.

But there are problems with the invasion theory. For one thing, archeologists have not (or not yet) been able to reconstruct the route of invasion with any great plausibility. That is, in certain areas which, on other grounds, one would have expected to lie directly in the path of the Doric advance, Mycenean civilization seems to have survived for a while, albeit in a somewhat less prosperous condition. The chronology also presents difficulties. It is still in a very controversial state (since the events are only a century or two apart, objective dating by radiocarbon is unfeasible) but there is considerable evidence that the destruction of Mycenae and Pylos occurred some time before the Dorians arrived.

Pondering these problems a few years ago, Carpenter evolved the theory that the Myceneans were not evicted or massacred by invaders but simply deserted their territory, forced out by a generations-long drought. This, he believes, was the result (I am using my own terminology rather than his) of a northward shift of the Desert Zone, so that the climate of the Mediterranean grew warmer and also dryer. Its long hot summers, now some four months long in Greece, stretched to eight months or more. Such a climatic shift would certainly have had a catastrophic effect on Mycenean civilization, but in trying to prove that it happened that way, Carpenter was hampered by the total lack of any direct evidence on the Grecian climate during the period. He was therefore forced to rely on indirect evidence, notably, the peculiar regional survivals of Mycenean culture already mentioned. These, he claimed, occurred in areas where the existence of mountain ranges produced "orographic" rainfall (from air forced upward by the mountains) which ameliorated the drought. The depopulated areas, on the other hand, lay either in the rain shadow of the mountains or in relatively flat country.

When I first read Carpenter's account of his reasoning, it struck me as pretty persuasive. It was true that (as he scrupulously noted) subsequent study of pollen deposits in a lagoon near Pylos had revealed no trace of a climatic shift. But this evidence he dismissed as at best inconclusive on grounds that seemed reasonably sound. Against it he cited the opinion of the distinguished English climatologist H. H. Lamb who, said Carpenter, "signified very nearly complete agreement with my chronology of the major phases of climatic change in Mycenean . . . times." Rereading the Carpenter theory, however, I began to be troubled about some of its details, to the point where I started wondering whether Lamb had really written the almost unqualified endorsement ascribed to him. When I looked up Lamb's original review, I found myself confronting, not an endorsement but a delightful academic comedy of manners, a masterpiece in the art of the reluctant and genteel put-down.

In reviewing Carpenter's work, Lamb was obviously hampered by a sense of gratitude that somebody outside the small and specialized field of paleoclimatology had at last begun to take it seriously. "There seems too little recognition amongst archeologists and historians," he wrote, "of the role climatic shifts must have played in history." Thus, "it is more than welcome to find an eminent classicist stepping

. . . over the cleft between our two cultures." In undertaking "necessary criticisms" of the Carpenter theory, "the last thing the reviewer wishes is to discourage . . . either the author or those who would follow in his footsteps." But having observed—and perhaps gone beyond—the necessary academic amenities, Lamb then had to buckle down to his "necessary criticisms." They were lulus. Carpenter's climatology, he was compelled to say, "relies far too heavily on [a] somewhat sweeping and oversimplified hypothesis." This reliance was "premature, excessively literal and leads to some misleading dating and interpretation." Noting that several centuries before 1000 B.C. were indeed unusually warm in *northern* Europe, he also noted Carpenter's seeming unawareness that a warmer northern Europe did not necessarily mean a dryer Mediterranean. Indeed "the next warm epoch [i.e., the Little Climatic Optimum, around 1000 A.D.] . . . is believed to have had a much moister climate in the Mediterranean. . . ." Moreover, in citing changes in sea level as evidence of climatic change, Carpenter had made "far too little allowance" for possible changes in land level, in a region where earth movements, including earthquakes, were and are common. To sweeten these distinctly unpalatable pills, however, Lamb coated them with the declaration that while Carpenter's knowledge of climatology lacked depth, "he admits as much with charming modesty." The outside observer might suspect that Lamb is a bit of a charmer himself.

When one examines Lamb's chronology in detail, it becomes apparent that the agreement which Carpenter found with his own theory is a great deal less than "very nearly complete." Lamb places the northern European warm period, which is also Carpenter's hypothetical drought period, in the fourteenth and thirteenth centuries B.C. But the first of these Carpenter himself describes as "the period of greatest cultural prosperity in Greece"; the second marks a phase of "scarcely slackened cultural activity and power," which only near the end "was rudely shaken by the destruction of the great [Mycenean] palaces. . . ." Whatever happened, it was certainly *not* the long period of decline and depopulation that Carpenter depicts. That, if it occurred at all, came later—at a time when, on Lamb's evidence, the climate of Northern Europe and, presumably, of Greece was returning to something like that of the present.

There are two other pieces of climatological evidence which Lamb does not cite, but which also weigh against Carpenter's theory. The first concerns what happened in the Levant. Just about the time the Myceneans were declining, the Levantine coast was invaded by the Peleset, apparently a mixed group of peoples from Greece, the Aegean, and Asia Minor (since Carpenter's alleged drought covered all these areas, it doesn't matter which). But if Greece was drought-striken, the Levant, which is hotter and farther south, would have been a howling desert. The Peleset would have found no sustenance there, but would have starved by the thousands. They didn't. In fact, they survived successfully enough to give their name to Palestine and to perennially oppress the Jews, whose version of their name has come down to us as "Philistines."

Even more puzzling, perhaps, is the case of Attica, the region of Greece around Athens. Carpenter and everyone else agree that Attica was one of the principal "refuge areas" for the Myceneans who streamed out of southern Greece (the Peleponnesus) when their civilization broke apart as a result of drought, invasion, or whatever. If drought, then Attica must have remained well watered. Carpenter of course explains this in terms of orographic rainfall, from moist westerly winds funneling up the Gulf of Corinth to discharge their moisture on "the cooling heights of Parnes, Pentelikon and Hymettos." When I looked at a relief map of Greece, however, I found that these hills were not as high as all that; moreover, the conformation of the land around the Isthmus of Corinth didn't seem quite right for Carpenter's funnel. It then occurred to me to examine a rainfall map of Europe. If orographic factors could ameliorate a drought in 1200 B.C., I reasoned, then the same factors would obviously produce local increases in rainfall today. And so it proved. The mountains of the northwest Peloponnesus, for example, are markedly wetter than the surrounding lowlands, as they should be. But Attica, far from being wetter, is dryer! Eastern Attica, in fact, is the only section of modern Greece with less than 20 inches of rain per year.

If civilization declined precipitously in the Peleponnesus because of drought, it should have been obliterated in Attica. It wasn't. If the Dorians, as Carpenter claims, more or less peacefully took over regions depopulated by crop failure and famine, then certainly they would have taken over Attica. They didn't.*

If we discount Carpenter's drought as the destroyer of the Myceneans (and it seems to me we must) and also discount the Dorians (and there are, as Carpenter says, reasons for doing so), then what *did* happen? My own theory (and it is just a theory, and probably not an original one) is that the Myceneans destroyed themselves.

By any standard, Mycenean civilization was intensely predatory. Archeologists have puzzled over the basis of its prosperity; Greece of that day had neither the intensely specialized agriculture of Palestine nor the mineral resources of Asia Minor. Certainly the Myceneans traded with the Levant and other regions—but what did they trade with? Some have suggested that their main manufacture was weapons, which is significant in itself. However, a great deal of the "trade" seems to have approached outright piracy, in which portable loot in the form of weapons and metals (precious

* Since writing this, I have been informed by Reid Bryson that the situation is not quite so clear cut. Statistical studies of weather patterns in Greece, he says, indicate that drought in the Peleponnesus *can* be combined with adequate rainfall in Attica—meaning that here Carpenter might have been right for the wrong reasons. Since the studies are not yet published, it is hard to evaluate them, and even if true, they do not dispose of the other objections to Carpenter's theory. Personally, I will believe in the drought explanation of Mycenean decline when, and only when, I see some convincing evidence that a drought occurred—not that it might have occurred.

and otherwise) was combined with human loot that could be sold profitably in slave markets to the east and south.

Our view of the so-called Heroic Age of Greece is biased by the fact that our sources are later Greeks recalling their own glorious past. With a less ethnocentric eye, we can see that famous adventurers like Jason, of Golden Fleece fame, were hooligans out for what they could grab; the "fleece" described what they did as well as what they got. The general level of morality among the heroes can be judged by the squalid wrangle in the *Iliad* between Agamemnon and Achilles (two of the top leaders) over a feminine piece of loot from their Trojan robbery-with-violence expedition.

The archeological evidence itself is significant. Sinclair Hood, head of the British School of Archeology in Athens, in contrasting Minoan (Cretan) civilization and its Mycenean successors, cites the decline in such refined arts as seal engraving, ivory carving, and fresco painting even while "the inventiveness of the armorer remorselessly advance[s]. . . . Superb weapons accompany the warriors in graves . . . in Crete and on the mainland. Swords are now better hafted and are shorter, adapted to cut as well as thrust; while elaborate bronze armor makes its appearance. . . . A war-minded and militaristic spirit pervades the age."

Now predation abroad makes for predation at home. While the Agamemnons and Odysseuses were off looting and squabbling, usurpers could make their way into the hero's palace—and, often enough, into his bed as well. The hero's return—if he did return—was then the signal for a long cycle of revengeful slaughter and blood-feud. The spectacular delinquencies of the House of Atreus may have been an extreme case, but surely not an atypical one. Consider, for instance, the massacre that ensued when Odysseus finally got back to Ithaca, even though Penelope was more chaste than Clytemnestra. (Homer had to call in the gods to stop a civil war in Ithaca between Odysseus and relatives of the slaughtered suitors.)

Again the archeological evidence is significant. Hood tells

us that in the fourteenth and thirteenth centuries B.C. (i.e., at the height of Mycenean power), "massive defense walls were built at the chief settlements on the Greek mainland. . . . At Athens and Mycenae remarkable covered stairways descended to underground springs which lay hidden outside the walls," the purpose being, of course, to ensure a supply of water in case of a siege. Against whom were these ramparts built, so mighty that later Greeks thought them the work of the Cyclopes? Not against the Dorians, who by even the most generous chronology were still far away. Not against any foreign power, for there was none in the region. We must conclude that the petty Mycenean kingdoms were arming and walling their cities against one another. They doubtless believed, to paraphrase a poet long after Homer, that good fortifications make good neighbors.

But not, I think, good enough. Experts in the heroic skills of assault with intent to rob are unlikely to be inhibited by the fact that a potential victim lives in the next valley rather than overseas, especially if, as seems possible, the cream of the most accessible overseas loot had already been skimmed off. Certainly the Myceneans of those days were ranging ever farther in search of booty. Egyptian records of the fourteenth and early thirteenth centuries B.C. tell us that at least three times the land was assaulted by the "people of the isles." These seem to have been a coalition of predatory folk among which we can distinguish names that seem to be the Egyptian versions of Achaeans and Danaeans, both Homeric names for the Heroic Age Greeks. Like imperial looters of a much later century, the Myceneans may have sung the equivalent of "Wider, yes and wider may thy bounds be set; / God who made thee mighty, make thee mightier yet!" But the gods did not always cooperate. The Egyptians repeatedly beat back the sea peoples who, beaten back, may well have turned their predacious energies against one another.

We must not neglect, either, the possible role of the home peasantry. Their taxes paid for the armaments, ships, and supplies of their larcenous masters, their labor erected the

Titanic fortifications. It was their sons who were induced, or compelled, to fill the ranks of the Mycenean armies and reap the usual reward of the common soldier: plenty of wounds, plenty of lice, and little glory (try to find a common soldier in the *Iliad*). Returning home, battered and bitter, from some unsuccessful foray, would they not have gazed thoughtfully at the great palaces, crammed (as we know from the Linear B documents) with stores of grain, wine, and oil? And, gazing, might they not have begun muttering the Mycenean equivalent of "Burn, baby, burn!"? *Somebody* burned Mycenae, Tiryns, and Pylos. . . .

It seems quite possible, then, that in one way or another the fate of Mycenean civilization paralleled that of the legendary cats of Kilkenny: "They fought and they fit, and they scratched and they bit/ 'Til instead of two cats, there weren't any." That is the sort of thing that can happen to civilizations which devote too much of their energy and resources to armaments and war.

If climatic change seems to have had little or nothing to do with the collapse of the first Greek civilization, there is reason to suspect that it helped appreciably in the renaissance of Greece a few centuries later. Beginning about 1000 B.C., the climate of northern Europe turned colder and wetter. This process, which seems to have continued until close to the start of the Christian era, produced serious problems for the Germans and other northern folk, but it may well have brought a cooler and moister climate to Greece, thereby making far more bountiful crops.

So far as I know, there is no direct evidence that this actually happened. Greece did become re-civilized soon after 800 B.C., but that by itself proves nothing. Rather more significant, however, is that Greece also experienced something of a population explosion. Beginning around 700 B.C., the emigration and colonization of the Greeks is one of the main features of Mediterranean history. Over the next few centuries, they pushed east to the coast of Asia Minor, northeast to the coasts of the Bosporus and Black Sea, south to Libya and the Egyptian delta, and west to southern Italy, Sicily, and the Franco-Spanish coastal region centered on the Greek city of Massilia (Marseilles).

Now population explosions, before the rise of modern medicine, had one absolutely fundamental requirement: an increased food supply. This may result, of course, from improved techniques—as in the case of the Indo-Europeans of the Danube—but I know of no comparable technical advance in early classical Greece. The other most likely cause is an improved climate. (It is worth noting that the Myceneans, while predatory, were not conspicuously expansionist, which would suggest that their food resources were inadequate to sustain a rapid population increase.)

The intensive agricultural exploitation of Greece, however, severely damaged the land, and thereby, of course, the "effective" climate. The process was described some twenty-four centuries ago by, of all people, Plato. The passage, from his *Critias,* is worth quoting for its extraordinarily modern understanding of the problem.

> The soil which kept breaking away from the highlands . . . keeps continually sliding away and disappearing into the sea. . . . What now remains, compared with what existed [earlier], is like the skeleton of a sick man, all the fat and soft earth having wasted away and only the bare framework of the land being left. . . . What are now mountains were lofty, soil-clad hills; the stony plains of the present day were full of rich soil, the mountains were heavily wooded—a fact of which there are still visible traces. There are mountains in Attica which can now support nothing but bees,* but which were clothed, not so very long ago, with fine trees suitable for roofing the largest buildings—and roofs hewn from the timber are still in existence. . . . The country produced boundless pasturage for cattle.
>
> *The annual supply of rainfall was not lost, as it is at present, through being allowed to flow over the denuded*

* I.e., are grassland or scrub.

> *surface into the sea, but was received by the country, in all its abundance, into her bosom, where she stored it in her impervious clay and so was able to discharge the drainage of the heights into the hollows in the form of springs and rivers with an abundant volume and a wide territorial distribution.* The shrines that survive to the present day on the sites of extinct water supplies are evidence for the correctness of my present hypothesis. [*My emphasis*—R.C.]

A modern climatologist could hardly improve on the accuracy of this account—and could almost certainly not equal its eloquence.

On the face of it, the deterioration in the soil and climate of Greece should have led to a second disintegration of Greek civilization. That it did not was due to the adaptability of Greek farmers, aided in some cases by the shrewdness of Greek politicians. The vine and especially the olive can yield a profitable crop on plains which are too stony, or hills that are too precipitous, for raising grain. Indeed it has been said of the grape vine that the worse the soil, the better the wine. The shift from essentially subsistence farming to commercial agriculture, specializing in the production of wine and oil, was hampered by the years that must elapse between planting a vineyard or olive grove and obtaining a crop from it (the olive tree may not bear until its fifteenth year), a period in which the farmer will have little or no income. To ease this transition, the Athenian political leader Peisistratus inaugurated in 560 B.C. what may well have been the first system of "farm subsidies," low interest loans to enable farmers to plant the slow-yielding crops so valuable to farmer and merchant alike. The Greeks could then trade wine and oil, as well as various sorts of luxury goods, for the grain and timber their land no longer yielded in abundance. One of their main sources for these products were the northern and western coasts of the Black Sea, and here, too, it seems likely that climate provided some assistance.

The cooler and moister climate known to have existed in

Europe around 500 B.C. and thereafter extended well into the eastern portion of the continent. Among other evidence of this is the fact that the boundary between forest and steppe grassland apparently moved well to the east. In 2000 B.C., the grassland extended as far west as parts of Romania and Hungary; by 500 B.C. forested or partly forested land had pushed almost to the Don River, some 1,000 miles eastward.

The moister climate along the Black Sea probably made little difference to grain production most of the time; even in today's dryer conditions, the region is part of the famous Ukrainian "black soil" wheat belt. But it is a fact that a greater quantity of rainfall means also more reliable rainfall from year to year. Thus while the wheat harvests were probably no better in an average year, the moister climate ensured that there would be fewer years of drought. Providentially, even as the effective climate of Greece was deteriorating, thanks to erosion and accelerated run-off, the improving climate to the northeast was providing her with a reliable source of bread.

It is also possible that the eastward shift of the forest benefitted the Black Sea coast (and therefore its Greek colonies and their customers) in another way: by cushioning the impact of the nomadic steppe peoples. These fierce warriors were a perennial plague to the Near East and eastern Europe, but during the period of Greek flowering the fiercest tribes seem to have kept well to the east, where the grassland still rolled to the horizon unbroken by forest. Thus their major impact during this time was not in Europe but in Iran, Mesopotamia, and Asia Minor, where they proved on occasion a considerable military threat to the Persian Empire. The Greek triumph over the Persians, which made possible much of what we think of as Greek civilization, may have occurred partly because the moister climate in the west shunted the nomads toward Persia rather than Greece.

The erosion-impelled shift from subsistence agriculture to commercial wine-and-oil farming had one more important consequence. In the Mediterranean world, as in most other places before and since, the subsistence-farming peasant was only marginally involved in civilization. So long as he could produce most of what he consumed, he preferred to ignore the outside world as much as possible, being concerned chiefly with preserving the maximum proportion of his goods from the tax collector, and sinking ever deeper into what Marx, in a famous phrase, called "the idiocy of rural life." A commercial farmer, however, is tied to the city by the need to sell and buy there, and will inevitably concern himself more with what happens beyond his own narrow acres. Thus in Greece, as opposed to the Near East, says one historian, "the rural population was vitally and actively concerned in the affairs of the state. Herein lay much of the secret of Greek military effectiveness against Oriental [i.e., Persian] foes. . . ." To the same source, surely, we can partly trace the growing Greek notion that politics and civic affairs were the business of the average citizen, not merely of kings and nobles. It is not being fanciful, perhaps, to suggest that this process may have advanced further and faster in Attica precisely because, being dryer than the rest of Greece, it was forced to rely more heavily on commercial farming—and, of course, on trade. (Wealthy Greek shipowners are by no means a modern invention.) It would be a gross oversimplification to trace the burgeoning of Athenian democracy purely and simply to the effects of land erosion aggravated by the accident of a dryer local climate. Yet I feel there is a connection.

The decline of Athens, and with it the Golden Age of Greek culture, had nothing to do with climate. Instead there is a rather hair-raising similarity to the decline of Mycenae—the same emphasis on fortifications (the famous Athenian "Long Walls") and on the defense establishment (the "wooden walls" of the Athenian fleet). There was also—and here we know of no Mycenean parallel—the matter of the Delian League: a sort of North Aegean Treaty Organization which, beginning as a mutual defense pact against Persian aggression, became, as the Persian menace receded, an apparatus for Athenian aggrandizement. As Athens became an ever richer and more powerful democracy, she began to believe that the gods must be on her side, permanently. Imprudent military adventures then dissipated much of the riches, and with them went much of the democracy—as shown, for example, by the condemnation of Socrates by a Committee on Un-Athenian Activities.

It is tempting to relate the rise of Rome, which occurred only a little later than the blossoming of classical Greece, to the same favorable turn of climate, but the analogy breaks down. Greece prospered and expanded as a whole—but the Romans were only one of half a dozen peoples in the Italian peninsula, and the climate there, good or bad, certainly can't explain how they alone managed to dominate first their Italian neighbors and eventually the entire Mediterranean. But if climate has no apparent connection with the rise of Rome, it may have had something to do with the checking of Roman expansion and, later, with the Roman decline.

South of the Mediterranean ("our sea," the Romans called it) the power of the Caesars stopped at the Sahara. On the north, there may have been another barrier, less obvious than the desert but equally conclusive. If you examine the northern boundary of the Roman Empire and then look at a modern climatological map of Europe, you will find a rather close relationship between the boundary and winter temperatures—in particular, the climatic isotherm marking the line beyond which the average January temperature today dips below freezing. In London (Londinium) and Paris (Lutetia) winters are actually warmer than in, say, Milan (Mediolanum), though summers are, of course, much cooler. But as you cross the Rhine, moving northeast, the winters grow more severe. The picture is much the same in eastern Europe. From time to time, indeed, Roman power pushed north of the Danube, which roughly demarcates the isotherm in question, but trans-Danubian Dacia was always a backward frontier province, with no settlements at all comparable to those in more southerly regions.

I have no idea whether this climatic boundary played any decisive role in halting Rome's northward expansion. Certainly the Caesars were considerably influenced by the defensible frontiers provided by the Rhine and Danube, which happen to lie close to the 32° F. isotherm. But consider also that Germania (it was partly inhabited by Gauls then) was distinctly colder and moister than Gaul, or than present-day Germany. It would have been densely forested, except where it was swampy or boggy. (Dacia, interestingly enough, lay in dryer territory, which today is partly grassland and even in the Roman era may well have been less densely wooded than Germany.)

Now the strength of the Roman army, vis-à-vis the barbarians it confronted, never lay in numbers but in organization, discipline, and maneuverability, and the German swamps and forests must have been hellish country to maneuver over. We know, in fact, that on one occasion no less than three Roman legions thrusting into Germany were cut off and almost annihilated in a forest ambush. It is perhaps not overimaginative to conclude that, in this indirect fashion, climate did set bounds to Rome's military expansion. And when we add the fact that northern Europe, impoverished by cold and damp, had little that the Romans or anyone else wanted, the case for climate becomes even more plausible.

The decline and fall of Rome is a subject on which the numbers of theories is exceeded only by those on the causes of the Glacial Epoch. I am not about to promulgate another one. But equally, I am not about to pass over one of *the* major events in world history without at least trying to drag climate in.

One cause cited by some historians is a decline in agricultural productivity in Italy. We know that Rome imported quantities of grain from other parts of the Mediterranean, notably Egypt, but in part this must simply have reflected the rise in the population of Rome itself, swollen by imperial bureaucrats, luxury tradesmen, and other immigrants propelled by more complex social causes. Agricultural productivity in classical times is hard to document because there are so few documents. (To take another area of history, it has been calculated that Roman army clerks must have made out a minimum of some quarter-million pay vouchers a *year,* over a period of at least 300 years. From these innumerable bits of bureaucratic paper, there have come down to us exactly six.)

Fortunately, we now have more direct methods for assaying the state of Italian agriculture. For instance, a mud core from a now-drained lake near Rome shows that the rate at which mud accumulated jumped sharply during the second century B.C., when Roman power was first showing its muscle. From this we can calculate that the amount of soil erosion in the lake's watershed must have risen from perhaps an inch and a half per thousand years to the same amount in a century. In places where the slope is greater (and a great deal of Italy ranges from hilly to mountainous) the loss has been estimated at up to four inches a century. Thus during the four centuries of Roman power (roughly, 200 B.C. to 200 A.D.), many parts of Italy would have lost from six to sixteen inches of top soil. The direct effect on fertility would have been serious, and when we add in the presumable decline in the effective rainfall, through increased runoff (the same process that Plato described in Greece), it is clear that Italian agriculture must indeed have fallen on bad times.

Whether this development was important enough to seriously weaken the Empire is a problem I will let the historians wrangle over. There is another factor, though, which everybody agrees had a lot to do with the Decline and Fall: the barbarians, who first weakened the empire by forcing it to engage in perennial frontier warfare and at last conquered large portions of it. And here the case for climate is persuasive if not positive.

Exactly when the climate of northern Europe began to improve is still less than clear. The process may have begun as early as the beginning of the Christian era; H. H. Lamb suspects that by the end of the third century the climate may, albeit temporarily, have become even a little warmer than the present.* Warmer and dryer times would have made far more bountiful northern harvests, as well as the expansion of cultivation into formerly swampy areas. Both imply an increasing population. This development was doubtless aided by the use of iron tools, already widespread in northern Europe—in particular, the iron ax, which facilitated clearing of the still-dense forests, and the iron plow, which made for easier and more effective cultivation of the sticky and often waterlogged lowland clay soils that are widespread in the region.

During the first century, then, we see the beginning of the gradual expansion of the Teutonic peoples of northern Europe. By the second century, they have overcome the remaining Gallic tribes beyond the Rhine; by the middle of the third century the Teutonic trickle is becoming a flood into central and eastern Europe, and in another fifty years the Goths have pushed as far east as the Don.

Inevitably there was a backlash, with climate as usual playing a useful role in the wings. For the same increasing warmth and dryness that helped set off the German expansion in northern Europe must also have thinned out the woodlands of the south Russian and Hungarian steppes. In A.D. 372 a confederation of fierce grassland nomads, whom we know as the Huns, attacked the eastern Goths and in three years had pushed them west of the Danube, thereby wiping out a century of Teutonic expansion. By 450, the Hun Empire (by now doubtless including many Teutonic vassal tribes) controlled most of Europe between the Baltic and the crumbling Roman frontier.

Their expansion blocked on the east, by a people much more at home in steppe climate and vegetation, the Teutons turned west and south. The Roman Empire in the west col-

* Such a development might well imply a drop in rainfall to the south, further aggravating the decline of agriculture around the Mediterranean.

lapsed; during the next century, the history of western Europe is the marching and countermarching of German armies: Ostrogoths, Visigoths, Vandals, Burgundians, Franks, and Lombards. The eastern half of the Empire, having barely beaten off the Huns (whose confederation collapsed with the death of Attila) was almost immediately faced with new incursions from the Slavs. These people, living farther east than the Germans and therefore in a colder climate, had been slower to benefit from any climatic improvement, but by the end of the fifth century their expansion had carried them to the Danube on the south and to the base of the Danish peninsula on the west.

I must conclude by emphasizing that climate was far from being the whole story of the collapse of the Roman Empire which, even in its palmiest days, was by any definition a sick society. A feverish material prosperity was coupled with a profound political and personal corruption. Representative institutions declined, as successive comanders-in-chief set at nought the opinions of the Senate. Alienated intellectuals succumbed to philosophical pessimism; alienated conquered peoples and slum populations burst forth in desperate revolts whose suppression decimated the young manhood of Italy.

Another symptom, perhaps, was the appearance of bearded, dirty, sandal-wearing groups who organized communes to share their bread and wine. They made subversive statements about the prospects for salvation of the rich, urged nonresistance to violence and aggression ("turning the other cheek" was the way they put it) and even suggested that in gaining the whole world one might lose one's own soul.

With all these things working against it, a society probably doesn't need much of a push from climate to go into a tailspin.

34. THE STEPPE WARRIORS

A Climatic Nuisance

Twice already I have mentioned the role of the nomadic warriors of the steppes. Just who were these people?

Broadly speaking, they were those who, for nearly three thousand years—from before 1500 B.C. to nearly 1500 A.D.—inhabited and dominated the dryer regions of northern Eurasia that stretch from the Danube on the west, around the northern Black Sea coast, and on north of the Caucasus and Caspian deep into Asia.

The Eurasian steppe belt is not unlike the grasslands of our own Great Plains. Instead of stretching north and south, however, it runs east and west, thanks to the rain shadow of the Caucasian and Himalayan systems which create a series of climatic belts to the north. These, enumerated very schematically, run from the cold dry mountain and plateau climate, through the milder, moister foothills, into lowland desert; farther north, with somewhat more rainfall and considerably less evaporation, we get steppe grassland, then wooded steppe, and at last the full forest. The climate throughout is, of course, Continental, the most Continental in the world, with warm to hot summers (depending on latitude) and cold to frigid winters.

Attempts to disentangle the origins of the nomads are confused by the fact that many of the peoples customarily so called were not nomads at all. The roots of nomadism lie in a number of sedentary, mixed-farming cultures which filtered into the steppe from the Danube, probably also from the Mesopotamian hill country via Iraq, and possibly from Asia Minor via the Caucasian passes.

The essential difference between a barbarian farmer and a barbarian nomad is, of course, mobility, involving both food supply and household goods. A mobile food supply (I am not talking, remember, about hunting-gathering savagery) means pastoralism: reliance on the meat, milk, butter, cheese, and yogurt furnished by the grass-cropping tribal herds. Mobile homes mean essentially what they still mean: a wheeled vehicle pulled by some power source.

Pastoralism is hard to establish conclusively at any early date. The bones of sheep, goats, and cattle prove nothing; just about all early farming communities kept them. The bones of pigs are a species of negative evidence, the animals cannot be herded over long distances. The absence of pig bones may not mean mobility, however, but merely that a particular sedentary community did not keep pigs. (Most people in today's Middle East do not.) Even less conclusive is the absence of traces of grain cultivation, for these may be due simply to gaps in the archeological record.

The need for wheeled vehicles helps us narrow things down. These, to our certain knowledge, were invented in Mesopotamia around 3500 B.C., or possibly a bit earlier. On the steppes, the first traces of them do not show up until nearly 2000 B.C., in southern Russia and Hungary; a clay model of a two-wheeled cart, turned up in the north Caucasus, has been dated a couple of centuries later. Significantly, it looks not unlike a cut-down version of an American prairie schooner.

At this point the story becomes embroiled in a dispute already touched on: the identity of the Indo-Europeans. A remarkably wide-spread collection of Indo-European languages have similar words for wheel and axle ("wheel" itself is akin to the Greek *kyklos* and the Sanskrit *cakra*), which would argue that the "original" Indo-Europeans knew about these things. This would immediately rule out the Danubians; there is no evidence that they knew about the wheel and every reason to believe that at the time they began their expansion (soon after 4500 B.C.) the device had not yet been invented. It would rule in the Battle-Ax folk, who originated on the steppes and whose migration into western and southern Europe began soon after 2500 B.C., at a time when wheels were very likely in use on the grasslands. But this neglects a third possibility; that the Danubians (for reasons stated in Chapter 31) *were* the original Indo-Europeans, but that their descendants, spread across Europe, learned about wheels and axles from their Indo-European cousins, the Battle-Ax invaders. When exotic institutions and technologies are introduced to a new people, they carry their special vocabularies with them much more often than not. Thus the skills of Arab alchemists taught the English about *al qaliy* and *al kuhul,* even as the stern French-speaking Norman barons had earlier taught their sullen Saxon churls the virtues of *rente, prisun* and *gibbet.**

While I doubt that the first Indo-Europeans were pastoral steppe dwellers, there is very little doubt indeed that the first steppe pastoralists were Indo-Europeans. Until the beginning of the Christian era, every one of these people of whom we have any record appears (often through linguistic analysis of their chieftains' names in historical records) to have spoken languages of the group.

Wheeled vehicles gave the Indo-European pastoralists mobility, but not, at the beginning, very much. The heavy, ox-drawn carts with solid wheels which they had taken over from Mesopotamia (almost identical carts are still used in such places as India and Turkey) travel at no more than a slow walking pace. A series of inventions and discoveries, however, converted these clumsy vehicles into a species of Bronze Age rapid transit—and something more.

The steppe dwellers may have domesticated horses as early

* Again I find Calvert Watkins in support. "Most [European] terms relating to wheeled vehicles," he notes, "seem to be metaphors formed from already existing words"—i.e., not primary elements of the original Indo-European vocabulary. Thus "the NAVE or hub of the wheel . . . is the same word as NAVEL." WHEEL itself he derives from a verbal root *kwel-,* meaning simply "to revolve or go around."

as 3000 B.C., but at first only as a food animal. We do not know who first conceived the notion—and the apparatus—for harnessing these fast-stepping creatures to a cart; a famous Sumerian frieze dating from shortly after 3000 B.C. shows horselike creatures pulling a four-wheeled cart, but these were probably asses of some type. We do know that the horse's utility as a draft animal was hampered by its inefficiency at pulling heavy weights. The first domesticated horses were by no means as strong as an ox (the heavy draft horse is a product of medieval breeding), and the animal is, moreover, not anatomically constructed to bear the yoke through which an ox transmits power to the load. Prior to the medieval invention of the horsecollar, the only way of harnessing a horse to a cart was by means of a strap around its lower neck, which, when the animal pulled too vigorously, half strangled it.

A partial solution came, however, with a new design for cartwheels, in which the massive slabs of wood were replaced by a circular rim joined to the axle by tough but light spokes. Again we do not know where the invention was made, but the steppe seems a reasonable candidate, if only because the chronic shortage of wood would have supplied an additional motive for lightening cart construction. The wood shortage may also have contributed to another invention: the "compound" bow, a weapon made of wood reinforced with horn and sinew, which because of its greater stiffness can be made considerably shorter than the all-wood type, and therefore usable in a confined space.

Some time before 1700 B.C., all these inventions had coalesced into a spoked, two-wheeled cart which could be drawn at high speed by a pair of horses and manned by a driver and a bowman. The ox-cart had evolved into the war chariot. These first armored divisions swept over the Near East and, as the now-decisive weapon in warfare, were quickly adopted by the local defense establishments. It was the chariotry of the Hyksos which first broke into Egyptian isolation—and, incidentally, at last convinced the conservative Nile dwellers that the wheel might really be a useful invention. The same steppe-born weapon also played a part in Mycenean warfare, as we know from the *Iliad,* but is portrayed there as little more than a taxicab for carrying the heroes into battle (archery is depicted as not really sporting). We do not know, however, whether Homer was telling it like it was or merely like men remembered it 400 years after the sack of Troy.

The general adoption of chariotry established a new balance of power between Mediterranean and Mideastern civilization and the barbarian steppe. But not for very long. Over the next several centuries—possibly thanks to the gradual breeding of heavier, stronger horses—the steppe peoples evolved another "ultimate" weapon: cavalry. Mounted bowmen from the steppes helped overthrow the Assyrian Empire, helped their own Indo-European cousins set up the Persian Empire, the largest the world had yet seen, and served as a perennial annoyance to that empire until it was absorbed into the still greater state of Alexander.

At the other end of Eurasia, the repercussions were equally notable. Eastward flowed the tide of the steppe warriors, impinging at last on the state of Ch'in in northwest China. They did not succeed in overwhelming it, but the Ch'in rulers were forced, in self-defense, to adopt the same military tactics—which, by no means incidentally, enabled them to set up shop as the first rulers of a unified China (ultimately known to some as the land of Ch'in).

Eventually the civilized states once more reestablished the balance of power. In barbarian territory, however, the steppe warriors' impact was more drastic. Agriculturists and peaceful pastoralists alike, unable to stand against the swift-striking steppe cavalry, were forced to become their tributaries—or in a few cases, perhaps, develop cavalry of their own. Either way, the relatively settled and probably relatively pacific barbarian societies became dominated by a warrior aristocracy, even in areas (such as southern Russia and Hungary) well outside the steppe proper. These military rulers took to themselves most of the profits of the region's burgeoning trade in grain, hides, dried fish, furs, and, no doubt, slaves. Their profits were taken in luxury goods from the Mediterranean—wine, fine pottery, and especially gold in the form of richly ornamented vessels, massive bracelets and plaques, and the like. Judging from the quantities of these objects found in the so-called royal tombs of the Scythians, Sarmatians, and other warrior steppe peoples (which, we can be sure, are only a small sample of the total, most of which must have been surreptitiously looted and melted down), it was a very rewarding business.

Yet the very process of adopting a relatively settled life somewhat tamed these rude warriors. The vices of civilization are pleasant, but tend to erode the military virtues—and a good thing, too. (Some steppe tribes developed vices of their own; Herodotus tells us with some puzzlement that after important funerals it was the habit of tribal notables to gather in a felt tent and inhale the smoke from hemp seeds strewn over hot stones. Recently archeologists have turned up actual evidence of these funereal pot parties.) Deep in Central Asia the less civilized tribes continued to hack away at one another, forming and reforming kingdoms and coalitions of which we know little but the names, but on the west and south peaceful coexistence based on trade was the rule rather than the exception.

Around the beginning of the Christian era, the steppe way of life had spread to the arid, bitterly cold region now known as Mongolia, where it drew in a quite different group of people. Physically they were Mongoloid in type, unlike the earlier steppe people (at least some of whom, according to Chinese accounts, had green eyes and red hair). Linguistically, they were part of the Altaic group of which the present language of Outer Mongolia is a representative.

Just what set the Altaic people to expanding is a real puzzle. An old theory has it that they were thrust into motion by the drying up of their pasture-lands, but even apart from the fact that there is (so far as I know) no evidence of such a climatic change, it seems to me improbable on other grounds. As noted, waves of conquest are most often generated by population explosions, and these result not from a deterioration of climate, but from an improvement. Now it is a fact that during the centuries the Altaic people were pushing westward from Mongolia, the climate of the northern lands was moving toward the Little Climatic Optimum. What effect this had on Mongolia we do not know; a warmer climate would presumably have made for less effective rainfall, but on the other hand, it might also have allowed more moist air to penetrate northward from the South China Sea, meaning more rainfall. Even the latter hypothesis does not explain how these rather sparse and scattered tribes, relatively new to the steppe way of life, could so rapidly come to dominate all of Central Asia and, on occasion, large parts of Europe and the Middle East.

But dominate it they did. Beginning with the Huns and continuing with the Khazars, Patzinaks, Turks, and Mongols (to name merely the main groups), wave after wave of Altaic peoples burst out of Central Asia, killing, pillaging, raping, and burning as they came. The high point of steppe conquest came in the thirteenth century, when the Mongol chieftain Chinghis Khan put together a warrior coalition which briefly dominated all of Central Asia, nearly all of Russia, much of eastern Europe and the Near East—and all of China. The final repercussion of this cataclysm occurred some three centuries later, when a motley horde of warriors pushed from Iran into northern India. They were by this time a cultural mixture of Turkish and Persian, but they still called themselves Mongols (in Persian, *mughul*) and their Indian Empire, which lasted until the beginning of the eighteenth century, was known as the Mogul Empire.

The waves of Altaic steppe warriors had waxed as the Little Climatic Optimum advanced, and waned as it retreated. Coincidence? Quite possibly. Certainly their subsequent decline had a lot less to do with climate than with the invention of gunpowder, a substance whose manufacture required a fairly complex and sedentary civilization (saltpeter, one of its chief ingredients, was obtained chiefly from barnyards). Firearms gave civilization a decisive advantage over a people with neither barns nor barnyards, and the history of nomadic Eurasia during the past five centuries is essentially a matter of its gradual takeover by the civilizations of China and, especially, Russia.

Yet there was one climatic aspect to the steppe way of life that bears emphasis: its stringent limitations on population. All the steppe peoples were pastoral to some extent, and the Altaic group almost entirely so. Economically, pastoralism simply cannot support anything like the population that can live by agriculture. Considered in terms of calories produced per acre, plant foods (such as grain) outweigh animal foods (which are, of course, ultimately derived from plants) by something like 10 to 1. To be sure, large areas of the steppe were too dry to have supported agriculture in any case, but that very factor reduced what modern ranchers call its "carrying capacity" for sheep, goats, horses—and people.

Thus from a population standpoint, the steppe pastoralists were not much better off than the Ice Age hunters of western Europe, who also had lived chiefly on meat. The steppe grasslands, even as they provided a base for the warrior populations, ensured that those populations would be, and remain, sparse. They could not support any elaborate state organization, nor, for that matter, was there any way in which such an organization could have persisted. It is hard to exercise long-term control over people who can literally pull up stakes and move on to greener pastures.

Steppe conquests did not depend on numbers, but on speed; the impressions of "hordes" that have come down to us reflect more the rapidity with which the warrior tribesmen could concentrate their forces than the size of the forces. And the really large-scale incursions were episodic and brief, depending on the happenstance of some charismatic leader who could temporarily unify the scattered and feuding tribes into a truly formidable military force. Attila was such a one, and so was Chinghis Khan—and in neither case did the coalitions they built up survive their deaths. For similar reasons, the steppe dwellers were almost invariably absorbed by the agricultural people they conquered. The Iranian steppe dwellers conquered Iran when that region was still largely semidesert; thanks to qanat irrigation, they later managed to build up a sizable population. Subsequently, Iran was conquered by Arabs, Turks, and Mongols (and invaded by a score of other "hordes"), but the Iranian languages are still dominant there, as they are in the relatively moist foothill country of Soviet Tadjikstan to the northeast.

The Altaic peoples of the steppe, which for nearly a thousand years terrorized their neighbors, survive linguistically almost nowhere but on the steppe itself, in areas such as Soviet Uzbekistan, Kazakhstan, Turkmenistan, Kirghizia, and in western China—as well as in Mongolia, of course. The chief exception is Turkish Asia Minor which, thanks to its dry climate and rugged terrain was never very heavily populated to begin with and was, we can believe, partially depopulated by the long, seesaw struggle between Turks and Byzantines. At that, many coastal regions of Turkey, with a less forbidding (Mediterranean) climate and therefore a denser population, remained Greek-speaking until the 1920s, when the Turks exterminated or expelled most of the infidels.

Thus the Steppe climate, even as it set the stage for the warrior nomads placed stringent limitations on the number of actors. For nearly 3,000 years the steppe warriors were a nuisance to civilization—frequently a bloody nuisance—but in the long run, no more than that.*

* As a final footnote on races, I might mention that Africa also suffered from the activities of (black-skinned) nomadic warriors, who were quite as destructive as their white- and yellow-skinned Eurasian counterparts.

35. THE VIKINGS

A Climatic Explosion

There has been an awful lot of nonsense written about the Vikings. Thanks to energetic propaganda by their Scandinavian descendants, they are commonly portrayed as virile warriors, intrepid voyagers, and hardy explorers. The truth is that their wars were frequently massacres of civilians without distinction as to age or sex, their voyages were chiefly quests for loot, while many of their discoveries were achieved by characters one jump ahead of what passed in those days for the cops.

For the full flavor of Viking life, there is no better source than the Norse and Icelandic sagas, carefully read; all you have to do is keep a careful body-count. Thus the tale of an early voyage to Greenland ticks off no less than twelve murders in six paragraphs; Eric the Red's father emigrates from Norway to Iceland "because he had committed manslaughters" (plural), and his son later abandons Iceland for Greenland because of equally pressing health reasons. The sagas are sown with individuals whose names, rendered into colloquial American, come out as Dirty Eyolf, Shoot-Out Hrafn (two of Eric's victims), Cockeye Ulf, Lowlife Steinholf and Loudmouth Thord—not to mention Bloody-Ax Eric, a Viking king of Norway.* A gathering of these worthies would have borne a marked resemblance to a Cosa Nostra convention at Apalachin or Las Vegas, the main difference being

* While on the subject of names, I cannot refrain from mentioning a lady with one of the most evocative in history. She was Eric the Red's mother-in-law, and she hight Ship-Bosom Thorbjorg. Picture the prow of a Viking ship. . . .

that the guests arrived in longships rather than Cadillacs. And Viking life as a whole comes across as not unlike the Chicago of Scarface Al Capone and the O'Banion boys.

Had the Vikings operated out of Sicily or Malaya, they would have come down to us as what they were: a crew of piratical cutthroats. To their good fortune, however, they had fair skins, blue eyes and blond or reddish hair—to north Europeans, the perennial earmarks of the Good Guys. (In fairness to the Scandinavians, I should point out that the Vikings were a minority among them—according to some accounts, the descendants of the Battle-Ax aristocracy who had conquered Scandinavia, along with most of western Europe, some 2,500 years earlier. Most Scandinavians were peaceful farmers and fishermen; the proportion of Vikings among them was probably not much greater than the proportion of gangsters and political crooks in today's Chicago, of muggers in Harlem, Mafiosi among Italo-Americans—or bankers among American WASPs.)

In a book on climate, however, the Vikings are bound to occupy a place of some honor because their explosive story is perhaps the only unambiguous case of a major historical event set off by climatic change. They were the cutting edge—in every sense—of a great wave of Scandinavian expansion which was turned on and, to some extent, turned off by the arrival and departure of the Little Climatic Optimum.

When we talk of Scandinavia, we mean here Denmark, Norway, and southern Sweden—which is to say, the areas in which maritime influences shape the climate. (Northern Sweden, cut off from oceanic winds by the mountainous spine of the peninsula, was almost certainly inhabited by reindeer-hunting Lapps, living a life little different from that of Ice Age Europe.) The ocean's benign influence on Scandinavia is intensified by the presence offshore of a branch of the warm Gulf Stream (called, in that part of the world, the North Atlantic Drift). Even in the dead of winter, ocean temperatures off the Arctic North Cape of Norway average better than 35° F.—not much colder than the waters off Boston in the same season.

Scandinavian summers are cool, of course, and winters are cold—but almost never bitterly cold. Thus in Tromsoe, Norway, well north of the Arctic Circle, the all-time low is a mere −1°; in Continental Omaha, nearly 1,700 miles closer to the Equator, the record is −32°! By eliminating the most extreme plummetings of the thermometer, the ocean makes for a longer frost-free Scandinavian growing season than one would expect from a mere inspection of temperature averages. But plant growth depends not merely on the length of the growing season, but also on temperatures during it, and here Scandinavia comes off poorly: Oslo's warmest month averages only 64° (Omaha's average is a torrid 78°). Thus if the inhabitants of northern Germany had a thin time of it in the chilly centuries just preceding the Christian era, we can believe that for the Scandinavians, times were even thinner. Gradually, however, things improved, and by the ninth century or earlier, all of northern Europe (in fact, all other northern regions for which we have evidence) had become warmer than it now is.

We know this from many lines of evidence. The Domesday Book, in which William the Conqueror's Norman clerks catalogued the riches of the land he and his barons had grabbed, lists no less than thirty-eight vineyards; English wines, other sources tell us, were almost as good as those of France. Even allowing for boosterism, this means that English summers must have been distinctly warmer and dryer than they are today. (In southern England at present, grapes ripen only in a specially favorable microclimate—in a greenhouse, or up against a south-facing wall.) Other records describe at least occasional winemaking all over northern Germany (today, Rhine wine marks the northern limit) and even in Latvia. It was probably not very good wine, but by modern climatic standards was still remarkable.

More significantly for our Vikings, we know that in Norway, for instance, farming was pushed higher up on the mountains—the altitude of the "crop line" went up 300 to 600 feet. In a country where so much land is mountainous, this made for no small addition to the arable acreage. And the spread of rye and barley fields to higher ground implies even better crops on the warmer, already cultivated lowlands; together, these imply an increase in population —plus a taller and more vigorous population, an important military consideration at a time when much fighting was hand to hand.

For some reason, many Scandinavian historians have simply ignored this climatic factor in their history, seeking out all sorts of other explanations for the Viking outburst, and in the process performing not a few feats of mental gymnastics. Thus Eric Oxenstierna, for example, cites as one important factor improvements in shipbuilding, whereby "about A.D. 600 sizable sailing vessels were being built." He does not explain why, in that case, the first recorded Viking raid (in which the monastery of Lindisfarne, near the English-Scottish border, was sacked and burned) took place nearly two centuries later.

Johannes Brønsted cites overpopulation as a "contributory cause" of the Viking raids, but ascribes it to "the widespread practice of polygamy among the Vikings, who rather prided themselves on the number of sons they could beget; it was common for them to have concubines, mistresses and underwives. . . ." I am quite ready to believe what he says about the Vikings' amatory propensities; sexual exploitation of women fits well with the rest of their obsessive *machismo.* I am not at all ready to believe that even the most wholesale polygamy would explain overpopulation. From a reproductive standpoint, six women impregnated by one man will produce no more children than the same six women impregnated by six men. Indeed, assuming that all six ladies are truly faithful to their one lord, they are likely to produce less; even the most compulsively virile Viking could hardly average better than one woman per night *—meaning that the chances of any given woman being taken during her fertile period would be proportionately reduced. But even if one gives Brønsted the best of this particular argument, he still ignores what every parent knows: in producing children, it's not the initial cost but the upkeep. From a population standpoint, the number of conceptions among the Vikings (or any other contemporary people) is scientifically trivial; what matters is the number of children which could be nourished to maturity. Which brings us back to food supplies—and climate.

Brønsted himself sees the chief cause of Viking activities as commercial, stemming from a shift in trade routes ultimately traceable to the Moslem conquest of North Africa. That a great deal of Scandinavian activity, especially to the southeast, was commercially motivated seems clear; but motive, as detective story fans know, is not means. To build and equip the hundreds of ships needed for either commerce or piracy (the largest Viking aggregation, the "Great Army," numbered some 700 keels), with hand-hewn timbers and handwoven sails, and to man them with the necessary fighting men (the Great Army totaled over 20,000 and they were drawn mostly from Denmark alone) argues a sizable surplus of labor that could be spared during the crop-raising season —which was also, of course, the raiding and trading season. Add to this Brønsted's own statement that "the biggest Viking campaigns in the west were . . . undoubtedly motivated by a colonizing impulse . . ." and the case for population as a main cause, perhaps *the* main cause, of the Viking explosion seems unanswerable. With colonies being set up in Holland, France, England, Ireland, the Scottish Isles, Iceland, much of the Baltic coast, and a sizable belt of Russia between modern Leningrad and Kiev, plus an added allowance for military casualties (fierce as they were, plenty of the

* Bearing in mind that the Vikings were away from home for several months of the year, and at home partook copiously of mead and beer in the evenings.

Vikings were sent to Valhalla), we have to assume either that Scandinavia became depopulated—which it certainly didn't—or that its population was expanding fast enough to take up the slack.

No question but that the Norwegians, Swedes and Danes were all over the place. In 845, Hairy-Pants Ragnar sailed a fleet of 120 ships up the Seine and captured Paris, discouraging further French attacks by sacrificing more than one hundred of his prisoners to Odin; he departed only after King Charles the Bald paid him 7,000 pounds of silver as protection money. A generation later, his three sons seized and colonized eastern England; a generation later still, Rolf the Norseman made himself so obnoxious to Charles the Simple of France that the king granted him and his followers the land that soon became known as Normandy. Other Viking fleets pillaged Lisbon and the Moslem-Spanish towns of Cadiz and Seville; ranging on into the Mediterranean, they set up a base in the Rhone delta, ravaged the French coast, and sacked several Italian towns including Pisa.

To the southeast the Scandinavians, especially the Swedes, were no less active. Sailing up-river from the Baltic and portaging to the headwaters of the great Russian rivers, they established a rich trade with Byzantium, via the Dnieper and Black Sea, and with the Moslem lands further east, via the Volga and Caspian. Tens of thousands of Arab coins have been dug up in Sweden and Denmark.

According to some accounts, it was the Scandinavians who laid the foundations of the Russian state. A monkish chronicle, said to have been compiled in Kiev soon after 1100, has it that because of dissension among the Finns and Slavs along the Baltic, "they went across the [Baltic] sea to the Varangians [Swedes], to the Rus . . . and they said . . . 'Our land is large and fruitful but it lacks order. Come and rule over us.' " Three brothers, headed by one Rurick, heeded the summons and founded settlements near the Baltic, whence they spread south. From other evidence, this process would have begun early in the ninth century. Russian historians, both Czarist and Soviet, violently dispute this account, and doubtless the truth was a good deal more complicated than the simple legend. At any rate we know from plentiful archeological evidence that the "Rus" (the word is Scandinavian and means "oarsmen") were a powerful influence in Russia for generations, at a period when the first Russian states were developing. On whether they were responsible for that development, as on whether they bequeathed their name to Rus-sia, we had best reserve judgment.

The ubiquity of the Vikings is summed up by the career of a Norwegian known as Hard-Boss Harald. He was the half-brother of Olaf, king of Norway, but when he was fifteen the king was slain in a battle against an army of peasants who disliked his tyrannical ways. Harald fled to Russia and thence made his way to Byzantium, where he entered the service of the Empress Zoë (did she, one wonders, prefer blonds?). Becoming commander of her fleet, he fought a naval engagement in 1042 near Naples against a Viking fleet from the west. A few years later he returned to Norway, rich and famous, and made himself king, ruling for twenty years or more in a manner which earned him his sobriquet. His final effort was an invasion of England, which he undertook in 1066 against the king of that land, another Harald. He was defeated and killed at Stamford Bridge, but the English casualties, plus the forced marches they were forced to undertake, probably provided the margin which a Viking descendant, Bastard William, needed to win the battle of Hastings a few days later.

To Americans, of course, the most interesting aspect of the Viking saga is their explorations to the west, to Greenland and the North American continent. By far the best account of this that I know of is *Westviking,* by the Canadian author Farley Mowat, and the story as he tells it is too good to spoil by summarizing. So I will confine myself to the climatic aspects—which, by the way, Mowat understands better than any other writer I have read.

The exploration of the North Atlantic did not, of course,

begin with the Vikings. The earliest such adventurer of whom we have any record was a Greek intellectual named Pytheas, who some time before 300 B.C. made a remarkable voyage from Massilia (Marseilles) to Scotland and thence to a land called Thule. Unfortunately, Pytheas' own account is lost, but from later references to it we know that Thule lay six days' sail from Scotland, and that there were "no nights at midsummer . . . and on the other hand no days at midwinter" (the quotation is from Pliny the Elder). Moreover, one day's sail from Thule lay the "frozen sea," a climatic detail which makes almost certain that the land referred to was Iceland. Lying just below the Arctic Circle, Iceland indeed has no true night in midsummer and no true day in winter.

Subsequent centuries produced a number of ambiguous tales of islands in the North Atlantic, but the next really circumstantial account dates from around 825 A.D., when an Irish monk, Dicuil, reported the experiences of some of his confreres who had visited Thule. They told him "that not only at the summer solstice but in the days around it, during the evening, the setting sun hides itself as if behind a small mound so that there is no darkness for even a little while;" a man can "even pick lice from his robe." The midsummer sun in Iceland behaves in precisely this manner. Dicuil then comments on earlier (lost) accounts which declared "that the sea surrounding Thule is frozen." The monks, he says, found the seas open, but frozen, as Pytheas had earlier reported, a day's sail toward the north. This, too, carries conviction. From much more recent records, we know that Iceland was almost surrounded by winter pack ice during the cold early years of the nineteenth century, and the same would doubtless have been true during the cold early centuries of the Christian era. By the time Dicuil wrote, however, the warming trend would have shoved the winter pack-ice line north to at least its present position—just about a day's sail (i.e., 100 miles) from Iceland—and quite possibly a good deal farther. This suggests that the louse-ridden Irish monks made their visit in an abnormally cold year for that time.

There is a fair amount of evidence that at least a few Irish actually settled Iceland before the Vikings arrived there; according to one saga, they departed because, being Christians, they did not wish to live with the pagan Vikings. Knowing what we do about Viking manners and morals, one wonders whether the Irish did not depart from this world at the same time. It is even possible that the Irish had reached Greenland before the Vikings, but certainly the first positively known settlement there was founded by Eric the Red—he having made Iceland, as it were, too hot to hold him.

Greenland in those days had an annual average temperature estimated at from two to four degrees higher than at present. Four degrees may not sound like much, but it is the difference between London and Edinburgh, or San Francisco and Seattle. Until one looks at a map, moreover, one can easily forget that southern Greenland is well to the south of Iceland; the milder climate of the latter is due not to latitude but to warmer ocean currents (another branch of the Gulf Stream) which bathe its southern shores, and to the absence of the enormous ice-cap that covers nearly all of Greenland. Even in the best years of the Little Climatic Optimum, southeast Greenland, with prevailing winds from the north and northeast off the Arctic Ocean, offered few attractions to settlers. Southwest Greenland, however, even today enjoys a fair amount of southwesterly winds from warmer waters, and must have had more of them then; moreover, a series of deep fjords (gouged by glaciers during the Ice Age) carry at least some Maritime influence as much as fifty miles inland. The result was a land which, while it supported little agriculture, provided good pasturage for sizable flocks of sheep and cattle. These, together with fish, seal, and whale, made up the bulk of the Viking Greenlanders' diet. Sealskins, furs, and walrus ivory were commodities that could be traded east to Norway for grain, iron, and similar items.

A major problem for the Greenlanders was the almost total absence of trees. The need for timber seems to have been a major motive for voyages farther west, to North

America, which appears to have been first sighted by one Bjarni Herjolfson after having been blown off course on a trading voyage from Norway to Greenland. Exactly what he sighted is debated; Mowat believes it was Newfoundland, and his argument sounds convincing. (Mowat, indeed, argues that Eric the Red had even earlier crossed the Davis Strait and explored the coast of Baffin Island, but this, while plausible, is unsupported by any direct evidence.)

About ten years after Bjarni had, with some difficulty, made his way back to Greenland, Eric's son Leif set out to find the land Bjarni had described as "well wooded," full of the very desirable timber. What route Leif followed is uncertain and so is the place he landed. The sagas make clear at any rate that he purchased his ship from Bjarni (who, with his father, had settled in Greenland) and therefore presumably had the benefit of whatever sailing directions the trader could give him. Also clear is that either he or Bjarni (Mowat thinks the latter) saw and described three distinct territories: Helluland ("boulder-land"), which was probably the rocky, mountainous coast of northern Labrador; Markland ("forest-land"), probably southern Labrador; and one other. And it is over the identity of this "other" that much of the debate over the Norse discovery of America has raged.

Interestingly, many of the most useful, and most controversial, clues are climatic. We do, to be sure, have one unambiguous statement which, if we could fully understand it, would give us a pretty close fix on Leif's discovery. "The length of day and night," says the saga, "was more nearly equal than in Greenland or Iceland"—meaning, of course, that the place must have been well to the south of those lands. "Sunset," the saga continues, "came after Eykarstad and sunrise before Dagmalastad on the shortest day of the year." This account of the day-length at the winter solstice would tell us the approximate latitude—except that nobody is wholly certain what "Eykarstad" or "Dagmalastad" are. Clearly they are times, but equally clearly not clock times, since the Norsemen had no clocks. The best guess is that they refer to about 3 P.M. and 9 A.M. respectively, which would put the latitude somewhere between 46° and 52° N, which is to say, between Cape Breton Island and the southern tip of Labrador. On topographical grounds, the latter can be ruled out, which leaves us with either Cape Breton Island or Newfoundland. The latter must receive first consideration, since it is the only place in North America where unquestionable Viking relics have been found: foundations of typical Norse houses and the site of a forge, charcoal from which gives a date around 1000 A.D. (the Indians of that time, of course, knew nothing of iron-work).

Here is where the climatic problem arises. The saga states that in that country "there was never any frost all winter and the grass hardly withered at all." Even more significant, Leif and his friends found wild vines and grapes, for which reason they named the land Vinland. On the face of it, neither feature sounds much like Newfoundland, which today has no grapevines. Some authorities have tried to get around this by claiming that the term *vin* in the sagas is actually a word meaning "pasture" or "fertile land." But the account of the grapes is thoroughly circumstantial, and with one detail that could hardly have been invented. The first man to come upon the grapes was one Tyrkir, a German, and when Leif doubted his report he retorted that "Where I was born there were plenty of vines and grapes." As noted earlier in this chapter, there *were* vinyards in most parts of Germany at that time.

Mowat, after considerable digging, has come up with historical evidence that grapevines existed in parts of Newfoundland much later than Viking times, when the climate was little different from what it now is. This region was the Avalon Peninsula, the extreme southeastern portion of Newfoundland on which its capital, St. John's, lies. Between 1662 and 1670, for example, one James Yonge made four trips to Newfoundland as surgeon to the English fishing fleet. He reported "wild grapes incredible"—and since, as Mowat notes, he had previously traveled in the Mediterranean, including a voyage on a ship in the wine and raisin trade, he must have known what grapes looked like. Grapevines are said to have survived in Newfoundland until amost the end of the eighteenth century, when they were destroyed by a series of bitter winters. If there were grapes on the Avalon Peninsula in the 1600s and 1700s, they would have been even more plentiful in the warmer 900s and 1000s, so that we can easily believe the saga's account of how the expedition filled the ship's dinghy with the fruit. During the Little Climatic Optimum, Mowat observes, grapes were cultivated in Europe at least 270 miles north of their present limit, and the Avalon Peninsula is less than 200 miles north of the present limit of wild grapes in America.

But can it really be Newfoundland where "there was never any frost all winter"? Living several hundred miles to the south, and knowing what even a New York winter can be like, this seemed to me impossible. But apparently not. For one thing, Newfoundland is an island, with the Gulf of St. Lawrence to the west to moderate the frigid Canadian winds. Mowat has checked out the winter climate at a locality charmingly called Tickle Cove where, for a variety of reasons, he places the most probable site of Leif's landing. The cove, though on the northern coast of Newfoundland, lies at the foot of a deep bay which shelters it from the effects of the bitter Labrador current. Across a narrow isthmus to the south are Placentia Bay and the Atlantic, with relatively warm winds off the Gulf Stream. Thus even today, says Mowat, "snow-fall is generally light. . . . Winter temperatures seldom go more than a few degrees below freezing point and during some winters there is very little frost at all." Around 1000 A.D., then, we can safely conclude that there would be little or no frost during *most* winters. Leif and his friends may have shaded the truth just a little, as discoverers of new lands often do, but in the light of present climatological knowledge, their account seems basically accurate, particularly when you remember that, having endured years of frigid Greenland winters, Newfoundland surely must have seemed to them even milder than it undoubtedly was.

Having discovered this climatic Eden, how is it that the Norsemen did not found any permanent settlements? The answer is that they tried to, but the attempt—a few years after Leif's return from Vinland the Good—was aborted, apparently by typical Viking trigger-happiness, or sword-happiness. Those who expect violence generally manage to find it, and the Norse were no exception. The colonizing expedition, led by one Thorfinn Karlsefni, became embroiled with the natives, whom they called "skroelings"—apparently meaning "wretches" (the equivalent of "gooks"). They began with trade (on the usual European terms, exchanging scraps of red cloth for valuable furs) but progressed rapidly to warfare. And while the Vikings had iron swords, the skroelings—Mowat thinks they were Dorset Eskimos—had slings, arrows, and harpoons, weapons quite sufficient for a military standoff so long as their possessors eschewed hand-to-hand combat. In any event, the colony was abandoned, though the Greenlanders may have continued voyaging to Markland for timber and to Baffin Island for furs.

In another couple of centuries, the Greenland settlements themselves were in trouble. Around 1200, the sagas begin to contain references to drift ice as a navigational hazard in the Denmark Strait between Greenland and Iceland, and by 1300 it was bad enough to severely hamper commerce between the westernmost Scandinavian outpost and the homelands to the east; the last recorded voyage from Iceland to Greenland was in 1410. By that time, Mowat believes, "the Greenland climate had become so bad that it was probably impossible for a pastoral people to survive there, at least in any numbers." Grim evidence of the climatic deterioration was turned up in the 1921 excavation of a Norse graveyard in Greenland. The earliest graves were at normal depth, indicating that the ground was unfrozen during the summer, but later ones grow increasingly shallow, marking the beginning of a layer of permafrost, which moved closer and closer to the surface. Permafrost, by interfering with underground drainage, makes for boggy ground and poorer pasturage; moreover, by accelerating surface freezing in autumn and retarding the spring thaw, it shortens the grazing season.

We do not, of course, have any written documents to give us a first-hand picture of the disintegrating Greenland settlements. Some idea of what it must have been like can be gathered from Icelandic documents of the early eighteenth century, not long after the beginning of an even colder era. A 1709 parish record, for example, describes two abandoned farms: "Breðamörk: Half King's ownership, half owned by the farmer. Derelict . . . a little woodland, now surrounded by glacier. . . . Fjall: Owned by the church. Derelict . . . Fourteen years ago had farmhouse and buildings, all now come under the glacier." Other documents tell us that Breðamörk had been abandoned since 1698. "There was some grass visible then, but the glacier has since covered all except the hillock on which the farmhouse . . . stood and that is surrounded by ice so that it is no use even for sheep." Later even the hillock was overrun.

Increasingly, therefore, the Greenland Norse must have been forced to rely on hunting and fishing (and the latter may well have been reduced by colder sea temperatures, which would have sent the great schools of cod farther south); this would necessarily imply a seminomadic existence (hunters must go where the game is). But here they would have soon found themselves in competition with the Eskimos, whose whole culture was geared to that way of life. These were not the Dorsets (who were probably extinct by that time), but the so-called Thule culture, ancestors of today's Greenland Eskimos. Originating probably in Alaska, the Thule people had migrated across northern Canada and by 1200 or so had reached northwest Greenland. There is reason to suppose that the increasingly bad climate then pushed them south toward the Norse settlements, with resulting open conflict. In the year 1540, a Dutch whaling vessel sought shelter from a storm in one of the south-Greenland fjords. On shore they found "a dead man lying face down on the ground. On his head was a hood, well made, and otherwise clothed in frieze cloth and sealskin. Near him lay a sheath knife, much worn and eaten away." This, to the best of our knowledge, was the last of the Greenland Norse.

The Norse settlements in Greenland, a great geographer has written, were "at the verge of the possible." The climatic deterioration of late medieval times thrust them over the verge. The rediscovery of America was made across warmer seas where ice could not penetrate, and where—by no means incidentally—the ever-blowing trades gave assurance of a rapid and certain passage.

Slackening Scandinavian pressure to the South and southeast was only secondarily related to climate. The primary cause, I would guess, was similar to that which had checked so many steppe assaults from the east: political and military changes in the "target" countries. For if commerce and loot provided the motives for the Viking outburst, and an expanding population the means, political disunity in Europe furnished the opportunity. Later, partly in reaction to the Vikings, partly as a result of Viking colonies such as Normandy (which rapidly became a major military power), western Europe became a far tougher target for looters. Russia, thanks to its own gradual unification by Varangian or other princes and to renewed incursions of the steppe peoples to the south, became equally unattractive. The brief moment of Scandinavian greatness had passed. Apart from occasional Swedish outbursts during the seventeenth and eighteenth centuries, it never returned, and for more than a century now those countries have been known more for pacifism than for militarism. Whether many Scandinavians regret the dissipation of their warlike Viking heritage I do not know. Considering the devastation wrought by various like-minded folk elsewhere in Europe over the centuries, I myself cannot see it as any great loss.

There may be, indeed, one major echo of the Viking way

of life that permeates western European and American civilization even today—and not, I would say, altogether for the better. The historian W. H. McNeill, in a remarkable passage, has focused attention on what he calls "the continuity of ethos between piratical raider of the ninth and European merchant of the tenth centuries." As with the Mycenean Greeks and not a few other peoples, the line between raider and trader was thoroughly blurred, with particular Viking gangs shifting from one role to the other as circumstances made expedient. Thus, says McNeill,

> even when trading became prudent in more and more circumstances, bands of itinerant merchants still expected, like their piratical predecessors, to manage their own affairs and look after their own defense. They also tended to treat peasants and [even] lords of the land as strangers—potential victims of sharp practice, if no longer often of the sword. Eventually, such merchant communities began to establish themselves permanently at some convenient location. . . . From such nuclei, the towns of northern Europe took form . . . the central and essential psychological and institutional forms of north European urban life were shaped in the chaotic age when piracy transmuted itself into trade.

In the other great centers of civilization—Byzantium, the Middle East, India, China—McNeill sees a sharply contrasting ethos. There the merchants "were primarily caterers to the tastes of their social superiors: officials, landlords, rulers. They were accustomed to regulation and taxation from above . . ." and therefore failed to develop the "aggressive, ruthless and self-reliant ethos of western European merchants."

What McNeill is saying, I think, is that the merchants of the older, more settled civilizations necessarily accommodated themselves to the institutions and power relationships which surrounded them. In western Europe, by contrast, civilization was still in the process of crystallizing, so that the rising class of merchants could to some extent shape it in their own piratical image.

Certainly when one contemplates some European and Euro-American commercial ventures of later centuries—the depredations of the East India Company, the ravaging of Africa by slavers and gold-seekers, the massive peculations of the American Gilded Age, and the innumerable instances of dollar-, pound-, franc-, and mark-diplomacy—it is hard to ignore their resemblance to Viking raids on an enormous scale. The "continuity of ethos" from pirate to merchant seems, for good or ill, to have lasted long after the Vikings, and the climatic shift that propelled them, had vanished into the past.

36. THE MILLS OF EUROPE

A Climatic Gift

The domination of the modern world by the culture of northwest Europe, while a familiar fact, is also a remarkable one. That the sparse population of one section of one small continent—not even a continent, a mere jagged bulge on the western shoulder of Eurasia—should have colonized most of North America, nearly all of Australia and parts of Africa is by itself extraordinary; that the same peoples should have achieved effective, if temporary, control over the rest of Africa and the teeming masses of Asia is still more extraordinary. How did they do it?

Not through their moral qualities. The Vikings may have been an extreme case among Europeans, but hardly a very extreme one. Yet not, I think, wholly from their immoral qualities either. Giving full weight to the Viking strain in European culture, one still must note that the Indian Ocean and southwest Pacific had their own Arab, Malay, and Chinese "Vikings"; that the Aztecs ravaged the Valley of Mexico before the Spaniards, and that the Arabs, under the first Caliphs, and the Zulus, under their great emperor Shaka, need defer to no European in their aptitude for predatory land-grabbing.

If Europeans and European ideas eventually managed to subdue all these, and many other, rough customers, the reason is not that their intentions were much worse. In numbers, the Europeans started out puny; in power, they became giants—and their power, political and military, was based on technological power: horsepower. They were the first to apply on any important scale the techniques by which man can multiply his own efforts a hundred or a thousand fold, freeing himself from the limitations of his natural strength as the first farmers freed themselves from the limitations of natural wild foods. Europe, and its American offshoot, became great chiefly through the mastery of powered machinery. Nowadays, we tend to think of mechanical power primarily in terms of the electric motor and the diesel and gasoline engines; earlier, there was of course steam. But centuries before Thomas Newcomen, forerunner of James Watt, cobbled together his first crude steam pumping engine, Europe had already embarked on the road to power by harnessing the energy of falling water.

The water wheel was not invented in Europe but probably in the hill country of Mesopotamia (and independently, perhaps, in China). Some time before 100 B.C., an ingenious artisan devised a way of lowering a horizontal paddlewheel into a swift stream and using it to turn a millstone set in a framework above it. Within a few generations, the Greek poet Antipater could sing, "cease from grinding, ye women who toil at the mill; sleep late, even if the crowing cocks announce the dawn. For Demeter has ordered the nymphs to perform the work of your hands, and they, leaping down on the top of the wheel, turn its axle which, with its revolving spokes, turns the heavy concave Nisyrian millstones." The poet's description of the nymphs "leaping down on the top of the wheel" suggests that the horizontal wheel had already been modified into the much more efficient vertical wheel. This would have been connected to the horizontal millstones through an axle "with its revolving spokes," a crude form of gearing.

Yet despite such poetical panegyrics, the new invention took a long time to catch on. The water wheel can, for example, be used to raise irrigation water above bank level, and in some countries is still so employed. In Mesopotamia and Egypt, however, the bulk of irrigation water even today is lifted by man-power, using the primitive shadoof (a scoop on the end of a counterweighted pole), or by an ox- or camel-powered "wheel of pots." The mills of Rome, which ground flour for the bread which, along with circuses, kept the population docile, were ox-powered right through the decline and fall.

A common explanation for this cultural lag is the widespread existence of slavery. Human labor, so one theory goes, was cheap; machinery was expensive, so why bother to use, or even invent, it? This theory—derived, I suspect, from a superficial glance at Marx's "class struggle" theory of history—is persuasive at first glance but becomes more dubious the more it is examined. Slaves may have been cheap at times, and so may oxen, used for the heaviest jobs. But they were not always cheap. And whether initial cost was cheap or dear, upkeep was considerable; human and bovine draft animals both had to be fed and housed, however crudely, and the humans had to be clothed as well, if only in rags. Moreover, you can't shut down a man or an ox; the expenses continue, whether you need their power or not. The water wheel, by contrast, needs next to no upkeep and can be shut off when not needed.

There is, I suggest, a much simpler explanation of why the water wheel found so limited a use in the Mediterranean world: not enough water. The entire North African coast has not a single important river, apart from the mighty Nile itself. The Levant is only slightly better off, with one, the Orontes. Asia Minor, Greece, Italy, and Spain are somewhat more copiously supplied with rivers, but their waters are unreliable. The hot dry Mediterranean summer shrinks most streams to a trickle. Many summers ago, I strolled along the Arno at Florence and, peering down from the 20-foot walls that flank the river, watched the shallow stream wandering amid gravel banks and weed-grown hillocks. Yet the walls themselves, plus the height of the bridges that spanned the river, were testimony that in wetter seasons the Arno could become a deep and powerful torrent. Some years later, indeed, I read of the dreadful week in which the river, overtopping even its high retaining walls, inundated much of Florence, wreaking terrible havoc on treasures of painting, woodcarving, and fresco.* Living along the Arno, where would one put a water mill? On the bank, 20 feet above the water's summer level, or in the river bed, to be swept away with the first floods of autumn?

In fact, it appears that in the Mediterranean zone the water mill was used mainly in hill country, where springs sometimes supplied a sparse but reasonably reliable flow. But the really big rivers, with the potential for producing major quantities of power, were simply too variable to be of much use. Significantly, the largest known Roman waterpower installation—a colossus consisting of eight pairs of wheels, arranged "in cascade" down a hillside—was at

* Florentine floods are by no means the product of the Mediterranean climate alone; an important role has been played by human improvidence—the usual story of deforestation, overgrazing and erosion, a legacy, says one authority, from "the greed of the church, the avarice of the aristocracy, and the ignorance of the people." Credit should also be given to the bureaucratic stupidity of various Italian governments. The destructive effects of land misuse in Tuscany have been pointed out repeatedly over the centuries—among others, by Leonardo—but nothing was ever done about it. The only constructive effect of these floods, it seems, was on George Perkins Marsh, first American Minister to Italy. His observations of flooding in Tuscany helped stimulate him to produce the classic work *Man and Nature, or Physical Geography as Modified by Human Action,* a book which a generation or so later sparked the adoption of the first American conservation laws.

Barbegal, near Arles, France, which lies right at the boundary between the Mediterranean and Stormy zones. I would guess, too, that the stream that turned these mills flows down from the Alps, where elevation (orography) makes for more, and more evenly distributed, rainfall. Another likely spot for water power was the Po Valley. For rather complicated meteorological reasons, this portion of Italy, even though lying south of the Alps, has no dry season, and is climatically much more like Yugoslavia, lying to the east, than to the Riviera, on the west. It is perhaps significant that in Caesar's day the Po Valley was not even considered part of Italy—which, climatically, it is not—but was known as Gaul-this-side-of-the-Alps, with France being Gaul-Across-the-Alps.

Northwest Europe (or northeast Europe, for that matter) is quite different from Mediterranean Europe. Normally there is no such thing as a really dry summer (average July rainfall: Rome, 0.9 inches; London, 2.4 inches); moreover, the effectiveness of summer rains is stepped up by the much lower temperatures (July average: Rome, 76°; London, 64°). Yet an American who examines the climatic norms for northwest Europe may find them not quite what he expected. London, which we think of as perpetually damp, averages only about 24 inches of rain a year, which is a little more than half the figure for Boston or New York. Say what you like about the weather in those cities (and a good deal has been said), perpetual dampness is certainly not a common complaint in either. Of course there must be much more evaporation in, say, Boston than in London. Or must there? When you look at average annual temperature, they are precisely the same: 51°. Something is wrong.

What is wrong is our reliance on averages which, I noted earlier, are misleading in matters of both race and climate. Evaporation is not just a matter of annual averages, but of annual *range*—and especially of summer temperatures. Thus in Boston the July average is 74 degrees; in London, a full 10 degrees cooler, meaning that Boston loses much more summer moisture to evaporation. It is true that January temperatures in Boston average 10 degrees cooler than London, but at those temperatures, evaporation is so low anyway that the difference has little practical effect.

There are two other things about the climate of Northwest Europe that help to make the most of its rather limited rainfall and keep its streams flowing evenly throughout the year. First, since the prevailing winds are off the ocean, relative humidity tends to average higher than along the U.S. east coast, for example, where they blow off the land. And if incoming air is more nearly saturated with moisture, it follows that it is bound to pick up less as it passes over the land (try drying wet clothes on a damp or foggy day). Second, much of the modest northwest European rainfall descends in the form of drizzle or gentle showers rather than cloudbursts, meaning that, other things being equal, it is a good deal more likely to soak into the ground rather than run off rapidly. For example, though I don't have exact figures, I would reckon that London must have only about half as many thunderstorms per year as Boston.

For all these reasons, then, a miller or millwright seeking to go into business in northwestern Europe could count on plentiful and reliable supplies of motive power except in the case of a prolonged and abnormal drought (in which case, of course, there would be little grain to grind anyway). Again we find invaluable information in the Domesday Book (sometimes, since it was compiled for tax purposes, misspelled "doomsday" by Americans all too familiar with April 15). We find that at the time of the Norman conquest, England (excluding Ireland, Wales, and Scotland) had no less than five *thousand* water mills, which meant there could hardly have been a single hamlet without one. And England was a small and rather "underdeveloped" country.

Given so accessible a source of power, it is not surprising that medieval artisans gradually worked out many other things for the water wheel to do. With appropriate attachments it could saw wood, operate trip-hammers to crush ore or forge iron, and eventually turn lathes, run looms, and perform a host of other once-burdensome tasks. Long before Watt stared at his apocryphal tea-kettle, European industry was the most mechanized and the most power driven in the world. Had it not been, it is questionable whether Watt or anyone else would have bothered to tinker with steam engines. Inventions, by and large, arise when a social need for them inspires people to ingenuity, and Europe had already demonstrated that it needed all the power it could get.

The flowing rivers of northwest Europe gave the region one more asset: cheap transport. Land transport was fiendishly expensive even in Roman times (it was estimated that the price of a load of grain doubled every hundred miles); with the decay of the Roman road system, matters got worse. The Mediterranean world had long learned to rely heavily on water transport but, for reasons already cited, that means of moving goods almost invariably stopped at the sea coast or a few miles up the unreliable rivers. To the northwest, however, the Seine and the Scheldt, the Rhine, Elbe, Oder, and Vistula, not to mention the Thames and the Severn (to mention only a few), provided waterways for trade and commerce into the interior. Cheaper ways of moving goods meant better markets for goods, which set up an incentive for producing more goods, by devising more ingenious machinery—which required ever more power.

Water transport and water power are, of course, not the whole story of Europe's rise to world supremacy. China, in particular, had water wheels at least as early as Europe and its water transport, thanks to an elaborate system of canals, was probably more advanced. It is true, to be sure, that rainfall in China is somewhat less even than in Europe; due to the influence of the Siberian Monsoon, most of that country tends to have too little rain in winter and often too much in summer. But I doubt that this really explains the difference. More plausible, I think, is the lack of the aggressive, Viking spirit among the Chinese commercial classes—which, in turn, can surely be laid in part to the extraordinary stability of Chinese society, influenced by its climatic isolation. It appears that even Vikings may have their uses—in the long, long run!

The average American has been estimated to enjoy the equivalent in horsepower of some 1,200 human slaves. This power, and the other kinds of power that flow from it, are admittedly ill-distributed—remember about averages! But it is still nice to know that this priceless heritage from the water mills of Europe is there. If a nation has power, there is always the hope that it may learn to divide and use it better. But nothing divided by a hundred million is still nothing.

37. HERRING AND HITLER

A Climatic Fantasy

I don't quite know what it is about the Germans. Admittedly, I am prejudiced against them—not in the sense that I think all Germans are bad, simply that I expect any German, in the absence of evidence to the contrary, to be unlikable. But precisely where this unlikability resides is rather hard to pin down.

As a nation, and on the record, they don't seem to me notably more bloody-minded than a number of other peoples I could name. Nor are their public (or, so far as I know, their private) morals conspicuously worse than those of the Americans, French, British, Russians, or what have you. If anything, indeed, they seem to me rather less prone to the vice of hypocrisy than most nations. Nearly every people in the world has on occasion massacred large numbers of "foreigners," but almost always in defense of Christianity, Democracy, Socialism, the Law of Allah, World Peace, or simply for the good of the victim's soul. The Germans have actually been known to declare publicly that they kill because they are stronger, or to grab territory, or because of the shape of the victim's nose.

There is an old saying that when one German meets another, they walk in step; when they meet a third German, they appoint him corporal. It is, I think, this compulsion toward order, this hang-up on discipline as a *ding an sich,* that is at the root of my dislike. The quality of rigidity, of "Stand at attention when you speak to your superior!" is something I have noted even among German refugees from Hitler—including one whose health had been wrecked by the attentions of some concentration-camp guards. During the war he had been, at very nearly the cost of his life, one of the minority of actively anti-Nazi Germans, yet he remained a German by temperament—and therefore, to my taste, an unlikable though thoroughly genuine martyr.

If order and discipline are central to the German character (and I think that even most Germans would concede, sometimes proudly, that they are), there must obviously be something in their history that made them that way, just as the predictable Nile and the protection of the desert made the Egyptians easy-going conservatives and other ecological situations engendered Indo-European expansionism and Chinese parochialism. Purely as an exercise in imaginative history, I have tried to trace the German character to some climatic happenings in the fifteenth century.

The most obvious immediate explanation of the German predilection for order is the preponderant influence and prestige of the military. In no modern country except, perhaps, prewar Japan were the professional soldier and his *Weltanschauung* as honored as in the land across the Rhine. And surely a major reason for the prestigious position of militarism is that militarism made Germany a nation. Not much more than a century ago, what we now call Germany was a fragmented collection of kingdoms, principalities, and dukedoms—Westphalia, Hesse, Bavaria, Hannover, and a half dozen others—with militarist Prussia as their leader but not yet their ruler.

This takes us back another step. Why, at a time when Britain, France, and Russia (to say nothing of Holland, Spain, and Portugal) had been for centuries both unified and, at one time or another, even powerful states, was Germany still fragmented and feeble? For the explanation we look back to the seventeenth century, and the Thirty Years' War. Between 1618 and 1648, the struggle between Catholics and Protestants in Germany became the excuse for intervention by almost every other European power. The country was fought over at various times by French, Swedish, Danish, and Spanish armies—not to mention German armies of varied denominations and allegiances. The line-up was not always logical; Catholic France was active on the Protestant side, while Protestant Sweden at times favored the German Catholics. But logic, as so often happens in war, didn't matter. What mattered to Germany was that it was devastated. Casualties, famine, and disease killed off between one-third and one-half of the population—and this appalling fact takes no account of the destruction of wealth. Along with these cataclysmic quantitative changes, what we now call the quality of life dropped precipitously (a pretty good picture of it can be found in Brecht's *Mother Courage*).

The generations required to recover from that war are a major reason for Germany's slow development toward national power. And I can believe that the weakness and fragmentation which were the war's aftermath had a good deal to do with Germany's subsequent obsession with strength and order.

We then look back even further: How was it that the religious wars in Germany (we might even be stylish and say ideological wars) managed to reach such a pitch? Germany was, after all, the heartland of the Reformation, where Luther first nailed his defiance to the Wittenberg church door. She saved others, why could she not save herself? Most countries in Europe, after all, were officially either Catholic or Protestant (though their national unity was seldom obtained without beheadings, tortures, and autos da fé). Why was Germany so conspicuous an exception?

Germany's failure to achieve religious unity was, in fact, a reflection of her inability to achieve political unity. And this, in turn, derived in large measure from the weakness of that great nationalistic, unifying force in European history, the commercial classes and the towns they controlled.

Feudal lords could and did struggle with their equals and presumed superiors for the knightly right to run their own domains in their own way. The town dwellers-burghers-bourgeoisie cared nothing for this nonsense. For the master trader or master artisan, feudal disunity (at times, verging on anarchy) meant too many customs dues on his goods, too many different weights and measures to cope with, too many of the bandits and pirates who inevitably thrive when central governments are weak. The European merchant may have been psychologically descended from pirates, but this did not increase his love for those among his contemporaries who harked back to Viking days. Not surprisingly, it was the towns and townsmen who were the prime support, especially in the all-important finance department, of such anti-feudal absolutists as Henry VII of England and, later, Louis XIV of France.

And here we reach the nub of the problem. For up to this point, we have not said much more than that Germany was weak and disunited because she had been weak and disunited. But the German commercial classes were not weak because they had always been weak. On the contrary, well into the fifteenth century they were among the richest and most powerful bourgeois in Europe. Germany was a center of commerce, manufacture, and invention. German craftsmen, taking over the weight-driven mechanical clock from its Italian inventors, modified it into the spring-driven portable time piece called the Nuremberg Egg, prototype of the pocket watch. It was, of course, a German who invented printing by movable type, and in all probability another who began the manufacture of gunpowder in Europe. German supremacy in mining and metallurgy was reflected in the sixteenth-century publication of *De Re Metallica* by Georg Bauer, called Agricola, the first modern treatise on these subjects; a German-Latin reader for children published in 1568 was cast in the form of verses celebrating the skills of such craftsmen as the printer, papermaker, pewterer, and turner. The flowers of the Renaissance, first nourished by the merchant princes of Italy, sent tendrils northward to blossom in the works of Dürer and Cranach. Yet even as Germany was achieving these commercial and cultural triumphs, the roots of her commercial strength were withering. It began with the herring.

During the thirteenth and fourteenth centuries, the strongest commercial force in northern Europe was the Hanse, an aggregation of German merchant towns on the Baltic. Their strategic location enabled them to control nearly all the trade of the region: honey and wax, furs and salt from Russian Novgorod; amber, timber, and grain from the south Baltic coast (amber had been traded from that region since at least 3000 B.C.); timber, tallow, and iron from Scandinavia; wool and wine from the west and south. But the foundation of Hanse wealth was fish: the herring which German, Swedish, and Danish fishermen netted by the ton in Baltic waters and which the Hanse merchants exported, packed in German or Russian salt, to meet the religious needs of Europe. The Church, of course, forbade meat on Fridays and during Lent—the latter being a season in which meat was scarce anyway, since livestock did not thrive on the withered and sometimes snow-covered vegetation of a northern winter. Some time around 1420, the herring virtually disappeared from the Baltic. At one time it was thought that they had migrated out of it into the North Sea; more recently, it has become apparent that the Baltic and North Sea herring are different races. But whether through migration or decimation, the great Baltic herring shoals on which the Hanse towns had waxed fat and powerful were gone. The decline of German commerce, and the simultaneous rise of Holland and later England, partly based on the teeming herring of the North Sea, dates from that time.

Why did the Baltic herring disappear? Overfishing could hardly have been the reason; despite intensive exploitation, the North Sea herring population did not decline until the twentieth century. Disease is even less plausible, for the Baltic herrings were radically reduced, not wiped out. If we assume that the survivors were those with natural resistance to the (hypothetical) marine plague, their hardy descendants would long since have repopulated the region.

A distinguished British zoologist, Sir Alister Hardy, in an otherwise illuminating discussion of the herring blight, has suggested that "some submarine earthquake disturbance altered the sea-floor sufficiently to deflect the flow of currents from their previous course and so either upset the drift of [herring] larvae from the spawning grounds or affected their supply of . . . food." An ecological catastrophe of this general type seems very likely—but that the initiating cause was a submarine earthquake seems highly unlikely, to put it mildly. A shift in the ocean bed sufficient to alter the current flow could hardly have happened without somebody noticing it; it would have generated enormous tidal waves that would have devasted the entire Baltic coast and come down to us in history as a catastrophe comparable to the Black Death.

The most plausible reason, I suggest, was a shift in climate—the same shift that first hampered and ultimately destroyed the Norse settlements in Greenland. After A.D. 1200, northern Europe gradually became colder and perhaps wetter as well; interestingly, the process reached its first peak in the early decades of the 1400s. This combination of lower temperatures (less evaporation) and more, or at any rate more effective, precipitation would necessarily have had one important effect on the Baltic: a reduction in its salinity. The Baltic is a peculiar sea. Because it lies in such a cold region, fresh water flowing into it much exceeds evaporation from its surface. As a result, almost the entire Baltic is brackish rather than salty. Its salinity decreases, in fact, the farther inland you go—from a normal 3.5 percent in the Kattegat to around 2.0 percent in the Danish sounds to 1.0 percent or less over most of its area. At the eastern end of the Gulf of

Finland, off Leningrad, the water is fresh enough to drink.

Ecologically, this decreasing salinity is reflected in a rapid drop off in animal species, from 1,500 or so in the North Sea to a few hundred in the Danish sounds to less than eighty in the central Baltic. What must have happened, I suggest, is that the decrease in salinity as the Little Climatic Optimum waned wiped out, or radically reduced, some of the species of tiny marine animals (zooplankton) on which the herring, especially in its younger stages, depends for food. Herring larvae are known to be quite choosy about their diet; during one month of their development, in fact, they seem to feed exclusively on a single species of microscopic crustacean. A shift in salinity that hit this, or some other, critical food species would quickly have cut the Baltic herring shoals to a fraction of their former size.

It works out very neatly: The climatic shift reduced the salinity, the lowered salinity reduced the herring, the lost herring weakened the Hanse towns, the weakened towns failed to unify Germany, German disunity led to religious war, which led to foreign intervention, which led to devastation and worse disunity, which led to militarism and the compulsion toward discipline and order, which led to Bismarck, "Blood and Iron," and, ultimately, Hitler.*

Do I believe this climatic fantasy? Well, a little. As a theory, it would be strengthened by some actual evidence of changes in the salinity and marine life of the Baltic around A.D. 1400—which, for obvious reasons, we are unlikely to get. Meanwhile, it can serve as a toothsome morsel for anyone with a taste for oversimplified theories of history, and also, perhaps, as a useful illustration of a famous observation by Mark Twain. "There is something fascinating about science," he remarked "One gets such wholesale returns of conjecture out of such a trifling investment of fact."

* I might add that whatever climate may have done to create Hitler, it certainly helped destroy him. "General Winter," which so hampered his invasion of Russia (as it did Napoleon's) might just as well be called "General Continental Climate."

38. EMPIRICISM AND AMERICA

A Climatic Ideology

It's always stimulating to read an outsider's comments on something you've always taken for granted. Having spent virtually all my life in North America (chiefly the northeast United States), it had never occurred to me that there was anything very remarkable about the climate there. Judge my delight, then, on reading the account of British climatologist F. K. Hare.

"In their quality of dramatic changeability," he writes, "the climates of North America have no rival; nor is there any continent in which greater differences exist between region and region. . . . With the exception of the Pacific coast . . . all these climates are united by their inconsistent, sometimes violently erratic, character." I felt rather like the man who suddenly discovered that he had been speaking prose all his life! I also felt like a man who had discovered an important clue to some of the temperamental and philosophical differences between Europeans and Americans.

The variety and changeability of the North American climates as compared with those of western Europe derive chiefly from a basic fact of physiography: The "grain" of North America runs roughly from north to south, in contrast with that of Europe (or, for that matter, Eurasia as a whole), which runs from east to west. In the Old World the Temperate Zone mountain belt runs, with few interruptions, from the Pyrenees to the China-Burma hump; our own Rocky Mountain system, however, stretches from Alaska into Mexico. (The Appalachians follow the same trend, but are low enough to have no very profound effects on climate.)

To understand the significance of this contrast, let us take one more look at Europe's climates, and the reasons for them. If we limit ourselves to western and southern Europe (which is where most North Americans originated) we find only two climates: Maritime and Mediterranean. Neither can fairly be described as particularly dramatic; both are certainly variable in winter, and the Maritime in summer as well (the Mediterranean is, of course, monotonously hot and dry during that season), but the variations are generally rather small and unspectacular.*

Climatic violence and spectacular change are essentially a matter of the amount of variety and contrast in the air to which a particular region is exposed. Significantly, we find that the predominant air flow over all of western Europe is of a single basic type, Maritime Polar, which is to say that it originates over cool or cold waters and is therefore consistently moist and of relatively low temperature. Maritime Polar air does, indeed, vary considerably in temperature (and also in moisture content) depending on where it originates; obviously air from the Greenland Sea will differ considerably from air flowing from the Azores. Contrasts between various sub-types of Maritime Polar air are in fact responsible for most of Western Europe's weather, but since the contrasts are seldom great, the weather they produce is seldom violent or sharply changing.

There are three other types of air that on occasion reach western Europe. One is the intensely cold and dry Continental Polar air coming from eastern Europe and Siberia, but (as noted in Chapter 21) its incursions are sharply limited by the prevailing westerly winds. The other two come from the south, and in summer: hot moist Maritime Tropical air

* Indeed, so subtle and elusive are the transitions from partly cloudy to partly clear, from fog to drizzle and from drizzle to rain, that European weather forecasters must be even more ambiguous than their American counterparts. A lovely example is the September 18, 1969 forecast for southwestern Scotland: "Mainly cloudy but some bright or sunny periods but also scattered showers or outbreaks of rain." Ye pay yer siller and ye tak' yer choice!

from the Mediterranean, and the intensely hot, dry Continental Tropical air from the Sahara. The latter brings the dreaded sirocco to southern Italy, but by the time it reaches the northern Mediterranean it has been considerably gentled down by giving up heat to and absorbing moisture from the ocean.

Most southerly air of these two types, however, does not reach northwest Europe at all, since it is partially blocked by the east-west mountain ranges. And for the same reason, as already noted, the more extreme forms of polar air, whether Maritime or Continental, seldom reach the Mediterranean. The likelihood of dramatic shifts in weather are correspondingly reduced.

North America, of course, knows both the Mediterranean and Maritime climates. But the North American "Mediterranean," though it covers most of the California lowlands, runs inland only 200 miles or less to the Sierra Nevada; our Maritime zone, found in Oregon, Washington, British Columbia, and the Alaska panhandle, is similarly blocked by the Cascades and Coast Ranges. (By contrast, the European Maritime reaches nearly 1,000 miles inland, from Ireland to Poland, and the Mediterranean over better than double that distance, from Portugal to Palestine.)

North America east of the Rockies—which is to say, better than two-thirds of the U.S. and southern Canada (for simplicity, I am ignoring thinly populated northern Canada and Alaska)—is dominated by two sharply, even violently, contrasting types of air. From the Polar tundra and Subpolar taiga of Canada comes Continental Polar air, almost as cold and dry as its Siberian counterpart; from the Caribbean, Gulf of Mexico, and adjacent portions of the Atlantic come Maritime Tropical air. And this is a good deal hotter and moister than its counterpart in Europe, since these seas are markedly warmer than the Mediterranean (better than 10° on the average, year round). It is also less "stable," a meteorological concept which we need not go into except to note that unstable air masses, like unstable people, are more than usually prone to violent behavior.

To these we can add two other types: One of them is Maritime air originating over the Pacific which, by its passage across the mountains, has lost most of its moisture in precipitation and thereby gained a good deal of "latent heat" as the moisture condensed, so that it descends on the Great Plains as the warm, dry "chinook." The other, found only in the northeast, is cool moist Maritime Polar air from the North Atlantic, essentially the same as the predominant air-flow in western Europe.*

Considering merely these four contrasting types of air—cold-dry, warm-dry, hot-moist, cool-moist—there is obviously plenty of scope for radical shifts in weather merely through their replacement one by another, not to speak of the violence induced by their intermingling and interaction. And their constant interaction is assured by *the absence of any barrier to their movement.* In summer, Maritime Tropical air can thrust up from the Gulf almost to the Canadian border, turning Chicago and New York into tropical torture-houses for their inhabitants; in winter, similar, though less violent, incursions can bring the "January thaw" that punctuates so many bleak American winters—and also the March or April thaws that set rivers to raging, bursting banks and levees and inundating bottomlands.

In winter, Continental Polar air can turn much of the same region into almost a province of the Arctic, and bring below-freezing temperatures even to subtropical Florida, as citrus growers know too well. In summer, its breath is more benign, replacing the Gulf's muggy Maritime effluvium with the bracing breath of the north woods—and bringing cloudless skies whose brilliant sun can roast the unwary beach-goer. Maritime Polar air can bring to the northeast a fine, gray English drizzle; modified Pacific air still other cli-

* In summer, the desert regions of the Southwest generate still another type of air—hot, dry Continental Tropical of almost Saharan quality. But thanks to the mountains of New Mexico, this seldom moves out of its desert birthplace, for which Midwesterners can be thankful.

matic variations. And when we imagine the further variations on these basic themes that result when the great airmasses intermingle, begetting line-squall or nor'easter, thunderhead or fog, tropical downpour or blinding blizzard, we can understand the old saw (often said in New York but no less applicable to a hundred other American cities), "If you don't like the weather, wait a minute." To which one might add, "And be ready to duck!"

It is tempting to speculate that this climatic maelstrom infused some of its violence into the Europeans who settled these shores, and into their American children. There can be no doubt, certainly, that "Violence is as American as cherry pie," a remark that drew opprobrium onto a Black leader who was merely echoing innumerable white sociologists, historians, and critics.* One might even imagine that Canadians are rather less violent than we are because their country lies to the north of the main arena of climatic conflict. But there are limits to speculation, and this one lies well beyond them. In the first place, I know of no medical, psychological, or sociological mechanism whereby climatic violence could transfer itself to human behavior. Climate, violent or otherwise, is not something that can be resisted, violently or otherwise; for the most part, we endure it as best we can. To the extent that its more catastrophic aberrations may require massive human response—as in the case of a major flood, tornado, or blizzard—the consequence is rather intense human cooperation than human conflict; when men are busy fighting nature they are generally too busy to fight much with one another. There are, moreover, too many other obvious historic reasons for violence in America: the need to first dislodge and then exterminate the Indians, the need to keep the Blacks enslaved, and the general weakness of official law and order at so many American times and places, so that power flowed directly from the barrel of a six-gun.

* Though one might note that neither cherry pie nor violence is uniquely American!

But if the American climates do not serve to explain violence in America, they certainly have a clear relationship—to my mind, at least—with an equally characteristic American trait: our addiction to an empirical, pragmatic, cut-and-try view of life. Again Hare has put his finger on the central point: "The European settlers in the North American frontier encountered not one climate but many, each with its special hazards, but a few without compensations to reward the settler who remained." To which we need only add that these climates were without exception markedly different from what the settlers were used to. Checking over climatic tables for the major cities of western Europe and North America, the only reasonably close temperature match I can find is Oslo with Halifax, Nova Scotia—and the latter has more than twice as much precipitation. Unfortunately, Nova Scotia was settled by Scots, not Scandinavians.

The English settlers at Jamestown and around Chesapeake Bay would have found the winters not much colder than they were used to, but the summers far hotter—hot enough to induce sunstroke in the unwary. The Chesapeake region, in fact, lies just within the Subtropical Zone of North America (summer residents of Washington, D.C. would probably drop the "sub.") The Puritans in New England, having foolishly abandoned their first anchorage off Provincetown, Cape Cod, for Plymouth (where, according to one account, they fell first upon their knees and then upon the Indians), found both hotter summers and much colder winters. Add to the latter the much greater precipitation and the result is snow in quantities which gave a bleak point to the stern Puritan dictum that "whom the Lord loveth He chasteneth." Quite a few of the early Massachusetts settlers were loved to death.*

Both Virginians and New Englanders were from time to time exposed to a spectacular climatic feature altogether beyond their experience: the tropical revolving storm. These

* Of the hundred or so souls who landed from the *Mayflower,* no less than half perished during the first winter.

catastrophic disturbances—which the English eventually learned to call by the name of a Caribbean storm-god, Huracan—would, then as now, every so often swing northward from the Caribbean or tropical Atlantic and ripsaw their way along the coast, leaving in their wake leveled trees and houses, wrecked harbors, and swamped ships. To a Japanese or Polynesian they would have come as no great shock, but to the European settlers they must have seemed the end of the world.

Hurricanes ("typhoons" in the Pacific) are among the most violent manifestations of the atmospheric heat engine at work. They are powered by the heat-energy stored up (chiefly as the latent heat of evaporated sea water) in air passing over tropical seas. Under appropriate circumstances, still not wholly understood, this heat-energy is transformed over a rather limited area into the violent motion-energy of the hurricane's winds, which can range up to 120 miles per hour. Fortunately, the abundant Maritime Tropical air south and southeast of North America rather seldom undergoes this peculiar evolution; most of its energy goes to fuel the great storms that periodically sweep across the American Midwest and East, especially during the cooler months. These mighty disturbances actually involve much greater energy than any hurricane, but its release is providentially spread over much more territory. The storminess of North America, looked at in a very simple way, is the direct consequence of its abundant supplies of "energetic" Maritime Tropical air.

Hurricane or "ordinary" storm, the settlers learned the hard way to cope as best they could. They also, perforce, learned to cope with a rather different agricultural calendar than the one they were used to. While the growing season was hotter than in the home country, it was also shorter—in New England, much shorter. The last spring frosts came later and the autumn frosts earlier; the dates of both had to be guessed at on the basis of inadequate experience, with privation or actual starvation the penalty for the farmer or community who guessed wrong.

The natural, inevitable, result of all this was a propensity to expect the unexpected, to improvise as best one could with the means at hand, and above all to not dwell too heavily on existing knowledge, whether written in books or passed down from ancestor to descendant. There were few intellectuals among the early colonists; they were literate but seldom learned—in the fullest sense, *practical* men. More accurately, perhaps, the practical ones were the only ones who survived; the others either returned home or died off. To make a go of it in America, they had not only to expect the unexpected but be ready to try anything that seemed likely to deal with it successfully. A new experiment might contradict both book larnin' and ancestral wisdom, but if it worked, they thanked God and got on about their business.

Sometimes, however, cut-and-try methods just weren't good enough. Immediate problems—when to put in the spring crop, where to locate your cabin for shelter from the worst Continental Polar blasts of winter—could be solved quickly, with ingenuity and a bit of luck. But there were more complex and long-range problems whose genesis extended over decades and generations, and these required more and deeper thought than the colonists were willing or able to give. The forest soils of the North American east coast were thin, as all forest soils are. Europeans had faced that problem in earlier centuries and had devised methods of fallowing and manuring that restored fertility. But American soils faced the additional assault of a rainfall up to twice the northwest European average, and distinctly more violent to boot. The more land that was cleared of its forest cover, the more acute the problem became—to the colonists, the forest was not merely a seemingly inexhaustible source of material for fuel, houses, and tools (metal goods were costly imports) but also an expendable lurking place for equally expendable Indians, wolves, and panthers.

When the Puritans first landed on Cape Cod, like much of New England it was covered with white pine, 4 and 5 feet thick at the butt. By 1800, hardly a stick remained, and the thinning soil was increasingly being washed away by thunderstorm, nor'easter, and an occasional wandering hurricane. By the mid-1800s, the situation was so serious that some Outer Cape towns adopted an ordinance which forbade the cutting of a tree unless another were planted in its place. This was America's first conservation legislation, but it was already too late. Dig down almost anywhere on the Cape nowadays and you will find the results: perhaps half an inch of black humus and beneath it an ash-gray sandy loam, its nutrients leached away into the porous subsoil or washed into the nearest pond or stream. In all my wood ramblings on the Cape, I do not recall having seen a single white pine; the predominant growth is the gnarled pitch pine, whose outstanding characteristic is its capacity to thrive on poor soils.

Everywhere that the country was hilly or even rolling—most of New England and New York—the story was much the same. The settlers had farmed the land by cut and try; they had laboriously hauled out the boulder souvenirs of the Wisconsin glaciation and had heaped them into the stone walls that draw gray, wavering lines across almost any New England landscape—but now the land was giving out.

Some took to the sea, building and manning the whalers and tea clippers that founded a hundred New England fortunes. Others founded factories along the rivers that supplied plentiful waterpower. Still others piled into their wagons and trekked west, into Ohio and the Northwest Territory. For after all, there was plenty of land. . . .

West of the Alleghenies the settlers at first found the same sort of climate they had left, though a bit more so—summers a little hotter, winters a little colder. As the frontier pushed ever westward, however, things began to change. The Gulf's masses of humid air seldom drive straight north; the prevailing westerlies and the earth's rotation combine to swing them to the right, so that the states from Ohio on east receive a disproportionate share of precipitation, while moisture drops off as you push toward the Rockies. Average annual precipitation: Boston, 43 inches; Chicago, 33 inches; Omaha, 27½ inches. Through Indiana, Illinois, Iowa, and Nebraska, the forest gave way to prairie-savanna, the savanna to plains of long buffalo grass, and the long grass to short grass.

Ecologists argue about the relationship between dropping moisture and vanishing trees on the Great Plains. Some believe that grass fires, set both by lightning and by the Indians, have played a part. But whatever the cause, the pioneers on the grasslands found that the timber they had relied on for generations was now almost as scarce as gold. "There's n'er a log to sit upon along the River Platte/So when you eat you've got to squat, or set down square and flat."

More improvisation. One learned to burn chips of dried buffalo manure for fuel and to pile up walls of sod for shelter—or freeze in a norther. "The Little Old Log Cabin in the Lane," sung at barn dances back in Illinois Territory, was sourly rewritten into "That Little Old Sod Shanty on My Claim."

Though dryer, the weather was even more unpredictable. The prairie and Plains states are the arena par excellence where Polar and Tropical air fight out their unending battle. In forest country (and more so east of the Appalachians) the thrust of the great air masses is blunted, if not stopped, by irregular topography and the surface roiling and turbulence generated by trees. But on the Plains, flat as a billiard table and almost as treeless, the land supplies no such moderating influence. "Between Texas and the North Pole," the plainsmen say, "there's nothing but a barbed wire fence." Others tell how to gauge wind speeds: "You hitch one end of a logging chain to a post. If the chain blows straight out from the post, you've got a pretty fair breeze, but when it starts to whip around and links snap off, you can figure the wind's getting ready to really blow for a spell!" A plains norther sweeping south from Canada can drop the temperature 20 degrees in as many minutes; the chinook's eastward thrust down the long slope from the Rockies can lift the mercury no less abruptly. A plains blizzard can whirl snow up into the blinding "white-out" usually associated with Antarctica, in which a man can get lost between the house and the barn. The pioneers learned about all these things and improvised ways of coping with them. The ones that didn't learn were buried.

They also confronted a climatic phenomenon almost unique to North America: the tornado. The twister compresses the violence of a sizable thunderstorm into a radius of perhaps 50 feet. While even hurricane winds seldom get much over 100 mph, the air around the black tornado funnel is thought to reach speeds of 500 mph (this is an estimate; nobody sets out to carry instruments into a tornado, and whenever instruments have chanced to be in a twister's path, they have invariably been smashed). Tornado tracks are mercifully narrow, but within them houses and trees are chewed up as by a Gargantuan buzzsaw, to the accompaniment of "a roar like a thousand freight-trains," according to one lucky survivor who actually looked up into the funnel. Tornadoes have struck at least once in every one of the fifty states, but their prime hunting ground is the Plains, especially "Tornado Alley" in and around Kansas and Oklahoma; each of these two states averages around twenty-five twisters a year. The pioneers, confronting this terrifying climatic hazard, learned to build cyclone cellars for protection. ("Cyclone" is a confusing word; scientifically, it means a Temperate Zone storm, but is applied to hurricanes in parts of the Indian Ocean and to tornadoes in the Plains. It was a tornado, not a "cyclone," which whirled Dorothy and her little dog Toto from Kansas into the Land of Oz—and considering where she might have ended up, she was lucky!)

A partial compensation for all these hazards of the Plains was the soil. Uncounted generations of prairie grasses, enriched by uncounted generations of buffalo droppings and preserved from leaching and erosion by flat topography and limited rainfall, had compacted into a black topsoil ten and twenty feet thick. It was ideal farming country—provided there was enough rain. Sometimes there was—but sometimes not.

Early in the 1880s, a wave of pioneers began pushing west into the short-grass prairie of the Dakotas. It was a region which many authorities held was more suited to ranching than to agriculture; land agents for the Northern Pacific Railroad, however, took a more optimistic view. In fact, when the immigrants reached the ground, they found what was apparently typical long-grass prairie, having "the appearance of wheat fields." They quickly replaced the appearance with the reality; they were "from Missouri"—some of them, no doubt, literally—and seeing was believing.

What they did not know, or perhaps chose not to know, as the American practical man has so often shrugged off expert advice, was that the early 1880s were abnormally high in precipitation. The later years of the decade were the reverse; as noted earlier in connection with the Russian grasslands, the lower the rainfall the less reliable. This is not because the absolute variations in precipitation from year to year are more pronounced than in more humid regions. But while a drop from 38 inches to 32 inches a year will probably do no more than inconvenience the farmer, a drop from 18 to 12 inches can wipe out his crop—especially if the drop is concentrated in the warm growing season.

At Mandan, North Dakota, the 1882–85 warm-season rainfall averaged a modest but adequate 15.9 inches; during the succeeding five years the average fell to 11.6, with individual years registering as low as 8.3. Readings over much of the Dakota plains were similar. Once again pure pragmatism, no doubt helped by railroad propaganda, had entrapped American settlers.* The same pattern was to be repeated at intervals in many places on the plains—most

* I owe this information to Reid Bryson and his associate M. R. Moss.

spectacularly in the 1930s. A series of dry years obliterated the crops, and with no vegetation to hold down the soil, the swirling winds hoisted it aloft by the thousand tons.

> The dust storm struck, and it struck like thunder;
> It dusted us over and dusted us under;
> It blocked out the traffic, it blacked out the sun
> And straight for home all the people did run,
> Singing "So long, it's been good to know you. . . ." *

So sang Woody Guthrie, whose family was one of thousands "dusted out" of Oklahoma and Colorado and Kansas and forced into the long trek to California on Highway 66.

Others, of course, had made the journey long before and in the process had learned, or fatally failed, to cope with still other climatic novelties and hazards. Emigrants along the California Trail traversed the Great Basin, a typical Temperate Zone rain-shadow steppe-desert, and learned the hazards of desert waterholes; these may be so impregnated with alkaline minerals as to produce a rather spectacular diarrhea, and at times to mortally poison man and beast. If they got through, they joked about it; Sweet Betsy from Pike, having reached the gold diggings with her lover, remarked to a miner who requested the pleasure of a dance, "Don't dance me too hard—do you want to know why?/Gol durn ye, I'm chuck full of strong alkali!"

The Donner Party didn't quite get through. Nothing in their experience had prepared them for the 10- and 20-foot autumn snow drifts that even today can halt traffic across the Sierras. They starved to death every one, though some of the last to go staved off hunger for a while with the flesh of their deceased fellows. The Mormons didn't even try to make it. Seeking a territory sufficiently forbidding to discourage competing immigrants, their stern leader Brigham Young gazed

* "So Long, It's Been Good to Know Yuh (Dusty Old Dust)." Words and music by Woody Guthrie, TRO-©Copyright 1940 (renewed 1968), 1950, and 1951. Folkways Music Publications, Inc., New York, N.Y.

at the bleak terrain near Great Salt Lake and declared "This is the place!" That they survived and even prospered is due not merely to the efficiency of Young's theocratic dictatorship, but to their ingenuity in reinventing irrigation agriculture—a technique which neither they nor their grandfathers' grandfathers could have known except by rumor.

Irrigation agriculture was hardly less essential to those California settlers who, as the gold fields declined, sought their salvation in farming. In coping with the unfamiliar Mediterranean climate, however, they could draw on the experience of the Spanish-Mexican settlers from whom America had seized the territory and who, in introducing irrigation to southern California, had merely applied techniques used in Mexico for thousands of years. Farther north, in Oregon and Washington, the problems were less precedented and more puzzling. Parts of the lowland areas in that region are an odd blend of Mediterranean and Maritime, with summers dry enough to require irrigation yet with winters so wet as to necessitate artificial drainage of farmlands. In some coastal areas, on the other hand, the mountain-generated rains are so intense year round that the country is covered with timber "thick as hair on the back of a dog"—one of the world's few temperate-zone rain forests. "For four years I chopped and I loggered," the Puget Sound settler sang, "but I never got down to the soil!"

Thus from generation to generation Americans spread to fill up the land, at almost every step meeting new climatic challenges and difficulties and mastering them—or being mastered by them, like the gold-seekers who, starting out with the slogan "California or Bust!" at last turned back from the desert with the laconic motto "Busted, by Gosh!" If empiricism, improvisation, the pragmatic determination to try anything once and keep trying as long as it works are hallmarks of the American character—and I am convinced they are—then the ever-new challenges of climate are surely an important part of the reason. Significantly, it was an American president who remarked, "It is a condition that confronts us, not a theory"; the words might well be adopted as our national motto.

At its best, American empiricism implies a flexibility, an impatience with dogma, and a receptivity to new techniques and ideas that, as an American, I find admirable and exciting. At its worst, however, it implies an infatuation with gadgets and gimmicks, a grasping at immediate and simple solutions, and a contempt for intelligence and learning which I, as an intelligent American, find appalling. When Barry Goldwater in 1964 inveighed against "complicated, theoretical" explanations of social problems—which were, he intimated, the invention of a sinister Camorra of Eastern intellectuals—he typified the dark side of American empiricism. An even more recent example is the currently stylish mini-philosopher Eric Hoffer, who when he hears the word "intellectual" reaches for his cargo-hook (Blacks, adolescents, and "foreigners" of all sorts—except, possibly, his German ex-countrymen—affect him in much the same way).*

We Americans have been compelled to solve many problems, climatic and otherwise, by cut-and-try methods. At times the solutions have worked brilliantly. But in other cases, fuller experience has demonstrated what deeper thought would have revealed beforehand: not all problems have simple solutions, and even the most immediately effective simple solutions can eventually generate far more severe problems. The overgrown farmlands of New England, the ravaged strip-mining areas of Appalachia, and the dusted-out Okies faring westward in their jalopies testify to the limitations of the strictly empirical approach to climate and to life.

If we Americans are to learn to live in reasonable harmony with the drama, excitement, and change—yes, and the violence—which infuse our social environment as they do our natural one, we will need all the flexibility and in-

* He has now been joined in his anti-intellectual crusade by our Vice-President, who combines the pragmatist's contempt for intellect with the politician's dislike of criticism.

ventiveness that we have acquired from our generations-long struggle with new climates and new challenges. But we will also need to recognize a fact for which our empirical, experimental history has ill prepared us: What is "practical" is not always what is right, and the most practical short-term solutions can turn out in the long run to be the grossest and most destructive impracticality.

39. HILLSIDES, SEASIDES, AND CITIES

Climate in Miniature

If the climate most interesting to most people is the 1/16 inch next to their skins, hardly less interesting is the less intimate but still proximate climate right around home. In fact, the influence of the macroclimate—climate in the large—is often significantly altered by what might be called the mesoclimate: local variations in climatic factors covering areas ranging from the equivalent of a few city blocks up to several tens or even hundreds of square miles. These are determined by local features of topography, by local relationships between land and water and, in some of the most remarkable cases, by the activities of man himself. An understanding of them can sometimes help us to regulate our personal microclimate more efficiently, and also give us a more acute awareness of what is happening around us—which is stimulating in itself.

I remember one blistering July day on Cape Cod, for instance. We were at Ballston Beach on the Atlantic side, and the temperature, rare for the Cape, must have been close to 100°. The ocean was cold, almost too cold for comfortable swimming, but once out of the water you quickly started to

Rapid heating of land generates afternoon sea breeze.

pant. The combination of icy water and stifling beach was sufficiently unusual for me to wonder what was going on. I then noted that the breeze was not blowing in from the ocean, as it often does, but from the southwest, that is, from the land side. Hot to begin with (almost the entire northeast was undergoing a heat wave), it had been further heated by passage across the sunbaked moors and pinewoods and especially over the last couple of hundred feet, in which it had crossed first a blacktop highway and then a stretch of sand dunes, either one hot enough to peel the skin from you feet. It was this superheated air that was making the beach as uncomfortable in one way as the icy water was in another.

That being the diagnosis, the remedy was clear. We climbed into the car, drove over to the other side of the Cape some three miles away, and decanted ourselves onto Corn Hill Beach. This faces southwest, so that the breeze now reached us not across overheated moors and dunes, but over the cool expanse of Cape Cod Bay. The air temperature was just perfect—and the water was warmer too. What we had done was move from a mini-Continental to a mini-Maritime climate. The contrast, while hardly comparable to what we would have experienced had we moved across a conti-

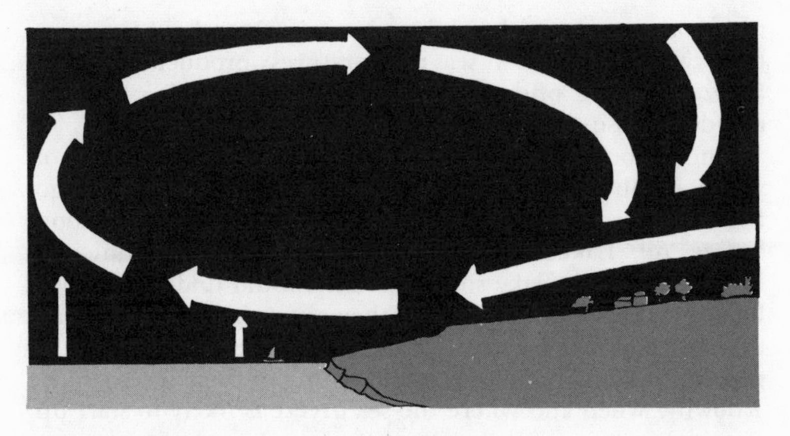

Slow cooling of water produces nighttime land breeze.

nent, was quite enough to make the difference between acute discomfort and relaxed delight.

An even commoner local Maritime effect is the afternoon sea breeze, observed from time to time in nearly all coastal areas that are hot at least part of the time. Land, of course, heats up much more rapidly than water, meaning that on a sunny summer day in a coastal area the ground temperature will within a few morning hours have risen much higher than the sea temperature (the difference may exceed 50 degrees). Air heated by the ground commences to rise, as in a hot-air furnace; to replace it, cooler air flows in from the adjoining water area (assuming there is no strong wind blowing from a contrary direction). A still, baking morning becomes a breezy, cool afternoon. Under favorable conditions (i.e., maximum contrast between land and sea temperatures), sea breezes can blow up to 20 mph and reach many miles inland. Ismailia, on the Suez Canal, is 40 miles from the Mediterranean but receives a sea breeze almost every summer day, beginning promptly at 3:30 P.M. Whenever people in coastal areas hear a summer forecast along the lines of "Afternoon high 90 except near 80 along the coast," a sea breeze is what the announcer is talking about. During the night, of course, the process reverses. Assuming skies are clear, land temperatures will drop below sea temperatures, producing a land breeze blowing offshore—and too often bringing with it clouds of mosquitoes from inland swamps and puddles.

Similar local winds are generated by any really big body of water, notably along the shores of the Great Lakes. Chicago friends have told me that were it not for the afternoon breeze off Lake Michigan, plus the cooling municipal beaches along the lake shore, the city would undergo a revolution every summer—which, they say, might not be an altogether bad thing at that. Crafty yachtsmen make a point of getting to know the pattern and timing of their local winds. Knowing when and where the sea breeze is likely to start up on a calm day can make the difference between winning a race and losing it.

Bodies of water, even relatively small ones, have other local climatic effects. Like the oceans, they serve as heat reservoirs during the colder months and as heat absorbers during the hot ones. The result is a slight moderating of extreme temperatures—in particular, and this is important to farmers, a somewhat extended growing season between the last frost of spring and the first one of fall. The wine industry in the eastern United States is, for this reason, concentrated almost entirely along the Hudson River, the Finger Lakes of upstate New York, and a few spots along Lake Erie.*

Lake Erie, and to some extent the other Great Lakes, also generate a troublesome local climatic problem for cities lying on their east or southeast shores. The lakes do not freeze until December, and in a mild winter may not freeze completely at all; as a result, a mass of Continental Polar air sweeping across the lake can pick up considerable heat and moisture in its lower levels. This makes the air mass unsta-

* However, while bodies of water help prevent late spring frosts, they also delay the spring temperature rise in shore areas because of their own slow warming plus the sea- (or lake-) breeze effect. Thus in the southeast corner of Wisconsin, along Lake Michigan, the lilacs normally flower some two weeks later than on the opposite, "continental," side of the state.

ble, meaning that when its relatively warm and moist lower levels are pushed upward—by being thrust against the rising ground of the shore—they will continue to surge upward, producing dense clouds and often a violent snowstorm. The high snow-removal bills of Buffalo, New York, lying at the eastern end of Lake Erie, are proof that its location, though superb from a commercial standpoint, could have been better chosen climatically.

The topography of the land itself can affect local climate. Hillsides will receive more or less sunlight depending on whether they face north or south; the same is true of valley sides. The Swiss, who are well equipped with steep-sided valleys (scooped out, in most cases, by Ice Age glaciers) have different words in several of their dialects for the sunny and shady sides.

Valleys, like coastal areas, can generate their own local winds. I learned about these the hard way many years ago when my wife and I were on a hitchhiking tour through France and Italy. In the course of it, we spent several days in the youth hostel at La Ciotat, a pleasant little Mediterranean town east of Marseilles. The hostel was distinctly cleaner and better appointed than many French hostels of that period, but possessed a major defect common to all hostels: the boys slept in one room, the girls in another. One night, therefore, we betook ourselves and our sleeping bags to a narrow ravine giving on to a beach. It seemed well sheltered from the wind, and for further shelter we bedded down in a thick clump of reeds. About 3:00 A.M., however, we were awakened by a distinctly chilly breeze blowing down the valley toward the beach, with sufficient force to penetrate both our reed shelter and our sleeping bags. We huddled together as best we could for warmth, but the latter part of the night was a great deal less pleasant than the beginning.

It was not until some years later, in leafing through a book on climatology, that I realized what had happened. Under the clear Mediterranean summer sky, the ground surface along the valley sides had cooled off during the night.

Slope winds move up a valley during the afternoon hours.

At night, sinking of cooler air reverses the wind's flow.

The air in contact with it, cooling and growing heavier, had naturally flowed down to the valley bottom and then along it to the mouth. It was this river of chilled air, its velocity no doubt increased by the nighttime land-breeze effect, that had given us such a chilly time. I also learned that during the afternoon hours, the effect would have been reversed; heated air, rising from the valley sides would have drawn cooler air through the valley mouth and along it to its upper end.

The downhill flow of cold air, and its "pooling" in lowland areas, explains why fog sometimes forms in the country on clear spring or summer nights. If the air is reasonably moist—perhaps as the result of an afternoon thunderstorm—the cooling may be sufficient to condense the moisture into clouds of water droplets in valleys and hollows while the hilltops remain brilliantly clear. The same mechanism explains why valley crops can be damaged by frost on spring or autumn nights when vegetation remains unscathed on higher, and seemingly more exposed, sites.

Sophisticated architects, provided their clients are sufficiently affluent, will take into account these and other local details of terrain in choosing a plot for a house and in siting the building upon it. In cold climates, for example, they will prefer a south-facing slope to a northward one, but will if possible build on the slope rather than at its bottom, thereby avoiding the pool of cold air that can form in the manner just described. Evergreen plantings on the north side can serve as a windbreak; deciduous trees to the south provide shade in summer, yet, stripped of their leaves, interfere little with the sparse sunlight of winter. Of course, if you are rich enough to be that choosy, you are rich enough to afford complete air-conditioning and to have no worries about your heating and power bills.

Even the best-paid architect, however, can do nothing about the most outstanding manifestation of local climate: the climatic peculiarities of cities. These are caused by the existence of the city itself and the human activities centered therein, which combine to make city climates noticeably different—in nearly all respects, worse—than those of the surrounding countryside. The main reason why city climates differ from country climates is that city air is markedly warmer than country air. We hear this dozens of times a year, whenever the weather announcer forecasts something like "Overnight low around 60 in the city, 50 to 55 in the suburbs." The reasons behind these forecasts are rather complex, however.

First, the brick and concrete, stone and asphalt of which cities are chiefly composed can store more heat, and more rapidly, than the grass, leaves, or even bare soil of the countryside. Because city materials are more dense, they require more heat energy to reach a given temperature, and they also conduct the heat inward much more efficiently.

Second, the structure of the city is such as to absorb more of the sunlight that strikes it. A countryside is largely composed of more or less horizontal surfaces (even a tree consists of hundreds of these surfaces); incoming sunlight, striking them, is likely to be reflected back immediately to the sky, though part of it will of course be absorbed and converted into heat. The city, however, has a jagged profile of horizontals and verticals, so that a beam of sunlight has an excellent chance of being reflected from vertical wall to horizontal street to vertical wall again, with more and more of its energy being absorbed at each step. "The walls, roofs and streets of a city," says the climatologist William F. Lowry, "function like a maze of reflectors, absorbing some of the energy they receive and directing much of the rest to other absorbing surfaces. . . . A city provides a highly efficient system for using sunlight to heat large volumes of air." *

Third, the city itself generates enormous quantities of heat, especially in winter. Furnace systems are a main source; since no building is constructed of perfect insulating materials, a heated building will lose considerable amounts

* On a hot summer night, I have experienced a noticeable and welcome fall in temperature merely by walking into a small city park, whose grass and trees absorb and store heat inefficiently.

of heat to the outside air. Year round, there are factories, power stations, and the exhausts and heated radiators of thousands of cars, trucks, and buses. In summer, air conditioning equipment partly makes up for the lack of furnaces. Heat removed from the interior of a building must obviously be gotten rid of outside it; ironically, the more extensively a city uses air-conditioning to improve interior microclimates, the more torrid will be the climate in "non-conditioned" areas.

Fourth, the city's all too efficient heating apparatus is matched by distinctly inefficient cooling mechanisms. In the country, most of the rainfall sinks into the ground, whence a great deal of it returns to the atmosphere by evaporation, and by transpiration through the leaves of growing plants —both being processes that absorb heat. In the city, however, much of the rainfall runs off immediately from roofs and streets, eventually passing into the sewer system where little if any of it is available for evaporative cooling. In addition, cooling by wind is hampered by the city's irregular surface; this stirs up turbulence, which often sets skirts delightfully flying but also cuts wind speeds by something like 25 percent, with a proportionate reduction in the amount of heat carried away.

The result of all this is that cities, especially their more densely built-up sections, form "heat islands" in the countryside. The size of the island, as well as its "height" (i.e., the amount of "excess" temperature at the center), will naturally depend on the size of the city. In large cities, daily minimum temperatures will average some 4° F. higher over a year in the city; daily maximum temperatures will only be a degree or so warmer, but this is considerable when averaged over a year.* The contrast between minimum (usually night) and maximum (usually day) temperatures occurs because during the daytime the city often, as it were, ventilates itself. As a

* Thus in Wisconsin, the delay in spring flowerings along the lake shore is noticeably less in the Milwaukee urban area.

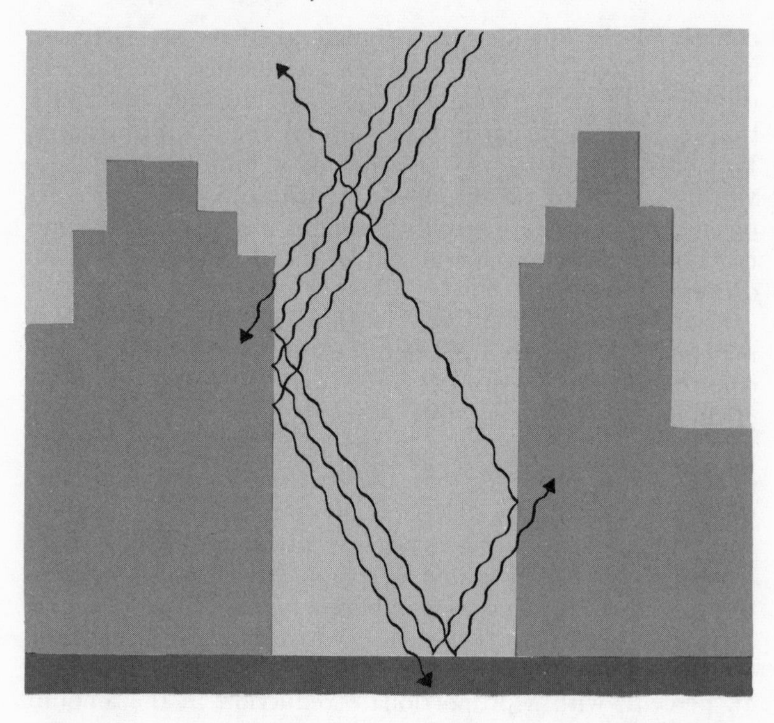

The jagged profile of a city traps major part of incoming sunlight.

heat island in a rural "sea" it behaves much like an actual island. Rising hot air over the city draws a "sea breeze" in from the countryside, while the hot air moves outward at higher levels, carrying with it a load of smoke and dust for the delectation of suburbanites.

Urban climatic peculiarities do not stop there. Because of lowered evaporation and higher temperatures, urban relative humidity averages appreciably lower than rural. Oddly enough, however, urban rainfall averages higher, by about 10 percent over the year. This, says Lowry, "builds up mostly as an accumulation of small increments on drizzly days, when not much precipitation falls anywhere in the area. On such days, the updrafts over the warm city provide

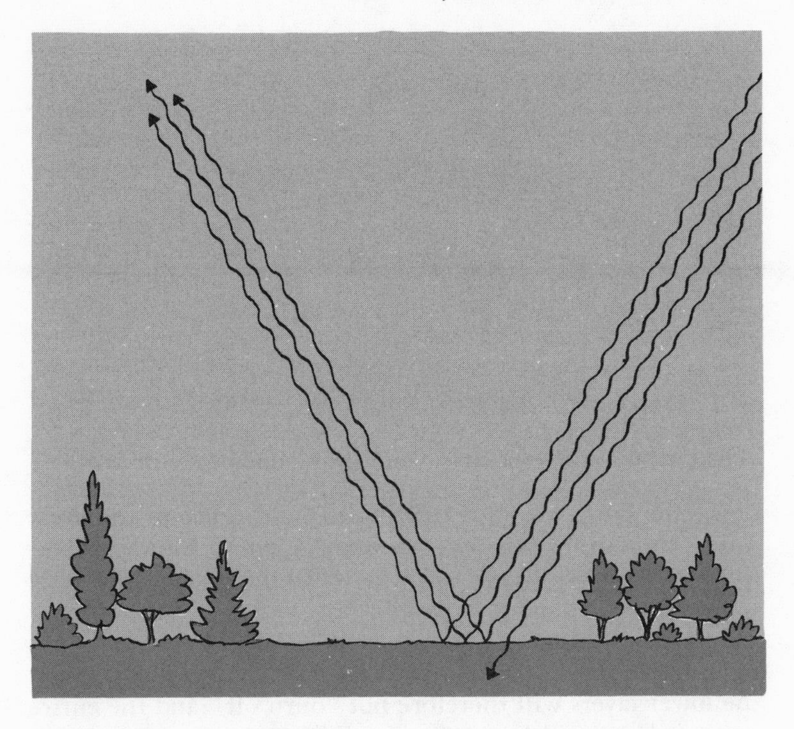

In rural areas, much more solar energy returns to space.

enough extra lift so that the clouds there produce a slightly higher amount of precipitation."

The most outstanding difference between city and country is the amount of fog—30 percent greater in summer and a full 100 percent in winter. This comes about through the copious emissions of dust and smoke already alluded to. Flying into Boston on a fine day, for instance, I have seen the dirty brown of the city's dust plume smudging the blue sky for twenty miles downwind. On calm days, city dust and gunk is not blown downwind but hangs over the city in a "dust dome." At night, the particles lose heat by radiation and, if conditions are right, will become nuclei on which water droplets condense. The droplets, fog, will by themselves weigh

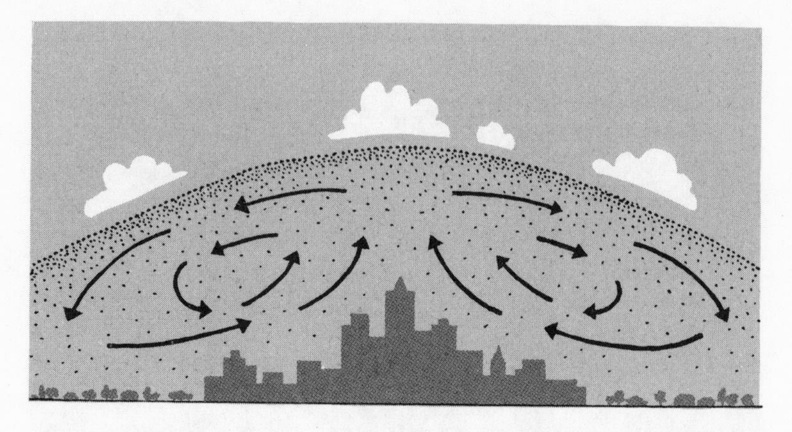

The dust dome over a city is most pronounced in still weather.

down the particles, thus tending to hold them in the area; unless something happens—a brisk wind to blow the particles away or a heavy rain to wash them out—the haze will build up from day to day.

In winter things are even worse: the feeble sun may not be able to penetrate beyond the topmost layers of the fog; the lower layers will therefore not "burn off," and the entire noxious brew will be further fortified by increased fuel consumption to warm the chilly, fogged-in city. This acrid smog reddens the eyes, catches at the throat, and, if it persists long enough, may build up to levels that can send scores or even hundreds of asthmatics and heart patients to their premature reward. These unpleasant-to-catastrophic episodes are especially likely in the late autumn and early winter, during periods of fair, still weather. At such times warm air from the south, hanging over the city, will have its lower layers cooled during the long nights. The result is a "temperature inversion," meaning that the air, instead of growing cooler with altitude as usually happens, turns warmer a thousand or two feet above the ground. The stagnant pool of chilled, smoggy surface air, too heavy to rise through the warmer, lighter layers above, hangs like a choking shroud over the city's reeking smokestacks and exhaust pipes.

Some cities on western coasts are unfortunate enough to possess a semipermanent inversion, the outstanding example being Los Angeles. The prevailing westerlies there, of course, blow in from the Pacific, whose surface waters along that coast are chilled by the cold California Current (summer ocean temperatures off California approximate those off Newfoundland). Overall, the Pacific air is relatively warm and dry, since the general atmospheric circulation has forced it down from higher altitudes. But as it crosses the California Current, its lowest layer is chilled and moistened by the sea surface, so that it must travel many miles over the warm land before acquiring enough heat to rise upward. On occasion, this chilled "marine layer" brings actual fog to parts of Los Angeles, as it frequently does to San Francisco. But always, unless winds are brisk—and the mountains around L.A. ensure that they often won't be—it traps auto emissions from the area's octopus of freeways, producing dimness if not darkness at noon. Every time I have landed at Los Angeles airport, I have noted the yellow-white smog layer below me and then, descending into it, had the sensation that some cosmic movie technician had thrown a dimmer switch on the sun.

Along with most of lowland California, Los Angeles also enjoys a much less frequent but equally unpleasant climatic feature: the "red winds," or Santa Anas. These occur when the normal westerlies are temporarily replaced by easterlies moving down from the Basin country of Nevada. The air, dry to begin with, is heated up and further dried by its 6,000-foot descent from plateau to sea level, and picks up still more heat from the hot valley lands and torrid desert just west of the mountains. A Santa Ana—one of several names it is known by—can crack lips, fray tempers, and fan a smouldering brush or forest fire into an inferno. It is said that in pioneer days, homicides committed during a red wind were automatically classed as manslaughter rather than murder.

The peculiarities of city climates are not wholly bad. Heating bills are lower than in the country, though air-conditioning bills, of course, are higher. And for those of us fortunate enough to possess city gardens, the growing season may be three or four weeks longer than for our neighbors in the suburbs. But these are rather feeble compensations for sauna-bath streets and subways in August and acrid smog in November. For the score of people who died during a 1948 "killer smog" in the little Pennsylvania town of Donora, as for the several thousands whose deaths were at least hastened by the catastrophic London smog of December 1952, such "compensations" are cold comfort indeed.

Our sultry, smoggy cities take their place with the Indus Valley deserts, the salt-choked soil of Sumeria, the ravaged, eroded Mediterranean lands, and the laterite-paved regions of southeast Asia, Africa, and Brazil. They are the latest and most conspicuous evidence that when man sets out to alter his environment in a big way, he is almost certain to end up with consequences he didn't bargain for and doesn't want. Especially when he barges ahead pragmatically, without giving thought to future costs as well as present profit!

40. AND NOW WHAT?

Climate Present and Future

We have come a long way: from the balmy, palmy days of the Oligocene Epoch, when the first apes roamed through forests in the now arid Fayum, through the majestic ice cycles of the Glacial Epoch and man's first fumbling efforts at climate control, into the event-packed Recent Epoch, with the challenges and problems of climate and climate control helping to shape the rise, spread, and decline of civilizations and cultures. Where have we come to—and where, climatically speaking, are we going?

The facts—we are dealing now with a period in which actual measurements of temperature and rainfall became first possible and then general—are quite clear. Between about 1600 and 1850, the earth passed through a period of worsened climate called, with some hyperbole, the Little Ice Age. Temperatures, at least outside the tropics, averaged lower than at any time since the centuries just before the Christian era, and perhaps even lower than that.

One clear indication is the persistence of sea-ice off the coast of Iceland. In the 1780s, the average ice season was nearly thirteen weeks a year, as against less than a week during the Viking Era. During this and the immediately following decades, measurements of water temperatures in the North Atlantic became sufficiently frequent to let present-day climatologists estimate the former position of the Gulf Stream. That great moderating influence on European climate, it appears, made a sharper right turn than it now does, so that its northern boundary lay south of its present position and its easternmost portions did not come as close to the coast of western Europe as they now do. The shift in the Gulf Stream is doubtless related to the extended sea-ice season off Iceland—but we do not know what shifted the Gulf Stream. On land, too, the effects were marked; mountain glaciers thrust farther down their valleys, in some cases several miles beyond earlier positions.

Some time after 1850, the chill began to slacken, though with frequent reversals; the ice began drawing back on both land and sea. This warming trend reached a peak in the 1930s, when the ice season off Iceland averaged less than a week and a half out of the year. Snow lines rose in many mountainous areas. Seal and walrus populations off southern Greenland declined, as the animals drew back toward their preferred colder waters—to the dismay of the Eskimos, who had for generations depended on them for food. In compensation, however, great schools of cod now made their appearance in the same waters. Birds began migrating and breeding farther north; in Iceland alone, seven new southerly species were reported, while some of the arctic species there dropped off in numbers since they were now breeding in more northerly regions.

Beginning around 1940, world temperatures leveled off and have since drifted downward, though thus far to nowhere near the levels of the Little Ice Age. How long this process will continue and how far it will go is anybody's guess—and the guesses depend on the guesser's own theory of climatic change. One point of view is that the climate is simply getting back to normal. H. H. Lamb, for instance, considers the first half of the twentieth century, especially its later decades, as "highly abnormal," so that the cooling trend of the 1950s and '60s would presumably be merely what one would expect. Of course this doesn't explain much—nor, to be fair, does Lamb suggest that it does. But aside from that, I wonder what "highly abnormal" really means. The recent warm spell was indeed "highly abnormal" compared with the preceding Little Ice Age. It was barely abnormal at all compared with the period 1400–1600. It was somewhat abnormal, though not markedly so, compared with the Little Climatic Optimum—but abnormally cool, not abnormally warm. It was distinctly abnormal in the same sense compared with the much earlier Climatic Optimum, which was itself abnormal in terms of the millennia before and after.

It does seem to be true that since the last retreat of the ice sheets, some 10,000 years ago, world temperatures have fluctuated around a norm somewhat below the 1900–50 averages. There is also reason to believe that the entire Recent Epoch is "abnormal" in comparison with at least some earlier interglacial (or even interstadial) periods. Between 100,000 and 80,000 B.P., for instance, sizable parts of the present Temperate Zone were subtropical, meaning that the world was a good deal warmer than during our own Climatic Optimum—for a period twice as long as the entire Recent

Epoch. And of course even this balmy spell was abnormally cool in comparison with conditions during most of the earth's history. The plain fact is that in climatology, as in most other areas of science, "normal" is a purely statistical concept, an averaging of a particular set of measurements over a given period of time. And the average figure you come up with depends on what set you pick.

It is also a fact that twentieth century climates may be abnormal in quite a different sense, in that they may be the result of human activities: specifically, the consequence of various substances that man has been tossing into the atmosphere by the billion tons.

The first of these is carbon dioxide. Man of course exhales this gas into the atmosphere with every breath drawn, but the output is trivial compared with the total mass of CO_2 on earth, which is better than 130 million *million* tons, most of it dissolved in the oceans. Moreover, man's output of CO_2 derives from his body's breakdown of carbon compounds, chiefly starches and sugars, in his foods. Since these are ultimately obtained from plants, which synthesize them from atmospheric CO_2, it follows that every pound of the compound added to the atmosphere metabolically by man or other animals merely restores a pound previously withdrawn by plants.

With the beginning of the industrial revolution, however, man began generating carbon dioxide that was not balanced by plant withdrawals. The coal, oil, and natural gas which power our machines were, to be sure, originally formed by plants—but tens and hundreds of millions of years ago. (They are being formed today in peat-bogs and on suitable parts of the ocean floor, but at a rate far smaller than they are being consumed.) As factories and power houses expanded and spread, their emission of CO_2 from fossil fuels accelerated; by 1960 it amounted to some seven billion tons a year. Even this is not very much—especially since nearly all of it is fairly rapidly soaked up by the oceans—but over the years it adds up. Actual measurements, in fact, indicate that between 1850 and 1950, the atmosphere's CO_2 content increased 13 percent.

Carbon dioxide is one of the atmospheric gasses responsible for the greenhouse effect, by absorbing heat radiation emitted by the earth rather than allowing it to pass into space. In fact, its contribution to the terrestrial greenhouse is out of proportion to its very sparse concentration in the atmosphere (a little over 0.03 percent, or about one part in 3,000) since it absorbs infrared radiation at wavelengths where the earth's emissions are particularly intense. It has been calculated that a 13 percent increase in atmospheric CO_2 should raise the temperature of the earth an average of 1°—and in fact this was just about the actual increase between 1850 and 1950.

So far, so good. But of course the increase in atmospheric CO_2 didn't stop in 1950; it continued, and even accelerated. Yet since 1950 world temperatures have slacked off, so that nearly half the earlier increase has been wiped out. What happened?

It may be, of course, that the downward temperature trend of the past twenty years is a temporary matter. Climatic changes, after all, happened before the Industrial Revolution and it is perfectly possible that the same forces, whatever they are, which produced earlier drops in temperature have, at least for the time being, superimposed themselves on the general upward trend set off by the burning of fossil fuels. On that assumption, the down-turn may well reverse itself before long. By the year 2000, world temperatures could be no less than 3.6 degrees above the 1850 average, as high as in the days of the Climatic Optimum. Other things being equal, of course. And with climate—since we are still not sure what all the "other things" are—we cannot be certain that they *are* equal.

Should such a development occur, however, it would be of immense scientific interest. One would like to know, for instance, whether the warmer temperatures would induce more rainfall (a "pluvial" period) in the tropics, or less; the answer would help to settle the long controversy over Ice Age climates in Africa. Again, H. H. Lamb believes that the Mediterranean Zone was not dryer during the Little Climatic Optimum, as one might expect, but wetter. This he explains by the appearance of a secondary storm belt over the Mediterranean, in addition to the "regular" one (which then lay well north of its present position). Presumably a similar pattern of circulation obtained during the Climatic Optimum; in another thirty years we may find out. We would also discover whether, as some have speculated, the Arctic Ocean was actually free of ice during the Climatic Optimum, which would conclusively dispose of Ewing and Donn's theory of ice ages. We might even find the Baltic once more teeming with herring.*

Should the warming trend continue for another few centuries, however, as the carbon dioxide theory indicates it must, temperatures will become distinctly sticky. It has been estimated that by the time all known fossil fuel deposits are exhausted, the earth will be more than 20 degrees warmer than at present—which would make New York tropical and London, Paris, and Berlin at least subtropical. Assuming, that is, that there is still a New York and a London. For a rise of this magnitude might well be enough to melt the Greenland and Antarctic ice caps, thereby adding enough water to the sea to raise its level by at least 150 feet. This would put my own fifth-floor New York apartment under some 16 fathoms of water—though by that time both I and the building would be long gone. It would flood a large part of the world's present inhabited areas, including much of its best agricultural lands—a climatic catastrophe far exceeding the various deserts and near-deserts created by human activity up to the present.

It is also possible, however, that the very recent downtrend in temperatures may be the result not of natural

* A warmer climate alone, however, would probably not be enough. Thanks to pollution of the rivers running into it, large sections of the Baltic are now described as being almost barren of life.

causes, but of other human activities whose impact on the atmosphere has first canceled and is now reversing the carbon dioxide effect. These have to do with dust, smoke, and similar substances which, instead of trapping outgoing radiation, block off incoming radiation.

Reid Bryson is one of those who has been pushing this theory. He points out that a large number of measurements indicate that "turbidity" in the atmosphere—its load of substances that reduce its transparency—has increased markedly over the past fifty years. One of the most striking findings in this area comes from the Soviet investigator F. F. Davitaia, who has studied glaciers in the high Caucasus. Glaciers often have a layered structure, whereby ice deposits of a particular year can be distinguished one from another. Measuring the amount of dust in these annual "growth rings," Davitaia found essentially no change up to 1930. Between then and 1960, however, the dust content increased no less than nineteen times. During much of this period the U.S.S.R. was industrializing at a breakneck pace; the chief exception was the war years, in which a large part of Soviet industry was destroyed and in which the dust content showed a drop. In fact, when Davitaia graphed the Caucasian dust fall and the Soviet capital investment in industrial expansion, the two curves showed a striking similarity.

There seems little doubt that increases in atmospheric dust would lower surface temperatures on earth, by blocking part of the incoming solar radiation. The question is, how much?

Part of the dust increase must certainly come from smoke and fine ash particles liberated by industrial combustion. But on the face of it, these quantities would not be sufficient to offset the carbon dioxide effect produced by the same processes. If the net effect of industrial CO_2 (warming) and industrial smoke (cooling) was to raise temperatures between 1900 and 1950, surely it should have continued to produce the same effect thereafter? Not necessarily. To assume this, it would be necessary to assume also that both processes were "linear," that adding twice as much CO_2 to the atmosphere would produce exactly twice the previous warming, and that twice as much smoke would produce twice the cooling. But we don't know this; conceivably, twice as much smoke could produce not twice but three times as much cooling, the more so in that the effects of smoke may be "amplified" by its stimulation of fog and cloud production, as noted in the last chapter.

Bryson also points out that there are a number of dust-producing processes that are certainly not being offset to any degree by CO_2 production. The first is mechanized agriculture. "Consider," he points out, "the dust stirred up by a tractor-drawn plow at 12 mph compared with the effects of a horse-drawn plow at 2 mph." Bulldozers and other mechanized earth-moving equipment are equally dusty. Another factor is mechanized traffic across deserts. A truck, unlike a camel, will break up the desert "pavement" of rocks and gravel which tend to hold down dust; the great North African tank battles of World War II generated dust clouds that reached the Caribbean (via the trade winds).

There is also the matter of contrails from high-flying jet aircraft. These are formed by the condensation of water vapor produced by combustion of the planes' fuel; before dissipating, they can spread to a width of half a mile or more. Bryson estimates that in the skies over the heavily traveled areas of North America, the North Atlantic, and Europe contrails may have increased the high-level cloud cover as much as 10 percent—which is "not negligible." If and when supersonic aircraft come into general use, this artificial cloud cover could increase even more. How much is impossible to say; most of the present information on supersonic flight is in the hands of the Pentagon, which is sitting on it as it has sat on so much other important information of public interest. Bryson believes that under some conditions contrails could almost cover the sky with their man-made "clouds."

What effect contrails are having, or may have, on climate

is impossible to say. Some meteorologists argue that they reduce temperatures by cutting down incoming radiation; others argue as vociferously that they raise temperatures by blocking outgoing infrared radiation. Both might be true. Even if their effect on climate is zero, contrails are blocking off part of the open sky, and may well block off more in the future. As Bryson puts it, "we would like our grandchildren to experience blue skies more often than on rare occasions!" Overall, he believes, atmospheric turbidity may very well be bringing on a new Little Ice Age: "The problem may not be how to keep New York from being flooded out but how to keep its harbor from freezing up in winter!"

Still another way in which we may be changing the climate in some regions is by felling tropical forests. The immense trees of the Equatorial Zone are enormously efficient mechanisms for returning moisture to the atmosphere; in fact, it is estimated that such a forest area will release practically as much water vapor as an equal area of open water, and far more than cropland or grassland. If large areas of Equatorial forest are cut down, air masses leaving them might be considerably dryer than they now are. The results could be serious for adjacent dryer areas that now obtain atmospheric moisture from Equatorial regions. For example, severe inroads in the great forests of the Amazon Basin could cut down rainfall in Venezuela and Colombia, which today derive a sizable percentage of their rain from air moving up from the south. In fact, weather records from at least two Colombian locations do show a serious decline in rainfall over the past quarter century. We don't know whether deforestation is the reason, but we don't know that it isn't, either.

The very problematical nature of these human modifications of climate indicates how little we still know about the subject—and how much we need to know. Man is no longer the rare species he was during the Ice Age. With 3.5 billion people on earth—and more coming—even our inadvertent actions can have potent and perhaps catastrophic effects. Those same population figures also mean that our margin for error is getting smaller every day.

Now that we have inadvertently modified the climates of our great cities, and perhaps of the entire world, where do we stand and where are we going as regards deliberate modification of climate?

Oddly enough, present practical techniques in this field are, with one exception, essentially the same as mankind has employed for thousands of years. Our devices for heating homes and workshops have grown bigger and more efficient, have moved from wood to coal to oil to natural gas and electricity, but in principle they are identical with the Neanderthal campfire in the family cave. Our irrigation systems, though they now cover millions of square miles, would be intelligible to any Sumerian or Egyptian priest-engineer.

The exception is the techniques developed over the past century or so for space cooling rather than space heating. The oldest and crudest of these is artificial refrigeration, including the artificial ice machine. As a ten-year-old, I watched men working one of the last of the old-time New England ice-ponds, sawing 6-foot slabs of ice from the frozen surface of Bantam Lake in Connecticut and laying them, packed in sawdust, in the great thick-walled icehouse, to be inserted into old-fashioned iceboxes when summer came round. Today, icehouse and icebox alike have vanished; mechanical refrigeration has brought ice manufacture into the home. At the same time, the deepfreeze has brought the Ice Age housewife's winter food-storage technique to all seasons and climes, while air-conditioning makes summer more tolerable to our overheated city dwellers.

We should not overemphasize the importance of this modern development in climate control. It has made life more convenient and comfortable for some, but not much more than that. Though we still have much to learn about the effects of environmental temperatures on human efficiency, there is no doubt that heat is a great deal less detrimental to it than cold. Man has survived successfully in the tropics from the very beginning, and, possibly for this reason, his biological cooling mechanisms are efficient enough so that he can function in the hottest climates as long as he has enough water. It may be, as some have said, that air conditioning will ultimately do as much for civilization in the tropics as central heating did for it in colder climes—but I wouldn't bet on it.

One need be no prophet to foresee a continued expansion of these established methods of climate control, especially, in light of our ever-increasing need for food, of irrigation. But here we are already running into a severe limitation: not enough water. In parts of the Southwest, underground water levels, thanks to the profuse expansion of irrigation, are sinking rapidly, meaning that shallow wells are running dry. In many coastal areas, and some inland ones, wells are turning brackish as salt water seeps in to replace what men have pumped out. This sort of thing is bound to happen in any region where more water is being extracted from the ground than is being put back in by precipitation; it is yet another illustration of a principle so often, and so destructively, ignored by men in dealing with their environment: You can't get something for nothing.

Of course we can, in principle, obtain unlimited quantities of water for irrigation and other purposes by desalting sea water. But you can't get something for nothing here either; desalinization requires energy, and energy costs money—according to present figures, a good deal more money than the water would yield if applied to agriculture. The difference would have to be made up by the taxpayer, who could well balk at loading himself with yet another subsidy to farmers.

One of the most costly, not to say grandiose, climate-control proposals now under consideration would seek water not in the sea, but in currently "useless" rivers. Thus it has been suggested that several Canadian rivers which now flow into the Arctic Ocean might be reversed and their waters sent south through aqueducts and tunnels into the Great Basin desert, most of which could thereby be converted into farmland. Similar proposals have been made for Siberia, where the Ob, Lena, and Yenesei carry millions of tons of water northward that would be a great deal more useful in Central Asia. Whatever its economic validity, climatic modification on this scale would unquestionably bring its own side effects. Kenneth Hickman, of Rochester Institute of Technology, has pointed out that these north-flowing rivers now carry heat into the Arctic Ocean from warmer latitudes; sending them south would make the polar region more polar, the tropics more tropical. He might have added that the increased contrast between tropics and northern polar region, by stepping up the pace of the atmospheric heat engine, would make the northern Stormy Zone more stormy. Hickman himself believes that the Canadian–Great Basin project would somewhat improve the climate of North America, except, presumably, for regions near the Arctic Ocean; winters would be warmer, but also snowier, and summers would be cooler. Evaporation from the irrigated Great Basin might increase rainfall in the Great Plains—and might also give the skiers of Aspen and Sun Valley far too much of a good thing.*

At present, neither the United States nor Soviet Russia can spare enough money from bombs, missiles, and war planes to undertake irrigation projects of this magnitude. I regard this as probably the only good result of the arms race; for the fact is, we still don't know enough about climate to be sure *what* the overall effects would be, whether a milder, moister climate in the wheat belt or a catastrophic increase in tornadoes in the same region. And before we, or the Russians, start in on any such massive intervention into climate, I hope to heaven that, for once, we will make sure we know exactly what we are doing.

Just how important this is can be seen from some recent studies in which far less ambitious manipulations of climate

* It might also, by increasing ice in the Arctic, spell farewell to the "Northwest Passage" to Alaska, recently opened with so much fanfare.

have, as usual, produced unexpected and unwanted side effects. In Israel, for instance, the shift from dry farming to irrigation and from furrow irrigation to overhead sprinklers, which use less, and less-skilled, labor, has led to an enormous upturn in half a dozen species of insect pests. (The Israeli entomologist E. Rivnay has titled his account of these changes "How to Provide a Nice, Wet Place Where Insects You Don't Want Thrive.") In Egypt, the shift in some areas from the seasonal irrigation pursued since the time of the pharaohs to perennial irrigation has led to a terrifying increase in the parasitic disease schistosomiasis. The wormlike parasite spends part of its complicated life cycle in water snails, which thrive in the quiet, warm water of perennial irrigation ditches. In such areas, schistosomasis has jumped from a nuisance affecting a few percent of the population to a plague that attacks the health and sometimes threatens the life of well over half the population—even nearly all of it.

One wonders, too, what will be the ultimate impact on Egypt of the Aswan High Dam. Certainly the waters impounded behind it will spread irrigation—and presumably schistosomasis—to tens of thousands of acres. But what of the silt that it will also impound, the Gift of the Nile that for five thousand years restored the fertility of Egypt's soil? *

But with all the problems attached to these conventional methods of climate control, surely our scientists and engineers are on the verge of breakthroughs which will make them obsolete. What, for instance, of the various experiments in rainmaking we have heard so much about?

At this point the conventional science-writer's approach would be to embark on a long description of current experiments in new techniques of climate and weather control,

* A mammoth flood-control project in the Mekong Valley, now in hibernation due to the Vietnam War, has been criticized on similar grounds. The critics have pointed out that blocking the flood waters would also block the revivifying, flood-borne silt that makes the region—or what is left of it after bombing and defoliation—one of the most fertile in southern Asia.

ending with a mouthwatering account of the progress that is just around the corner and a paean to science, technology, and American know-how. I would estimate that nine out of ten newspaper and magazine science stories could be summed up in these terms—and something like four out of five professional scientific papers, though their language is a lot more restrained. As applied to climate control, to take only one area of science, this approach is pure four-letter-word. That it is so prevalent is due to neither stupidity nor malice on the part of writers and scientists, but to a necessary symbiosis between the two groups.

Many scientists dislike reporting negative results—unless, of course, they are being negative about the results of some scientific rival. Negative findings are no less important in the development of science than positive ones, but they don't win their discoverers fame, fortune, or Nobel Prizes. Science writers are even more negative about negatives for the excellent reason that newspaper and magazine editors dislike "downbeat" stories. I don't pretend to know the reason for this, but suspect that it is partly theological: Progress is basic to our American secular religion, and to cast doubt on the reality of progress is rather worse, nowadays, than suggesting that God is dead.

Since my publisher does not share these religious scruples, I can say flatly that the prospect of major progress in climate control between now and the year 2000 (my crystal ball clouds up beyond that date) is nil. Research now underway will very probably chalk up certain achievements in local climate control in restricted areas, generating a little more rain, a little less snow or fog, and, perhaps, occasionally moderating some of climate's extreme violences. But these achievements, though they would be nothing to sneeze at, will only nibble around the edges of climate control—which is to say that thirty years from now the pattern of the world's climates will be, in every significant respect, precisely that described in the first chapters of this book.

Having made this somewhat blasphemous statement, I had

better explain my reasons. The first is our lack of knowledge. There is still an enormous amount we don't know about why weather patterns develop and change in the ways they do. If you have any doubt about this, clip your newspaper weather forecasts over the next month and check them against the actual weather in your neighborhood. I am not for a moment suggesting that the U.S. Weather Bureau is manned by a bunch of dimwits. On the contrary, its personnel are as able and dedicated a group of public servants as you are likely to find. But the phenomena they are dealing with are almost inconceivably complex. I believe it was the late mathematical genius John Van Neumann who described weather as not merely the most complicated natural phenomenon we know of, but the most complicated it is possible to imagine. The modern high-speed computer was originally developed as a means of dealing with meteorological data; yet though it has solved many simpler problems in other areas (meanwhile, of course, creating new ones), it has not yet enabled us to predict with any precision what the weather is going to do.

In a paper appropriately titled "One Finger on the Throttle of Nature's Weather Machine," Roscoe Braham, Jr., a fine negative thinker from the University of Chicago, summed up the state of knowledge as of 1968.

> Our knowledge of the inner workings of the weather machine is so fragmentary that we may not yet have identified the throttle handle and distinguished it from associated interactions and feed-back loops; even to the extent to which we have identified the throttle, in many situations we do not know which way to push it for an overall beneficial effect; moreover in the best situations our grip on the throttle is a very tenuous one.

The second, and far more serious, barrier to widespread climate control is the amount of energy that would be required. It has been calculated that 0.1 inch of rain (a sprinkling) falling over an area one hundred miles square (roughly the size of Connecticut) represents an energy release equivalent to the total U.S. output of electric power for six years. Enough rain to save crops after a two months' drought would represent ten times that already massive figure, and combatting a really widespread drought would multiply it by ten to a hundred times again. Climate control, like everything else, has its price—and the price in energy is one that not even the most developed, energy-rich civilization can afford today or in the foreseeable future. It is essentially for this reason that experiments in climate control have necessarily limited themselves either to very small target areas or to the triggering of energy release which natural forces have already brought to the point where it is almost ready to let go. Nearly all these experiments concentrate on what is called cloud seeding.

I do not want to get into a disquisition on cloud physics. It is a fiendishly difficult subject which even meteorologists do not understand very well and any attempt by me to explain it would at best be confusing—as well as full of errors in five years' time. I will only say that there are certain types of clouds in which the condensation of water vapor to water droplets can, sometimes, be artificially stimulated, so that the cloud increases its size, and in which the coalescence of water droplets into raindrops or snow crystals can, sometimes, be set going artificially—provided that one can identify the proper types of cloud and know when and where to apply what kind of stimulus. The stimulus is a mass of "seeds" of some sort—commonly a "smoke" of silver iodide crystals, or pellets of dry ice—which either serve as "nuclei" on which droplets can condense if other conditions are right (just as dust and smoke particles can help form fog over a city) or which can set off condensation processes by chilling the adjacent air.

The most extensive of these cloud-seeding experiments were carried out in 1960–64 under the name of Project Whitetop. Braham, who was in charge of them, reports that a careful analysis of the results indicates that on the average the seeding *decreased* the rainfall instead of increasing it. "Present day cloud-seeding techniques," he says dryly, "do not always produce the desired effects."

Taking the most optimistic view of progress in weather control over the next generation or so, here are some of the things that we might see:

1. Seeding of fog on airport runways to convert it into drizzle. This is perhaps the most likely because it is the simplest; the main problem at present seems to be cost.

2. Treatment of snow clouds in some problem areas, such as the shores of the Great Lakes. The procedure here is called "overseeding," meaning that by pumping enough particles into the cloud its conversion into snow crystals might be slowed. The effect would be to spread the snow over a relatively wide area, instead of letting the cloud dump it all within a few miles. However, efforts of this sort will immediately bring in the third major barrier to weather and climate control: conflicting legal interests. Overseeding which reduced the fall in a particular Buffalo snowstorm from 3 feet to 1 foot might delight the city fathers there, but if the spread-out of snow meant an extra foot of it in Rochester or Syracuse, those municipalities might well feel disposed to go to court. And quite possibly win.

3. The moderation of some thunderstorms and conceivably the prevention of some tornadoes. This would be done either by overseeding or by setting up "competing" storm clouds. Collectively, these would release the same total amount of energy as would be released without seeding, or even more, but dispersed over a much wider area.

4. The seeding of storm systems to extract more rain from them over a particular piece of territory. Eugene Bollay, a positive-thinking meteorologist, believes that such a system, set up in the basin of the Upper Colorado River, could add as much as 10 percent to the river's annual runoff, an amount which he believes would more than pay for the system's estimated cost. He may be right. But while such a system would doubtless draw firm support from the truck farm-

ers of Arizona, who depend heavily on Colorado River water, it could evoke a much more jaundiced reaction from the wheat farmers of Colorado and Kansas, who might well claim that they were being done out of rain that would otherwise have moistened their fields.*

5. The seeding of tropical storms to prevent them from contracting into the compact fury of a hurricane, or the seeding of hurricanes to "loosen" their structure, thereby diminishing their destructiveness by spreading their furious energies over a larger area. Winds of 100 mph will do far more damage in an area of 1,000 square miles than will winds of half the velocity spread over twice, or even ten times, the area. Here, fortunately, nobody is likely to object very strenuously. The courses of hurricanes are unpredictable enough so that almost any coastal city in the general vicinity is under threat, and the certainty of a gale would doubtless seem a quite acceptable payment to ward off the possibility of a hurricane.**

These, as I say, sum up the most optimistic prospects of climate control. You will note that every one of them is essentially a matter of working with whatever natural forces are available at a particular time and place; the most they

* At least as promising as this approach are various more conventional techniques which focus, not on increasing rainfall but on making more efficient use of such rain as falls naturally. Recent experiments at Michigan State University, for example, involve implanting a layer of asphalt, by means of a special plowlike machine, some two feet below the surface of the soil. This layer greatly reduces seepage of rainfall into the subsoil, where little of it would be available for plant growth; the technique is reported to have raised output of a variety of crops from 50 to 100 percent. Israeli agronomists, working in the near-desert Negev, have adapted and improved techniques originally devised by the Nabateans, who inhabited that region some two thousand years ago. Water from the brief but violent desert showers, which normally would be lost in destructive flash floods, is shunted into catchment areas where it has a chance to soak into the soil. Legally, such approaches are on far firmer ground than cloud seeding, since nobody disputes a man's right to conserve water on his own land.

** Recent experiments in hurricane seeding ("Project Stormfury") have produced encouraging, if still tentative, results. On one occasion, the wind fell from 99 to 68 knots for several hours after seeding. Let's hope it was cause and effect!

envision is shifting the release of those forces slightly in space or spreading it over a somewhat greater portion of space and time. But the forces themselves are as far beyond our capacities to alter as their nature was once beyond our comprehension.

We can, perhaps, arrange for a little more rain in western Colorado, provided eastern Colorado, Kansas, and Oklahoma don't object. We cannot possibly arrange for much more rain in any of these areas, let alone all of them together. The general circulation of the atmosphere, plus the land-forms of North America, have arranged matters so that the westerly winds give up nearly all their moisture to the Sierra Nevada long before they reach the Rockies or the Plains beyond; to arrange matters otherwise would be a task comparable to making the earth turn backward on its axis. We can, perhaps, reduce the frequency of tornadoes, somewhat moderating the most extreme violence of the Great Plains climate. But short of building a mile-high barrier along the U.S.–Canadian border, we cannot possibly prevent those polar outbreaks which help generate the violent climates of both the plains and of North America generally.

We can, perhaps, restore the grass cover of the Indian Desert and thereby raise its rainfall, converting it from an arid waste into a Steppe pastureland, assuming that the poverty-stricken Pakistani peasants can be persuaded or compelled to temporarily reduce their already exiguous flocks of sheep and goats. But we cannot convert that region into a moist, tropical paradise. Barring a mammoth expansion of conventional irrigation, using sea water desalted by the trillion tons, the Sahara will remain the Sahara.*

It was the very hostility of the earth's climate that first forced our ancestors down from the trees and then stimu-

* In fact, thanks to human misuse of the lands bordering on it, the Sahara is currently expanding by 40,000 acres a year! And in the Indus Valley, more agricultural land is currently being lost through salinization than is gained by expanded irrigation.

lated them to cope with natural environments of ever-increasing difficulty. The world-wide benign climates, mild even at the poles, in which our remotest primate ancestors throve were a long time going—and will be a long time returning. If it took fifty million years to build the great mountain systems which, it seems, have done so much to make the earth's climate what it is today, it will take not much less than that for frost, wind, and water to erode them. Significantly, the penultimate Glacial Epoch (the Permo-Carboniferous) seems to have lasted, all told, something like fifty million years. If that is any guide, the two million years of Pleistocene glaciations against which Homo erectus, the Neanderthals, and the first sapient men successively contended are but the beginning of the overture. In another 15,000 or 20,000 years, if Milankovich was right—or perhaps even if he wasn't—the ice sheets will once more grind down from the north. By that time, indeed, our technology may have progressed sufficiently to stop their advance—assuming that earlier technological "progress" has not already obliterated man and machine together.

Meantime, however, the very limitations of what we can do to climate, now or in the near future, as compared with what it can do to us may serve as a reminder of a truth too often concealed by our celebrations of technological triumphs: For all our ingenuity and intelligence, for all the resources of energy we dispose of, we are still small and feeble animals in an enormous universe which, if not actively hostile to life, is light years away from being hospitable to it. Our planet, insignificant as it is in the cosmic scheme of things, is yet what one of the astronauts so perceptively called it: an oasis. It is the only oasis we know of, or are likely to know of, within the universal desert. It is our survival capsule in the illimitable bleakness of space.

As we complain of conditions within our little craft, of suns too hot, of frosts too cold, of rains too violent or too sparse; as we seek, with often insufficient knowledge and reflection, to modify these conditions for our own convenience and prosperity, let us also appreciate the uniqueness of what we have and the importance of preserving it. For though we cannot yet, or perhaps ever, transform the earth into an Eden, we can certainly convert it into an inferno, by design or by sheer carelessness. The world we live on, climates and all, is the only one we've got; if we do not keep it fit to live in, nobody else will.

SELECTED BIBLIOGRAPHY

Asterisked entries are available in paperback.

Allison, Anthony C., "Sickle Cells and Evolution," *Scientific American,* August, 1956.

Anderson, Douglas A., "A Stone-Age Campsite at the Gateway to America," *Scientific American,* June, 1968.

Armstrong, Richard Lee, and others, "Glaciation in Taylor Valley, Antarctica, Older than 2.7 Million Years," *Science,* January 12, 1968.

*Battan, Louis J., *Cloud Physics and Cloud Seeding,* New York, Doubleday, 1962.

———"Some Problems in Changing the Weather" (mimeographed), American Meteorological Society, 1968.

Bibby, Geoffrey, *The Testimony of the Spade,* New York, Knopf, 1956.

Bollay, Eugene, "Rainmaking Now" (mimeographed), American Meteorological Society, 1968.

Braham, Roscoe R., Jr., "One Finger on the Throttle of Nature's Weather Machine" (mimeographed), American Meteorological Society, 1968.

Braidwood, Robert J., ed., *Courses Toward Urban Life,* Chicago, Aldine, 1962.

Broecker, Wallace S., "Absolute Dating and the Astronomical Theory of Glaciation," *Science,* January 21, 1966.

*Brøndsted, Johannes, *The Vikings,* Baltimore, Penguin, 1965.

Bryson, Reid, and Baerreis, David, "Possibilities of Major Climatic Modification and their Implications," *Bulletin of the American Meteorological Society,* March, 1967.

——— and Wendland, Wayne M., "Climatic Effects of Atmospheric Pollution" (mimeographed), American Association for the Advancement of Science, 1968.

*Carpenter, Rhys, *Discontinuity in Greek Civilization,* New York, Norton, 1968.

Cassidy, Vincent H., *The Sea Around Them,* Louisiana State Univeristy Press, 1968.

Chang, Kwang-chih, "Archeology of Ancient China," *Science* November 1, 1968.

Changes of Climate, UNESCO, 1963.

*Clark, J. Desmond, *The Prehistory of Southern Africa,* Baltimore, Penguin, 1959.

Coe, Michael D., *America's First Civilization,* New York, American Heritage, 1968.

*Cole, Sonia, *The Prehistory of East Africa,* New York, New American Library, 1963.

Coon, Carlton S., *The Living Races of Man,* New York, Knopf, 1963.

Coursey, D. G., and Alexander, J., "African Agricultural Patterns and the Sickle Cell," *Science,* June 28, 1968.

Cox, Allan, and others, "Reversals of the Earth's Magnetic Field," *Scientific American,* February, 1967.

Critchfield, Howard J., *General Climatology,* Englewood Cliffs, N.J., Prentice-Hall, 1966.

Dales, George F., "The Decline of the Harappans," *Scientific American,* May, 1966.

*Davidson, Basil, *The Lost Cities of Africa,* Boston, Atlantic, Little, Brown, 1959.

Deevey, Edward S., "Pleistocene Nonmarine Environments," *The Quaternary of the United States,* Princeton, 1965.

Donn, William L., and Ewing, Maurice, "A Theory of Ice Ages III," *Science,* June 24, 1966.

*Edinger, James G., *Watching for the Wind,* New York, Doubleday, 1967.

Eimerl, Sarel, and DeVore, Irven, *The Primates,* New York, Time Inc., 1965.

*Emery, W. B., *Archaic Egypt,* Baltimore, Penguin, 1961.

Emiliani, Cesare, "Isotopic Paleotemperatures," *Science,* November 18, 1966. (*See also* subsequent discussion by Donn and Shaw, and reply by Emiliani, *Science,* August 11, 1967.)

——— "Ancient Temperatures," *Scientific American,* February, 1958.

——— and Geiss, J., "On Glaciations and their Causes," *Geologischer Rundschau,* Vol. 46, Pt. 2 (1957).

Ericson, David B., and Wollin, Goesta, *The Deep and the Past,* New York, Knopf, 1964.

——— "Pleistocene Climates and Chronology in Deep-Sea Sediments," *Science,* December 13, 1968.

Evinari, Michael, and Koller, Dov, "Ancient Masters of the Desert," *Scientific American,* April, 1956.

Fairbridge, Rhodes, "The Changing Level of the Sea," *Scientific American,* May, 1960.

——— "New Radiocarbon Dates of Nile Sediments," *Nature,* October 13, 1962.

——— "Nile Sedimentation Above Wadi Halfa During the Last 20,000 Years," *Kush,* Vol. XI (1963), p. 96.

——— ed., *Solar Variations, Climatic Change and Related Geophysical Problems,* New York Academy of Sciences, 1961.

Feininger, Tomas, "Less Rain in Latin America" (letter), *Science,* April 5, 1968.

Flannery, Kent V., "The Ecology of Early Food Production in Mesopotamia," *Science,* March 12, 1965.

——— and others, "Farming Systems and Political Growth in Ancient Oaxaca," *Science,* October 27, 1967.

Frenzel, Burckhard, "The Pleistocene Vegetation of Northern Eurasia," *Science,* August 16, 1968.

Frerichs, William E., "Pleistocene-Recent Boundary and Wisconsin Biostratigraphy in the Northern Indian Ocean," *Science,* March 29, 1968.

Fritts, Harold C., "Growth Rings of Trees: Their Correlation with Climate," *Science,* November 25, 1966.

Glass, B., and others, "Geomagnetic Reversals and Pleistocene Chronology," *Nature,* November 4, 1967.

Hardy, Alister, *The Open Sea,* Boston, Houghton Mifflin, 1965.

*Hare, F. K., *The Restless Atmosphere,* New York, Harper & Row, 1963.

*Hawkes, Jacquetta, *Prehistory,* New York, New American Library, 1965.

Hole, Frank, "Investigating the Origins of Mesopotamian Civilization," *Science,* August 5, 1968.

* Hood, Sinclair, *The Home of the Heroes,* New York, McGraw-Hill, 1967.

Howell, F. Clark, *Early Man,* New York, Time Inc., 1965.

"The Human Species," various authors, *Scientific American,* September, 1960.

Judson, Sheldon, "Erosion Rates Near Rome, Italy," *Science,* June 28, 1968.

Kenyon, Kathleen M., "Jericho," *Archeology,* Vol. 20, No. 4 (1967).

*Kramer, Samuel Noah, *History Begins at Sumer,* New York, Doubleday, 1959.

Lamb, H. H., *The Changing Climate,* London, Methuen, 1966.

Leopold, Luna B., and Davis, Kenneth S., *Water,* New York, Time Inc., 1966.

Lowry, William R., "The Climate of Cities," *Scientific American,* August, 1967.

McCormick, Robert A., and Ludwig, John H., "Climate Modification by Atmospheric Aerosols," *Science,* June 9, 1967.

McDowell, R. E., "Climate Versus Man and His Animals," *Nature,* May 18, 1968.

* McEvedy, Colin, *The Penguin Atlas of Ancient History,* Baltimore, Penguin, 1967.

——— *The Penguin Atlas of Medieval History,* Baltimore, Penguin, 1961.

*Macgowan, Kenneth, and Hester, Joseph A., Jr., *Early Man in the New World,* American Museum of Natural History, 1962.

McNeil, Mary, "Lateritic Soils," *Scientific American,* November, 1964.

*McNeill, W. H., *The Rise of the West,* New York, New American Library, 1963.

MacNeish, Richard S., "Ancient Mesoamerican Civilization," *Science,* February 7, 1964.

* Mallowan, M. E. L., *Early Mesopotamia and Iran,* New York, McGraw-Hill, 1965.

Mangelsdorf, Paul C., and others, "Domestication of Corn," *Science,* February 7, 1964.

*Mellaart, James, *Earliest Civilizations of the Near East,* New York, McGraw-Hill, 1965.

Millon, René, "Teotihuacan," *Scientific American,* June, 1967.

* Mowat, Farley, *Westviking,* New York, Minerva Press, 1968.

Müller-Beck, Hansjürgen, "Paleohunters in America: Origins and Diffusion," *Science,* May 27, 1966.

Nairn, A. E. M., ed., *Descriptive Paleoclimatology,* Interscience, 1961.

——— *Problems in Paleoclimatology,* Interscience, 1964.

Öpik, Ernst J., "Climate and the Changing Sun," *Scientific American,* June, 1958.

Oxenstierna, Eric, "The Vikings," *Scientific American,* May, 1967.

*Phillips, E. D., *The Royal Hordes,* New York, McGraw-Hill, 1965.

Plass, Gilbert N., "Carbon Dioxide and Climate," *Scientific American,* July, 1959.

Portig, Wilfried Helmut, "Latin America: Danger to Rainfall" (letter), *Science,* June 26, 1968.

Proceedings of the Conference on the Climate of the Eleventh and Sixteenth Centuries, National Center for Atmospheric Research, 1963.

Roberts, Walter Orr, "Climate Control," *Physics Today,* August, 1967.

Ross, M. R., "Climatic Change and Pioneer Settlement" (mimeographed), University of Minnesota, 1966.

*Samuel, Alan E., *The Mycenaeans in History,* Englewood Cliffs, N.J., Prentice-Hall, 1966.

Schlauch, Margaret, *The Gift of Tongues,* New York, Viking, 1942.

Shapley, Harlow, ed., *Climatic Change,* Cambridge, Mass., Harvard University Press, 1956.

Shaw, David, "Sunspots and Temperature," *Journal of Geophysical Research,* October 15, 1965.

Simons, E. L., "The Significance of Primate Paleontology for Anthropological Studies," *American Journal of Physical Anthropology,* November, 1967.

——— "The Earliest Apes," *Scientific American,* December, 1967.

Solecki, Ralph S., "Shanidar Cave," *Scientific American,* November, 1957.

*Strahler, Arthur N., *A Geologist's View of Cape Cod,* Garden City, N.Y., Natural History Press, 1966.

Sutton, Graham, "Micrometeorology," *Scientific American,* October, 1964.

Thompson, Philip D., and O'Brien, Robert, *Weather,* New York, Time Inc., 1965.

* *The Vinland Saga,* Baltimore, Penguin, 1965.

Waterbolk, H. J., "Food Production in Prehistoric Europe," *Science,* December 6, 1968.

Watkins, Calvert, "Indo-European and the Indo-Europeans," *American Heritage Dictionary of the English Language,*

New York, American Heritage, 1969. (*See also* following Appendix of Indo-European roots.)

Wexler, Harry, "Volcanoes and World Climate," *Scientific American*, April, 1952.

* Wheeler, Mortimer, *Civilizations of the Indus Valley and Beyond*, New York, McGraw-Hill, 1966.

* Wilson, John A., *The Culture of Ancient Egypt*, Chicago, University of Chicago Press, 1951.

Wisenfeld, Stephen L., "Sickle-Cell Trait in Human Biological and Cultural Evolution," *Science*, September 8, 1967.

*Woolley, Leonard, *The Beginnings of Civilization*, New York, New American Library, 1965.

* *World Climate from 8000 to 0 B.C.*, Royal Meteorological Society, 1966.

Wright, H. E., Jr., "Natural Environment of Early Food Production North of Mesopotamia," *Science*, July 28, 1968.

Wulff, H. E., "The Qanats of Iran," *Scientific American*, April, 1968.

Zeuner, Frederick R., *Dating the Past*, New York, Hafner, 1958.

INDEX

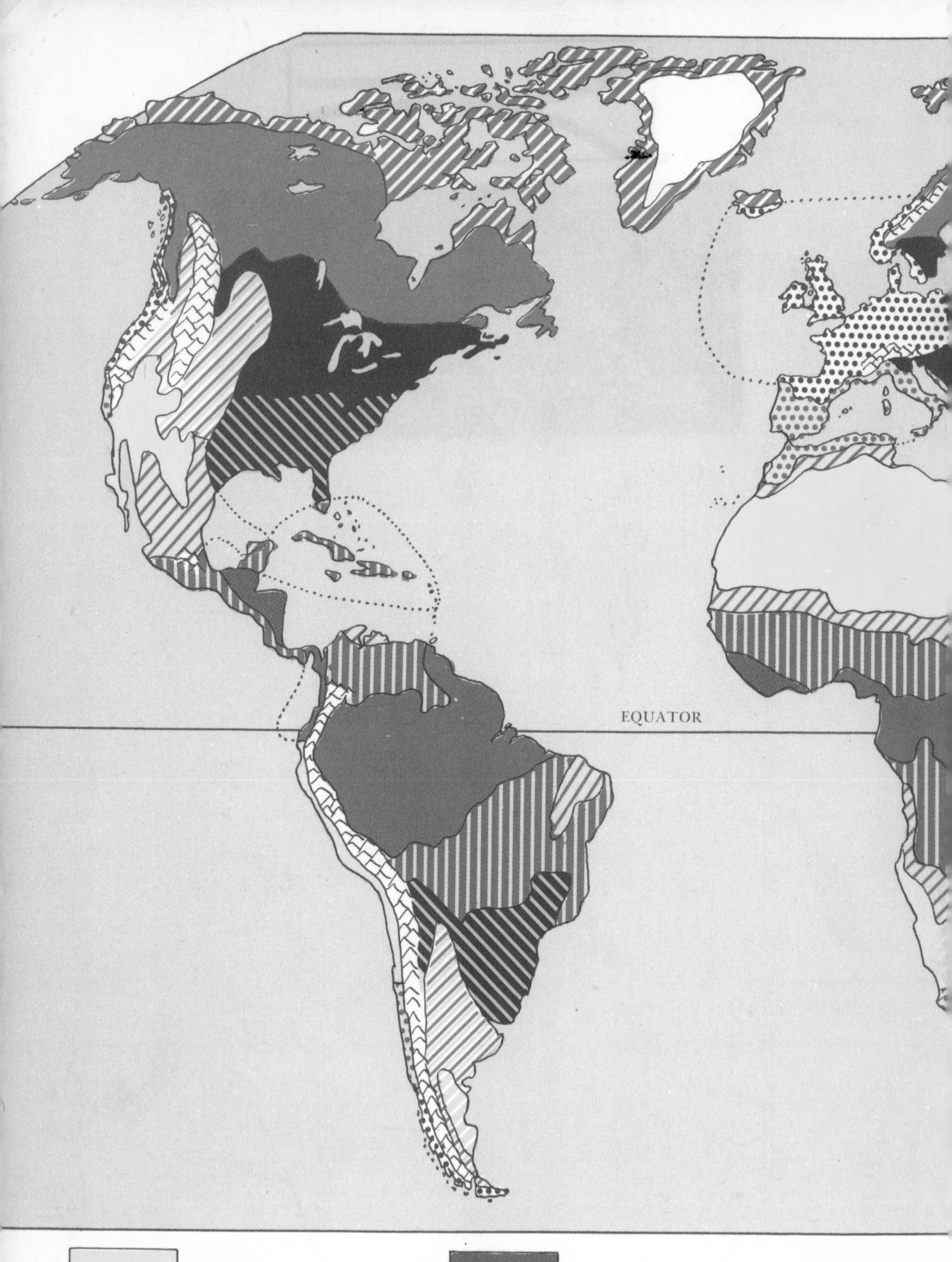

Desert

Steppe (tropical)

Savanna

Equatorial

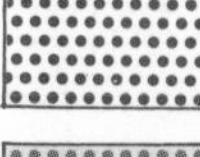
Maritime

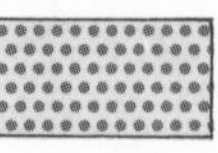
Mediterranean